33

Topics in Organometallic Chemistry

Topics in Organometallic Chemistry
Recently Published Volumes

Medicinal Organometallic Chemistry
Volume Editors: G. Jaouen, N. Metzler-Nolte
Vol. 32, 2010

C-X Bond Formation
Volume Editor: A. Vigalok
Vol. 31, 2010

Transition Metal Complexes of Neutral η^1-Carbon Ligands
Volume Editors: R. Chauvin, Y. Canac
Vol. 30, 2010

Photophysics of Organometallics
Volume Editor: A. J. Lees
Vol. 29, 2010

Molecular Organometallic Materials for Optics
Volume Editors: H. Le Bozec, V. Guerchais
Vol. 28, 2010

Conducting and Magnetic Organometallic Molecular Materials
Volume Editors: M. Fourmigué, L. Ouahab
Vol. 27, 2009

Metal Catalysts in Olefin Polymerization
Volume Editor: Z. Guan
Vol. 26, 2009

Bio-inspired Catalysts
Volume Editor: T. R. Ward
Vol. 25, 2009

Directed Metallation
Volume Editor: N. Chatani
Vol. 24, 2007

Regulated Systems for Multiphase Catalysis
Volume Editors: W. Leitner, M. Hölscher
Vol. 23, 2008

Organometallic Oxidation Catalysis
Volume Editors: F. Meyer, C. Limberg
Vol. 22, 2007

N-Heterocyclic Carbenes in Transition Metal Catalysis
Volume Editor: F. Glorius
Vol. 21, 2006

Dendrimer Catalysis
Volume Editor: L. H. Gade
Vol. 20, 2006

Metal Catalyzed Cascade Reactions
Volume Editor: T. J. J. Müller
Vol. 19, 2006

Catalytic Carbonylation Reactions
Volume Editor: M. Beller
Vol. 18, 2006

Bioorganometallic Chemistry
Volume Editor: G. Simonneaux
Vol. 17, 2006

Surface and Interfacial Organometallic Chemistry and Catalysis
Volume Editors: C. Copéret, B. Chaudret
Vol. 16, 2005

Chiral Diazaligands for Asymmetric Synthesis
Volume Editors: M. Lemaire, P. Mangeney
Vol. 15, 2005

Palladium in Organic Synthesis
Volume Editor: J. Tsuji
Vol. 14, 2005

Metal Carbenes in Organic Synthesis
Volume Editor: K. H. Dötz
Vol. 13, 2004

Theoretical Aspects of Transition Metal Catalysis
Volume Editor: G. Frenking
Vol. 12, 2005

Iron Catalysis

Fundamentals and Applications

Volume Editor: Bernd Plietker

With Contributions by

Matthias Beller · Bianca Bitterlich · Chi-Ming Che ·
Daniel F. Fischer · Masumi Itazaki · Sascha Jautze ·
Markus Jegelka · Kathrin Junge · Víctor S. Martín ·
Hiroshi Nakazawa · Juan I. Padrón· René Peters ·
Bernd Plietker · Kristin Schröder ·
Ella Lai-Ming Wong · Cong-Ying Zhou

Springer

Editor
Prof. Dr. Bernd Plietker
Institut für Organische Chemie
Universität Stuttgart
Pfaffenwaldring 55
70569 Stuttgart
Germany
bernd.plietker@oc.uni-stuttgart.de

ISSN 1436-6002 e-ISSN 1616-8534
ISBN 978-3-642-14669-5 e-ISBN 978-3-642-14670-1
DOI 10.1007/978-3-642-14670-1
Springer Heidelberg Dordrecht London New York

Cover design: KünkelLopka GmbH

Printed on acid-free paper

Springer is part of Springer Science+Business Media (www.springer.com)

Volume Editor

Prof. Bernd Plietker

Institut für Organische Chemie
Universität Stuttgart
Pfaffenwaldring 55
70569 Stuttgart
Germany
bernd.plietker@oc.uni-stuttgart.de

Editorial Board

Topics in Organometallic Chemistry Also Available Electronically

Topics in Organometallic Chemistry is included in Springer's eBook package *Chemistry and Materials Science*. If a library does not opt for the whole package the book series may be bought on a subscription basis. Also, all back volumes are available electronically.

For all customers who have a standing order to the print version of *Topics in Organometallic Chemistry*, we offer free access to the electronic volumes of the Series published in the current year via SpringerLink.

If you do not have access, you can still view the table of contents of each volume and the abstract of each article by going to the SpringerLink homepage, clicking on "Chemistry and Materials Science," under Subject Collection, then "Book Series," under Content Type and finally by selecting *Topics in Organometallic Chemistry*.

You will find information about the

- Editorial Board
- Aims and Scope
- Instructions for Authors
- Sample Contribution

at springer.com using the search function by typing in *Topics in Organometallic Chemistry*.
Color figures are published in full color in the electronic version on SpringerLink.

Aims and Scope

The series *Topics in Organometallic Chemistry* presents critical overviews of research results in organometallic chemistry. As our understanding of organometallic structures, properties and mechanisms grows, new paths are opened for the design of organometallic compounds and reactions tailored to the needs of such diverse areas as organic synthesis, medical research, biology and materials science. Thus the scope of coverage includes a broad range of topics of pure and applied organometallic chemistry, where new breakthroughs are being made that are of significance to a larger scientific audience.

The individual volumes of *Topics in Organometallic Chemistry* are thematic. Review articles are generally invited by the volume editors.

In references *Topics in Organometallic Chemistry* is abbreviated Top Organomet Chem and is cited as a journal.

Preface

Global energy consumption has increased tremendously in past decades due to intensified industrialization and growing world population. Apart from inventing new energy producing processes, the development of energy saving procedures offers the greatest chance to meet our energy needs in the long run. In this context, catalysis is certainly one of the key technologies for the coming decade. In order for a catalytic process to really make an impact, it has to be:

- Broadly applicable
- Selective
- Waste free
- Affordable
- Based on abundant materials

Modern research in catalysis tries to fulfill most of these criteria and a variety of powerful concepts have been developed within the past 30 years. However, the price of the catalyst is another variable that needs to be optimized in this equation. Iron offers some advantages with regard to this problem. It is abundant in sufficient amounts, nontoxic, and even nonnatural complexes can be metabolized into nontoxic products. However, iron complexes as catalysts have only recently been rediscovered after a period of almost 30 years, which was dominated by late transition metal complexes based on, for example, Ru, Rh, or Pd. This is even more surprising if one considers iron complexes to be the origin of modern organometallic chemistry. However, with regard to stability and reactivity, these complexes oppose severe synthetic challenges like, for example, spin ground state, oxidation state, etc. The advent of new spin-independent analytical techniques, plus the progress in the field of well-established techniques like Mössbauer spectroscopy, ESR, NMR, EXAF, SQUID, etc., have helped to address some of the fundamental problems in modern iron-based organometallic chemistry. Based upon this technological progress and the aforementioned general interest in identifying catalytic transformation by using readily available metal sources, catalysis by iron complexes is witnessing a tremendous comeback.

In line with these current developments, publishing a book dealing with the most recent achievements in this field is particularly timely. The volume is structured in chapters according to the type of metal complex. In every chapter, a brief introduction on the general chemical properties of the respective class of Fe-complexes will be given. Subsequently, representative examples for different catalytic transformations with a special emphasis on the various reaction manifolds will be presented. This structure implies that the reviews are not comprehensive but are meant to improve the understanding of the catalytic role a certain iron complex plays within the mechanism.

This book will complement existing reviews and monographs in this field and will be a source of inspiration for chemists working in various field of chemistry.

Stuttgart, Germany Bernd Plietker

Contents

Top Organomet Chem (2011) 33: 1–26
DOI: 10.1007/978-3-642-14670-1_1

Catalysis by Means of Fe-Based Lewis Acids

Juan I. Padrón and Víctor S. Martín

Abstract The interest and use of iron salts as catalysts in organic chemistry has shown an exponential growth in the past years. There has been an increasing demand for environmentally friendly and sustainable chemical methods, distinguished by the low cost and environmentally benign character of the iron salts used. This chapter focuses on reactions in which iron salts produce activation on unsaturated functional groups provided by the Lewis-acid character of these salts.

Keywords Alkene · Alkyne · Catalysts · Fe · Lewis acids

Contents

J.I. Padrón and V.S. Martín (✉)
Instituto Universitario de Bio-Orgánica "Antonio González", University of La Laguna, Avda. Astrofísico Fco Sánchez 2, 38206 La Laguna, Spain
Instituto de Productos Naturales y Agrobiología CSIC, Avda. Astrofísico Fco Sánchez 3, 38206 La Laguna, Spain
e-mail: jipadron@ipna.csic.es, vmartin@ull.es

Abbreviations

Ac	Acetyl
acac	Acetylacetonate
Alk	Alkyl
Ar	Aryl
bmim	Butylmethylimidazolium
Bn	Benzyl
cat	Catalytic
DCE	1,2-Dichloroethane
DMF	*N,N*-dimethylformamide
equiv	Equivalent(s)
Et	Ethyl
h	Hour(s)
HSAB	Hard and soft acids and bases
i-Bu	*iso*-Butyl
i-Pr	Isopropyl
Me	Methyl
min	Minute(s)
mol	Mole(s)
MS	Molecular sieves
NIS	*N*-iodosuccinimide
n-Pr	*n*-Propyl
Ph	Phenyl
rt	Room temperature
t-Bu	*tert*-Butyl
TfO	Trifloromethanesulfonate
TfOH	Trifluoromethane sulfonic acid
TMS	Trimethylsilyl
Ts	4-Toluensulfonyl

1 Introduction

Iron is the second most abundant metal on earth. It is a group 8 and period 4 element with [Ar] $3d^6 4s^2$ as electronic configuration. Iron as a metal is rarely found because it oxidizes readily in the presence of oxygen and moisture. Hence, it forms salts in its preferred oxidation state +2 and +3.

Iron halides are the most common of these salts [more specifically as a chloride: iron(II) chloride (ferrous chloride or $FeCl_2$) and iron(III) chloride (ferric chloride or $FeCl_3$)].

In 1963, Pearson introduced the hard and soft acids and bases (HSAB) principle, ordering Fe^{3+} as hard acid and Fe^{2+} as borderline acid [1]. Although the HSAB theory is qualitative, it has found widespread use in chemistry.[1] However, as a qualitative description it does not allow for the quantification of hardness or softness. Several attempts have been made to quantify hardness–softness on a normalized scale associated with physical properties, quantum mechanical principles using perturbation molecular orbital theory, and density functional theory [2–6]. According to this, Parr and Pearson obtained values of η (the absolute hardness) for molecules and ions relative to Al^{3+} as the metal with the highest value of 45.8 [2]. The expected increased hardness with increased oxidation state can be seen by comparing Fe^0 ($\eta = 3.9$), Fe^{2+} ($\eta = 7.2$), and Fe^{3+} ($\eta = 13.1$). Hence, Fe^0 is a soft Lewis acid. It should be noted that when transition-metal atoms act as Lewis acids it is usually in an excited valence state. Accordingly, iron atoms are $3d^8$ and not $3d^6\,4s^2$ when acting as a Lewis acid. A correction of the η value for this effect will lower the corresponding numbers. Furthermore, the addition of three chloride ions to Fe^{3+} lowers the hardness considerably since the η value of Fe^{3+} must be considered as a hypothetical upper limit in discussing the Lewis-acid catalyzed chemistry of iron salts. As a matter of fact, any Lewis base lowers the η value of an ionic acid.

Despite these studies, the effect of Lewis acidity strongly depends on the type of the reaction that is promoted or catalyzed. In 1972, Olah et al. classified Lewis acids in Friedel–Crafts alkylation reactions [7]. Iron(III) and iron(II) chloride were classified as moderately active and weak, respectively. In 2000, Kobayashi et al. reported a further classification of Lewis acids based on the activity and selectivity in addition reactions of a silyl enolate to an aldehyde and an aldimine [8]. $FeCl_3$ was classified in the active group as aldimine selective, while $FeCl_2$ kept the same type of selectivity but in the weak group. According to Olah, $SnCl_4$ and $FeCl_3$ have strong affinities for Cl atoms (compared with $AlCl_3$) especially in carbonyl-containing solvents [9]. In addition, Sn^{4+} and Fe^{3+} are considered softer than Al^{3+}, although they are still classified as hard acids [10].

Apart from the hardness and softness, two reactivity-related features need to be pointed out. First, iron salts (like most transition metal salts) can operate as bifunctional Lewis acids activating either (or both) carbon–carbon multiple bonds via π-binding or (and) heteroatoms via σ-complexes. However, a lower oxidation state of the catalyst increases the relative strength of coordination to the carbon–carbon multiple bonds (Scheme 1).

Second, apart from a different mode of activation, different modes of reactivity are observed. Hence, for $FeCl_3$ reactions like oxidative C–C but also nonoxidative C–C couplings are as well observed as bond formation via Lewis-acid activation (Scheme 2) [11–16]. Depending on the reaction type, one or more mechanistic pathways are accessible at the same time, which makes it difficult to shed light into mechanistic details.

---

[1]Once acids and bases have been classified as hard or soft, a simple rule of the HSAB principle can be given: hard acids prefer to bond to hard bases, and soft acids prefer to bond to soft bases.

Scheme 1 σ- and π-binding of bifunctional Lewis acid

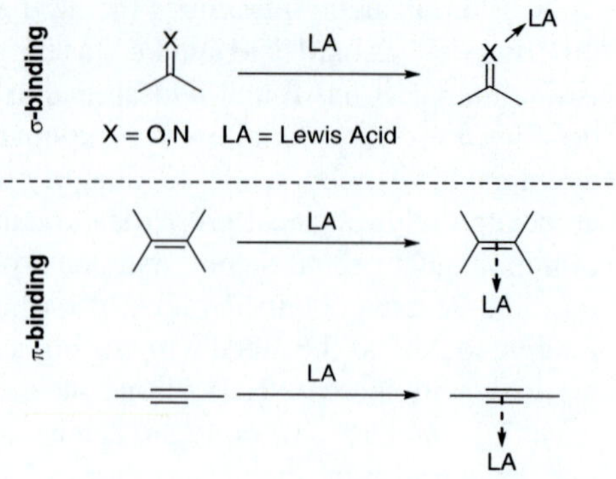

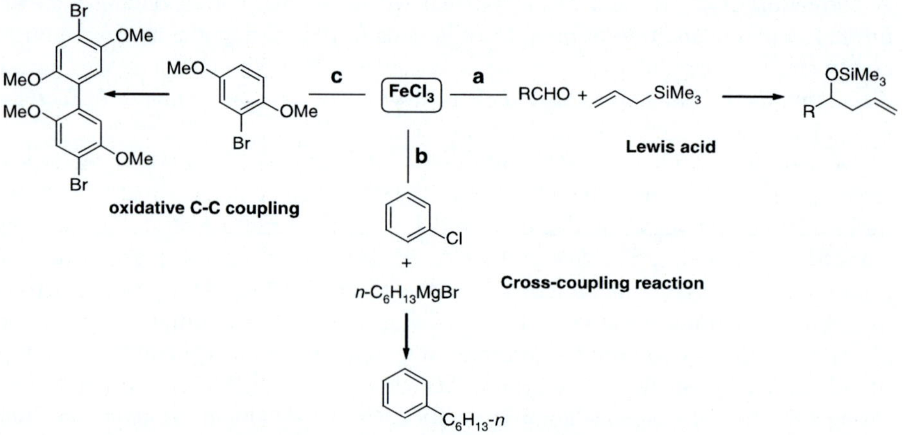

Scheme 2 Iron can work as Lewis acid, organometallic and oxidation catalyst

Since aromatic substitutions, aliphatic substitutions, additions and conjugate additions to carbonyl compounds, cycloadditions, and ring expansion reactions catalyzed by Fe salts have recently been summarized [17], this section will focus on reactions in which iron salts produce a critical activation on unsaturated functional groups provided by the Lewis-acid character of these salts.

2 C–C Bond Formation

2.1 *Aromatic C–C Bond Formations*

In the early 1970s, Olah et al. investigated the Friedel–Crafts benzylation of benzene and toluene with benzyl chloride derivatives in the presence of various

Lewis acids [7]. Friedel–Crafts catalysts were classified according to their activity in four categories: A, very active catalyst which also bring about intra- and intermolecular isomerizations; B_1 and B_2, moderately active and weak catalysts, respectively, which bring about no isomerization; and C, very weak or inactive catalysts. $FeCl_3$ was classified in the moderately active group B_1 providing high yields of benzylation with no extensive accompanying side reactions, whereas $FeCl_2$ was classified as a weak catalysts (group B_2) giving low yields of benzylation with no side reactions.

The toxic and/or hazardous properties of many solvents, notably chlorinated hydrocarbons, are making their use unadvisable, thus promoting the quest for novel reaction media. Since the pioneering work of Seddon [18], ionic liquids have emerged as a new class of stable, inert solvents with unique properties. Friedel–Crafts acetylations have been reported by Horváth using ionic liquids prepared from $FeCl_3$ and 1-*n*-butyl-3-methylimidazolium chloride ([bmim]Cl) [19]. This group has shown that the mechanism of the Friedel–Crafts acetylation of benzene in the presence of MCl_3 (M = Al or Fe) is exactly the same in [bmim]Cl as in 1,2-dichloroethane (Scheme 3).

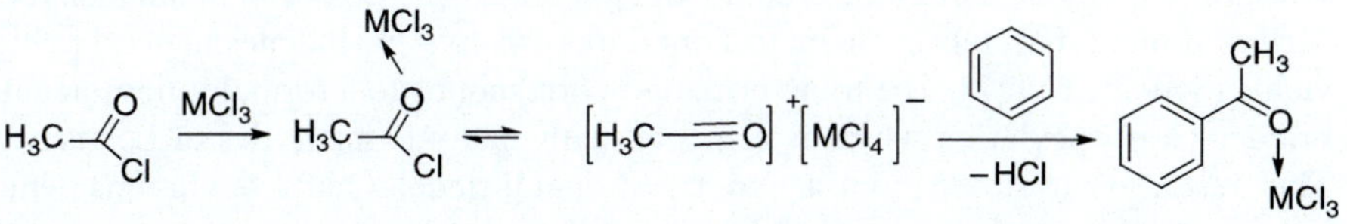

Scheme 3 Proposed mechanism of the Friedel–Crafts acetylation in the presence of MCl_3 (M = Al or Fe)

Furthermore, Jana et al. developed a $FeCl_3$-catalyzed C3-selective Friedel–Crafts alkylation of indoles, using allylic, benzylic, and propargylic alcohols in nitromethane as solvent at room temperature. This method can also be used for the alkylation of pyrrole (Scheme 4). The reactions were complete within 2–3 h without the need of an inert gas atmosphere leading to the C-3-substitution product exclusively in moderate to good yields [20].

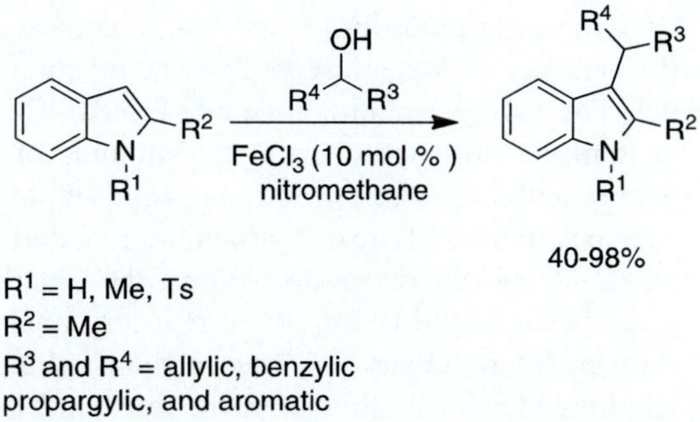

Scheme 4 $FeCl_3$ catalyzed Friedel–Crafts alkylations of indoles with alcohols

The $FeCl_3$-catalyzed Friedel–Crafts reactions of electron-rich arenes with imines or aziridines provide a facile and convenient route for the synthesis of β-aryl amines. Wu et al. found that the reactions of imines were highly substrate-dependent generating mono- or double-substitution products. The reaction of arenes with aziridines, however, led to a regioselective formation of the desired ring-opening products in moderate to good yields (Scheme 5) [21].

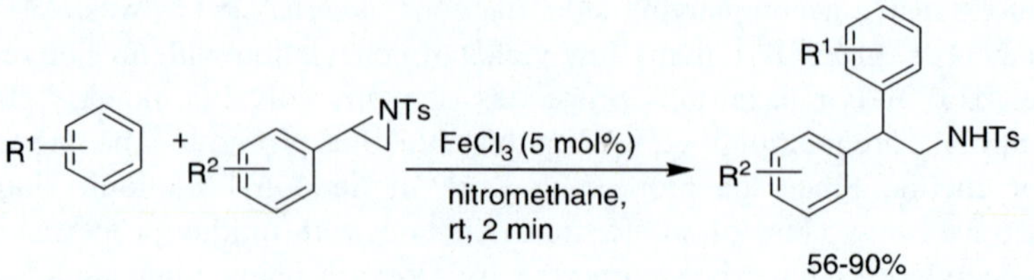

Scheme 5 $FeCl_3$ as efficient catalyst for reactions of electron-rich arenes with imines or aziridines

In 2007, Womack et al. published the conversion of 2-alkylcinnamyldehydes to 2-alkylindanones via a catalytic intramolecular Friedel–Crafts reaction. In the presence of 5–10 mol% $FeCl_3$ different in situ generated (*E*)-2-alkylcinnamaldehydes-derived dimethyl acetals cyclized to 1-methoxy-2-alkyl-*1H*-indenes in good to high yields (Scheme 6) [22]. The transformation corresponds to a formal intramolecular Friedel–Crafts acylation which is achieved with catalytic quantities of Lewis acid. This result is in strong contrast to traditional Friedel–Crafts acylations which require stoichiometric amounts of Lewis acid.

Scheme 6 Iron(III)-catalyzed conversion of 2-alkylcinnamaldehydes to 2-alkylindanones via a catalytic intramolecular Friedel–Craft reaction

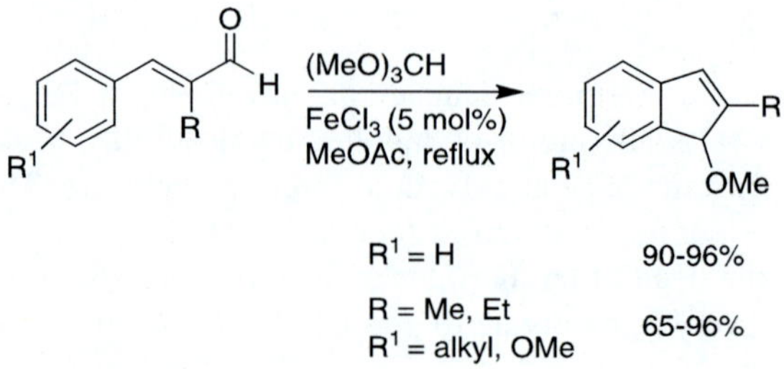

Womack's group improved the procedure using Ac_2O at room temperature in the presence of 4–6 mol% $FeCl_3$. The in situ generated acylals of 2-alkylcinnamaldehydes undergo an intramolecular Friedel–Crafts reaction to yield indenyl acetates at rt in the presence of catalytic amounts of Fe(III) (Scheme 7) [23]. Based on experimental results, the authors proposed an intramolecular Friedel–Crafts reaction in which an allylic oxocarbenium ion, derived from the in situ generated acylal, alkylates the phenyl group to form the indane skeleton (Scheme 7). A potential mechanism would be an allylic rearrangement of the acylal to a 1,3-diacetoxy-1-propene intermediate. This intermediate could regenerate either the *E*- or *Z*-acylal, resulting in isomerization of the double bond, or it could lead directly to the cationic species needed for the intramolecular Friedel–Crafts reaction.

As described above, propargylic alcohols can serve as electrophilic alkyl equivalents in intermolecular Friedel–Crafts reactions. However, the related intramolecular

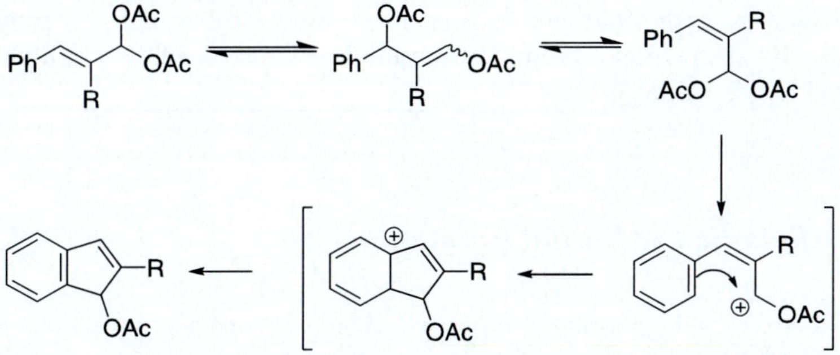

Scheme 7 Proposed mechanism to iron(III)-catalyzed indenyl acetate formation

Friedel–Crafts reaction remains unexplored, possibly due to the difficulty of the cycloalkyne formation. A mild, versatile, and efficient method for the one-step synthesis of substituted dihydro- and tetrahydroisoquinolines has been developed by the $FeCl_3{\cdot}6H_2O$-catalyzed intramolecular allenylation/cyclization reaction of benzylamino-substituted propargylic alcohols, representing the first example of the intramolecular Friedel–Crafts reaction of propargylic alcohols (Scheme 8) [24, 25]. $FeCl_3$, $InCl_3$, and $Yb(OTf)_3$ also exhibit good catalytic activity for the reaction.

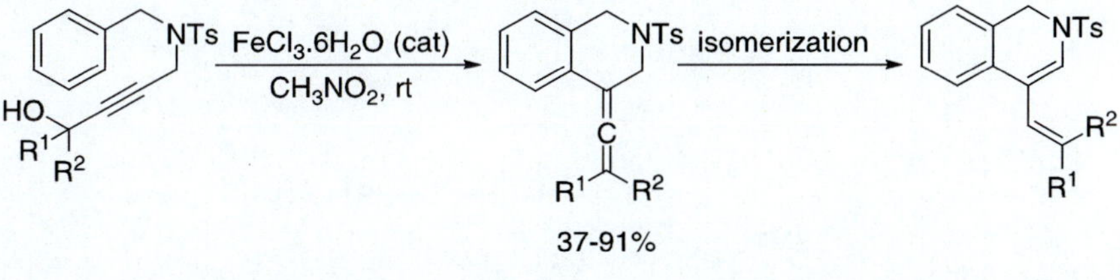

Scheme 8 One-step synthesis of substituted dihydro- and tetrahydroisoquinolines by $FeCl_3{\cdot}6H_2O$-catalyzed intramolecular Friedel–Crafts

An interesting intramolecular Friedel–Crafts-type cyclization was developed by Pericàs et al. Thus, aryl glycidyl ethers reacted to 3-chromanols as the only reaction product when treated with a catalytic amount of $FeBr_3$ in the presence of AgOTf in CH_2Cl_2 at room temperature (Scheme 9) [26]. The addition of a silver salt proved to

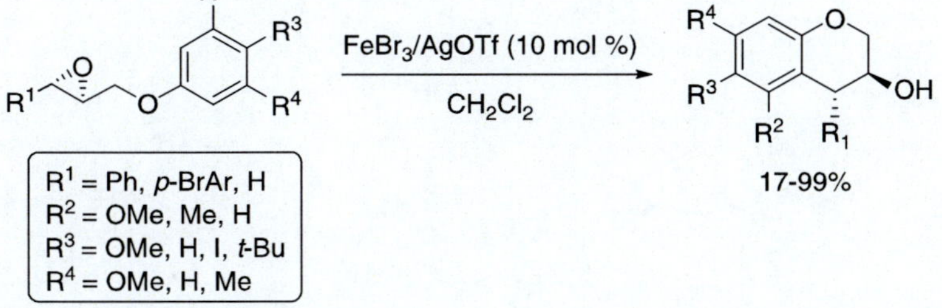

Scheme 9 Iron(III) bromide-mediated cyclization of aryl glycidyl ethers to 3-chromanols

be beneficial to both yield and reaction times most likely due to an activation of the $FeBr_3$ by abstracting a bromide from the iron center leading to a highly active Fe-based Lewis-acidic species.

2.2 Aliphatic C–C Bond Formation

Martín, Padrón, and coworkers have reported on the scope and limitations of the use of iron(III) halides as effective catalysts in the coupling of alkenes or acetylenes with aldehydes to achieve a wide variety of useful synthetic transformations. All these reactions are shown in Scheme 10, which serves as a guide through the aliphatic C–C bond formation section [27].

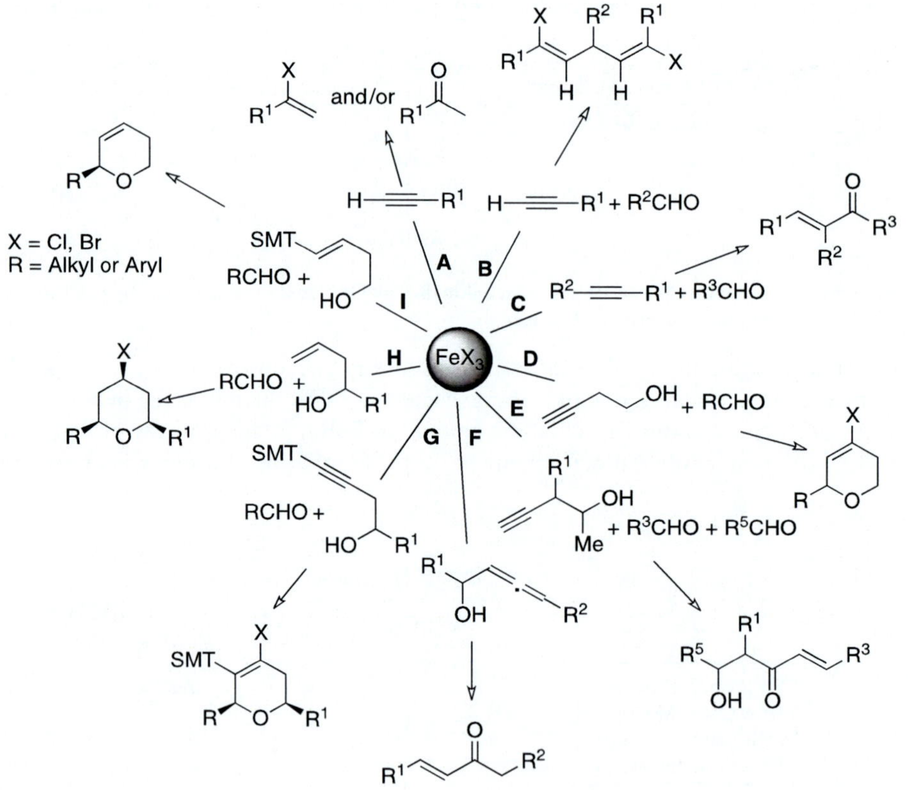

Scheme 10 Reactions of FeX_3 with different unsaturated substrates

2.2.1 Aliphatic C–C Bond Formation Using Acetylenes

Use of Nonfunctionalized Acetylenes

Terminal aliphatic alkynes (e.g., 1-octyne) react with iron(III) halides ($FeCl_3$ and $FeBr_3$) to give the corresponding 2-halovinyl derivatives (route A, Scheme 10). The moderate yields were remarkably improved upon addition of stoichiometric amounts of carboxylic acids.

On the basis of these results and Damiano's report [28], Darcel et al. described an iron-catalyzed hydration of terminal alkynes using catalytic amounts of iron(III) chloride (10 mol%). The reaction selectively leads to the corresponding methyl ketone derivatives (Scheme 11) [29]. Iron(II) species such as $FeCl_2$ or $Fe(OAc)_2$ were not able to promote the reaction, the starting phenylacetylene remained unchanged after several days at 75°C.

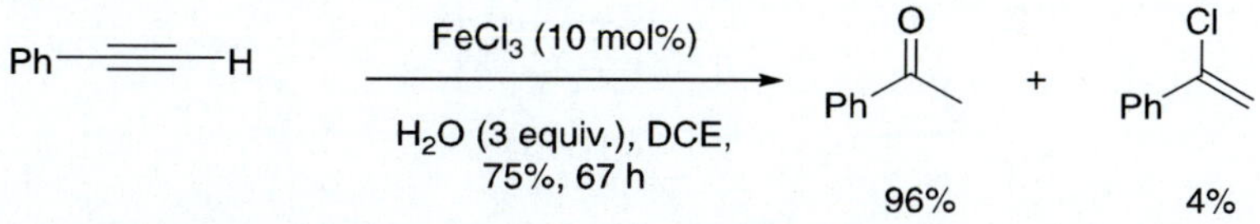

Scheme 11 Iron chloride-catalyzed hydration of terminal alkynes

The influence of iron(III) salts on coupling reactions of alkynes and aldehydes (Scheme 10, routes B and C) was also explored. In these routes, a new stereoselective coupling of alkynes and aldehydes was unmasked, which led to (*E*,*Z*)-1,5-dihalo-1,4-dienes (route B, Scheme 10) and/or (*E*)-α,β-unsaturated ketones (route C, Scheme 10) [27].

Both aliphatic and aromatic terminal alkynes reacted with aliphatic aldehydes giving exclusively a mixture of (*E*,*Z*)-1,5-dihalo-1,4-dienes and disubstituted (*E*)-α,β-unsaturated ketones, the former being the major products in all cases. When nonterminal aromatic acetylenes were used, the trisubstituted (*E*)-α,β-unsaturated ketones were the exclusive compounds obtained. The procedure was not valid for aliphatic and unsaturated alkynes. However, the catalytic system was found to be compatible with alcohols and their corresponding acetates although limited yields were obtained.

Experiments with terminal acetylenes, isolation of an intermediate acetal, alkyne hydratation studies, and ab initio calculations provide substantiation of a unified mechanism that rationalizes the reactions in which the complex formation between the alkyne and the iron(III) halides is the activating step (Scheme 12) [27].

The above-postulated overall mechanism considers two alternative pathways depending on the nature of the acetylene derivative. Region **A** outlines a proposal in which the formation of the σ-complex intermediate is supported by the fact that the treatment of aliphatic terminal acetylenes with $FeCl_3$ led to 2-chloro-1-alkenes or methyl ketones (Scheme 12). The catalytic cycle outlined in region **B** invoked the formation of the oxetene. Any attempt to control the final balance of the obtained

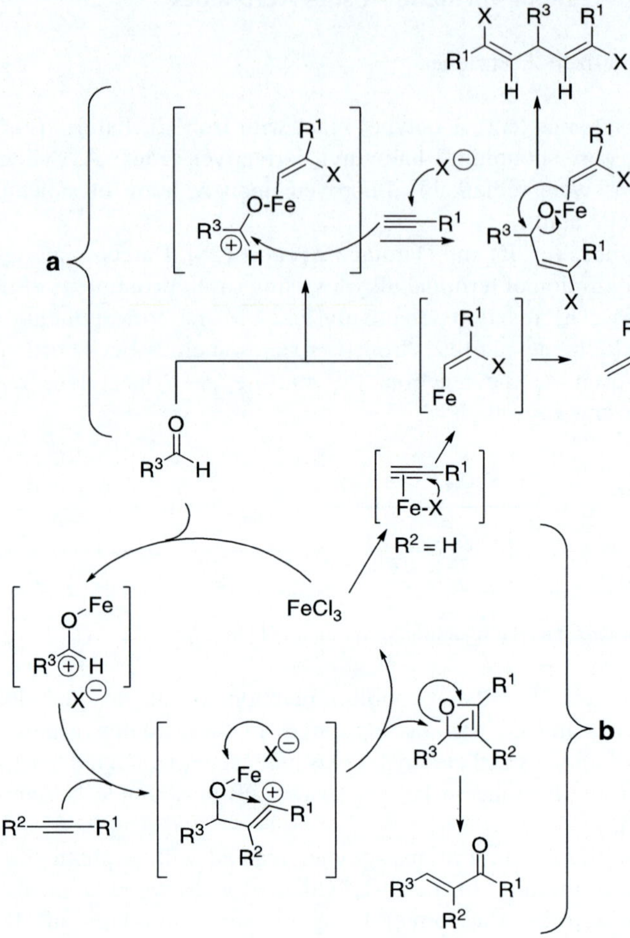

Scheme 12 Lewis acid role of FeX_3 in the coupling of acetylenes and aldehydes

products under experimental conditions (temperature, addition order, solvent, etc.) proved fruitless (Scheme 12). Thus, the tendency to follow either the catalytic (**B**) or the stoichiometric pathway (**A**) depends basically on the substitution pattern of the alkyne used.

A direct C–C bond formation by utilization of benzyl alcohols as alkylating agent is very attractive as it would be environmentally benign, water being the only by-product. Recently, Jana and coworkers described $FeCl_3$-catalyzed, sequential C–C and C–O bond-forming reactions of benzylic alcohols with aryl acetylenes for the regioselective synthesis of aryl ketones (Scheme 13) [30]. The reaction provides a simple method for the synthesis of substituted arylketones under mild conditions. Electron-rich alkynes reacted more efficiently and gave higher yields than did the neutral or electron-deficient alkynes.

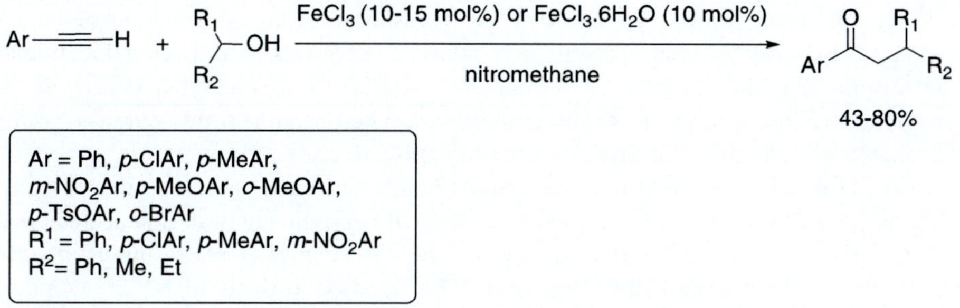

Scheme 13 Iron(III)-catalyzed addition of benzylic alcohols to aryl alkynes

$FeCl_3{\cdot}6H_2O$ turned out to be the catalyst of choice for this reaction, since the presence of water improved the yield. However, high yields of the desired ketones were obtained for electron-rich alkynes with anhydrous $FeCl_3$ at room temperature. Alcohols that are sensitive to acid-catalyzed dehydration were also tolerated under the present conditions (R^1 = Me or Et). Based upon experimental observations a mechanism for this reaction was proposed (Scheme 14).

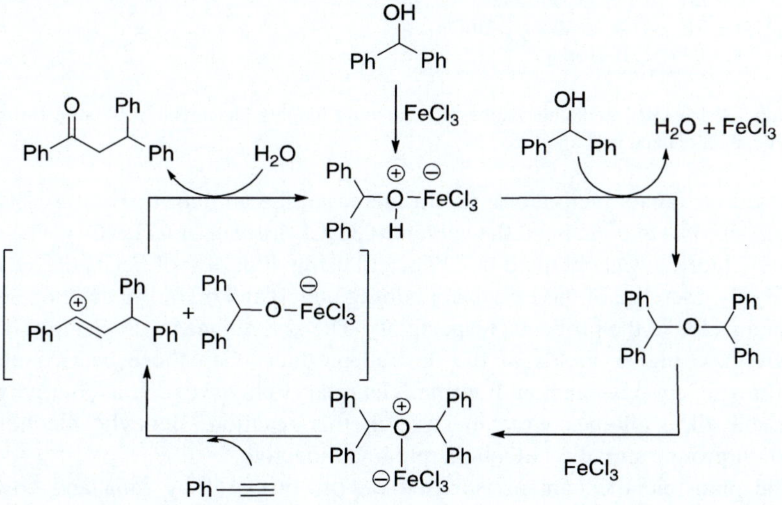

Scheme 14 Probable mechanism of iron(III)-catalyzed addition of benzylic alcohols to aryl alkynes

It was previously observed that with a catalytic amount of $FeCl_3$, benzylic alcohols were rapidly converted to dimeric ethers by eliminating water (Scheme 14). In the presence of an alkyne this ether is polarized by $FeCl_3$ and generates an incipient benzylic carbocation. The nucleophilic attack of the alkyne moiety onto the resulting benzyl carbocation generated a stable alkenyl cation, which suffer the nucleophilic attack of water (generated in the process and/or from the hydrated

$FeCl_3$ whenever used). To confirm the formation of the ether intermediate a separate experiment was performed between dimeric ether and phenylacetylene, observing that the dimeric ether reacted smoothly with phenylacetylene in the presence of $FeCl_3{\cdot}6H_2O$ to produce the desired ketone in a similar yield (≈42%) to that obtained when starting from the benzylic alcohol.

In 2009, Liu and coworkers accomplished an efficient method for preparing alkenyl halides via direct C–C bond formation of benzylic alcohols and aryl alkynes in CH_2Cl_2 by using 50 mol% of $FeCl_3{\cdot}6H_2O$ or 50 mol% $FeBr_3$ modifying the reaction conditions described by Jana. This method proved to be an excellent alternative to established procedures involving an excess boron trihalides and stoichiometric amounts of *n*-BuLi (Scheme 15) [31].

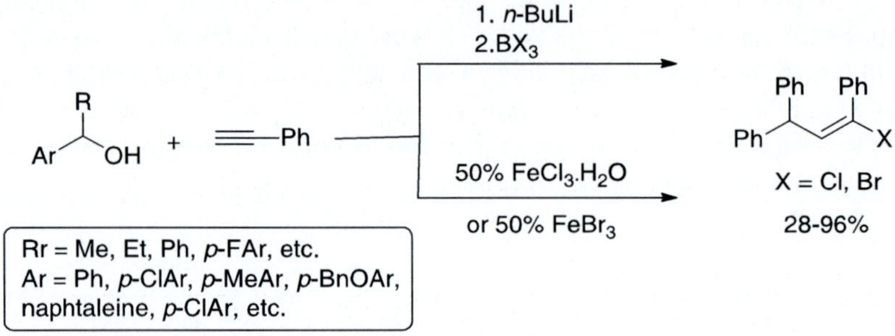

Scheme 15 Iron(III)-mediated synthesis of alkenyl halides via direct C–C bond formation of benzylic alcohols and aryl alkynes

Based on Jana's methodology, Liu increased the amount of $FeCl_3{\cdot}6H_2O$ (from 15 to 50 mol%) and changed the solvent (CH_2Cl_2 instead of CH_3NO_2). The desired alkenyl chloride was obtained in 82% yield using 40 mol% of $FeCl_3{\cdot}6H_2O$ at 50°C in CH_2Cl_2. Increase of the Lewis-acid dosage to 50 and 60 mol% gave 85 and 80% isolated yields of the products, respectively. The aryl alkynes with electron-donating groups gave higher yields of the desired products than those bearing electron-withdrawing groups, such as fluorine. Heteroaryl alkynes such as 3-ethynyl pyridine and alkyl alkynes were inactive in this reaction. Benzylic alcohols with carboxyl groups and alkyl alcohols remained inactive.

The plausible mechanism is based on the proposal by Jana and coworkers (Scheme 14). In this case, the *sp*-hybridized vinyl cation can be attacked by halide, instead of water, to give the *E*/*Z* isomer of the alkenyl halide. Compared with the systems using stoichiometric Lewis acid and strong base to prepare substituted alkenyl halides, the present method would provide an excellent alternative due to the environmentally benign system and atom efficiency.

Related to these strategies, $Fe(OTf)_3$/TfOH cocatalyzed the coupling reaction of terminal alkynes with benzylic alcohols in the absence of base by means of a sp–sp^3 C–C bond formation (Scheme 16) [32].

Although TfOH can also catalyze this reaction, the collaboration of $Fe(OTf)_3$ and TfOH was found to give higher yields and less side-reactions. The reaction was

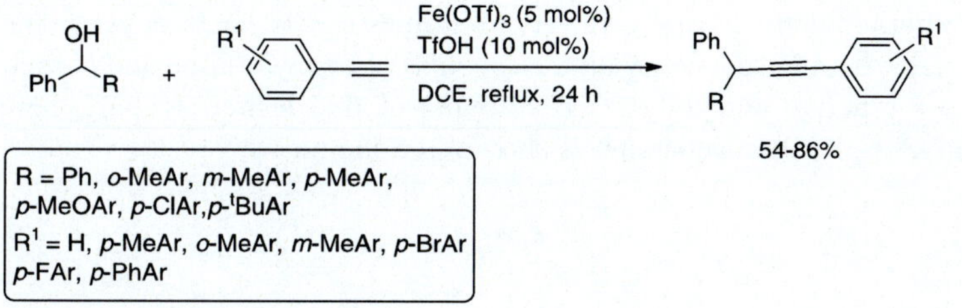

Scheme 16 *sp–sp*^{3} C–C bond formation via $Fe(OTf)_3$/TfOH-cocatalyzed coupling reaction of terminal aryl alkynes with benzylic alcohols

strongly influenced by the nature of the solvent. Polar solvents (DMF and $MeNO_2$) were not suitable for this transformation.

The coupling reactions of alkynes and aldehydes catalyzed by iron(III) salts have been discussed above (Scheme 10, routes B and C). The three-component coupling of aldehydes, alkynes, and amines is equivalent to the coupling reaction between alkynes and imines. Wang et al. reported such a three-component coupling catalyzed by $FeCl_3$ in the absence of any ligand (Scheme 17) [33].

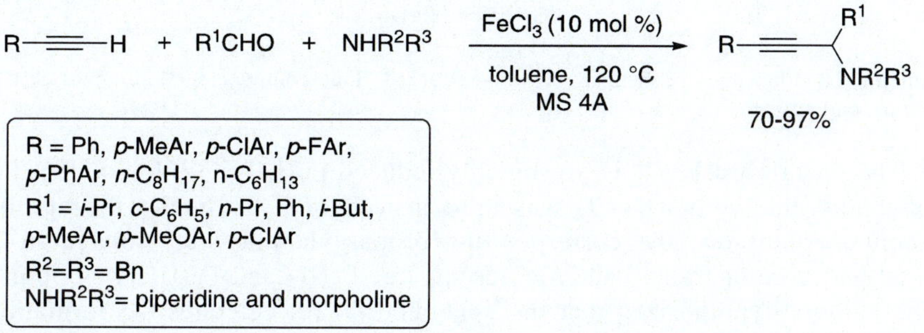

Scheme 17 Iron-catalyzed three-component coupling reactions of aldehydes, terminal alkynes, and amines

Several iron sources were screened in a model reaction of phenylacetylene, *iso*-butyraldehyde, and dibenzylamine. The three-component coupling reaction could be catalyzed by Fe(III) salts, Fe(III) oxide, or Fe(II) salts, such as $FeCl_3{\cdot}6H_2O$, Fe $(NO_3){\cdot}9H_2O$, $Fe_2(SO_4)_3$, Fe_2O_3, $Fe(acac)_3$, $FeCl_3$, and $FeCl_2$ in the absence of any ligand in toluene. The most effective catalyst for the reaction was anhydrous $FeCl_3$. In general, iron-catalyzed carbon–carbon and carbon–heteroatom bond formation reactions could be enhanced by a suitable ligand containing nitrogen or oxygen atoms. However, *N*-ligands and *O*-ligands did not assist in this three-component coupling reaction.

To check the scope of this coupling reaction, a study with different combinations of aldehydes, amines, and terminal alkynes was performed. Aromatic alkynes turned out to be more reactive than aliphatic ones. This study included aliphatic and aromatic aldehydes and a variety of amines (cyclic, heterocyclic, and acyclic). However, aromatic secondary amines such as *N*-benzylaniline and *N*-methylaniline were less reactive and only trace amounts of the desired propargylamines were isolated. A possible mechanism is shown in Scheme 18.

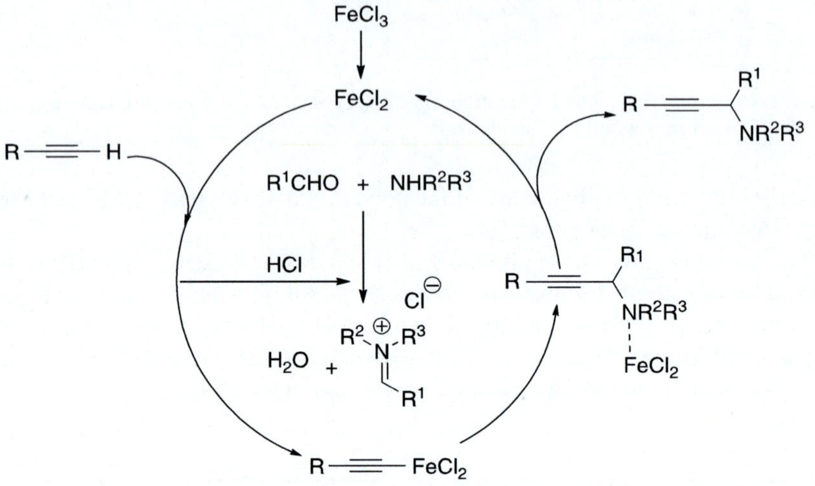

Scheme 18 Plausible mechanism of iron-catalyzed three-component coupling of aldehyde, alkyne, and amine

The precatalyst $FeCl_3$ is presumably reduced initially to a low-valent Fe(II) oxidation state. When $FeCl_2$ was used instead of $FeCl_3$ under the optimized reaction conditions, the corresponding propargylamine was isolated in 91% yield and a comparable catalytic activity for Fe(II) and Fe(III) was observed. The authors hypothesized that the generated HCl accelerates the formation of the immonium salt from the aldehyde and the secondary amines. $FeCl_3$ as a Lewis acid increases the electrophilic character of the starting aldehyde while stabilizing the immonium salt. The resulting Fe(III) alkynylate complex subsequently reacts with the immonium salt generated in situ to give the corresponding propargylamine and regenerates the Fe(II) catalyst for further reactions (Scheme 18).

Tu found that when aniline was used instead of the secondary amine under otherwise identical conditions 2,4-diphenyl-substituted quinoline was formed in 56% yield. Phenylacetylene and aniline were initially used as model substrates for exploring the aldehyde scope. With aromatic aldehydes the reactions proceeded smoothly to give the corresponding quinolines in moderate to good yields. A heteroaromatic aldehyde is also compatible with this transformation and the expected product was afforded in 83% yield. However, when aliphatic aldehydes were subjected to the reaction, the desired product was obtained in low yield (Scheme 19) [34].

Subsequently, the scope of amines and alkynes in this reaction was investigated, and it was found that substituted anilines (R^2 = *p*-Me, *p*-MeO, *p*-Cl and *m*-Br) were

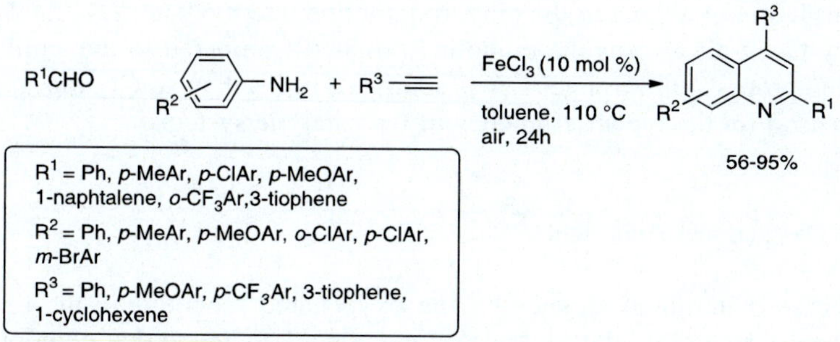

Scheme 19 Synthesis of quinolines by iron(III)-catalyzed three-component coupling/hydroarylation of aldehydes, alkynes, and amines

good substrates for this transformation. With regard to the alkyne scope phenylacetylenes, heteroaromatic alkynes, and aliphatic alkynes are suitable substrates.

Based on the experimental results above and together with Wang's previous work a tentative mechanism was proposed (Scheme 20) [33]. Thus, intermediate **A** is formed initially through coordination of imine and alkyne to the Lewis acid. This coordination sets the stage for an addition of the alkyne to the imine leading to the propargylamine intermediate **B**, which then undergoes an intramolecular

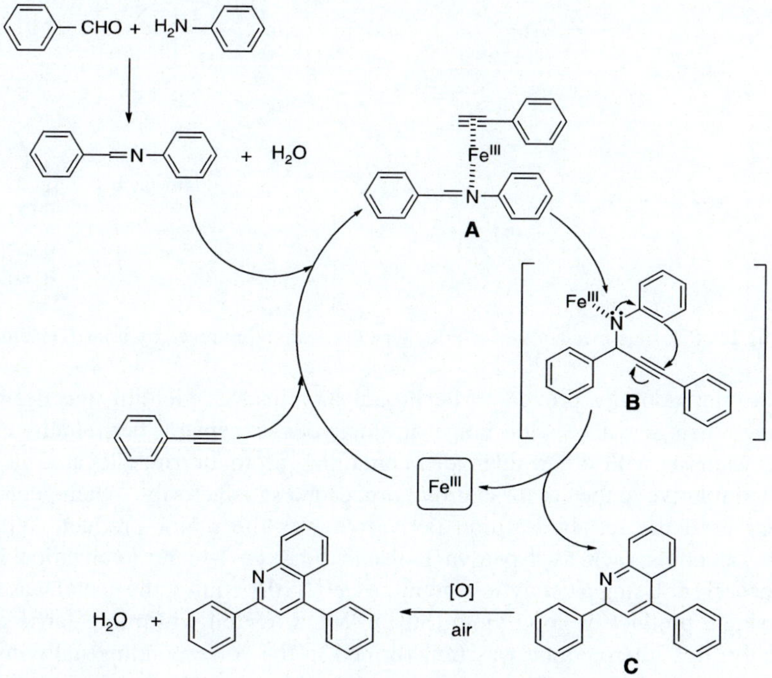

Scheme 20 Mechanism for iron-catalyzed tandem coupling/hydroarylation of aldehyde, aromatic alkyne, and anilines

hydroarylation of alkyne to the dihydroquinoline intermediate **C**. A final oxidation of **C** by O_2 in air affords the quinoline product. Compared to the similar $AuCl_3$-mediated process, the stronger Lewis acidity of $FeCl_3$ than $AuCl_3$ appears to be the main reason for the higher efficiency of the catalytic system.

Use of Oxygenated Acetylenes

In the case of terminal alkynes having oxygenated functions in the linear chain (Scheme 10, route D), Martín, Padrón, and coworkers found that homopropargylic alcohols reacted properly, yielding 2-substituted dihydropyrans as sole products, probably via a Prins-type cyclization. This cyclization provides a new approach toward 2-alkyl-4-halo-5,6-dihydro-2*H*-pyrans through a concomitant C–C and C–O bond formation (Scheme 21) [35].

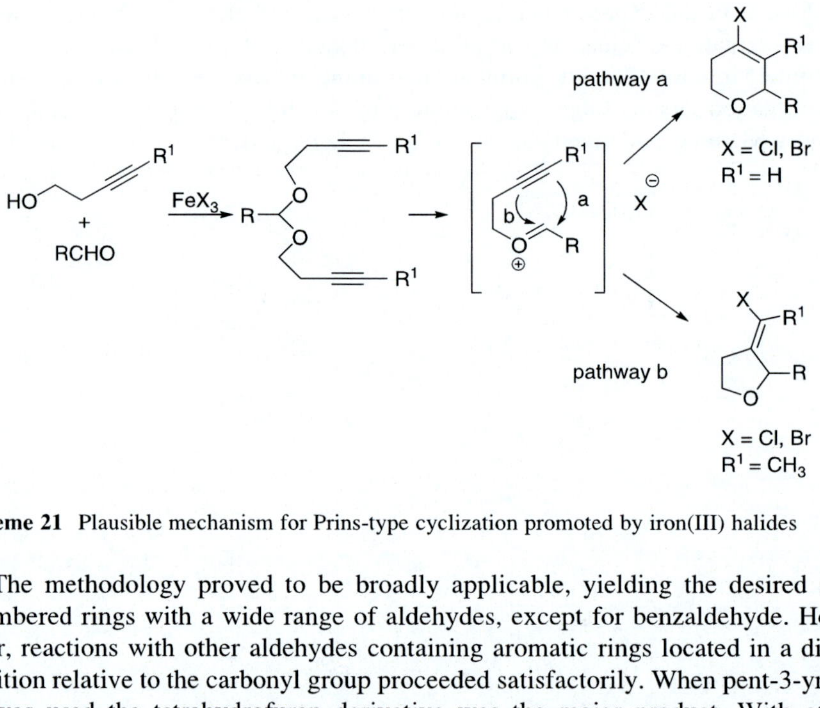

Scheme 21 Plausible mechanism for Prins-type cyclization promoted by iron(III) halides

The methodology proved to be broadly applicable, yielding the desired six-membered rings with a wide range of aldehydes, except for benzaldehyde. However, reactions with other aldehydes containing aromatic rings located in a distal position relative to the carbonyl group proceeded satisfactorily. When pent-3-yn-1-ol was used the tetrahydrofuran derivative was the major product. With other alkyne alcohols such as 4-pentyn-1-ol and 5-hexyn-1-ol no cyclization reaction was observed. Using a catalytic amount of $FeCl_3$ (0.1 equiv), the acetal was isolated as a single product in good yield (80%). When treated with more ferric chloride (1 equiv) this intermediate was transformed to the corresponding halovinyl tetrahydropyran. In light of this evidence, a plausible mechanism involves the acetal that, via FeX_3-mediated ionization, generates the oxonium ion that is intramolecularly trapped by the triple bond with further attack of the halide (Scheme 21).

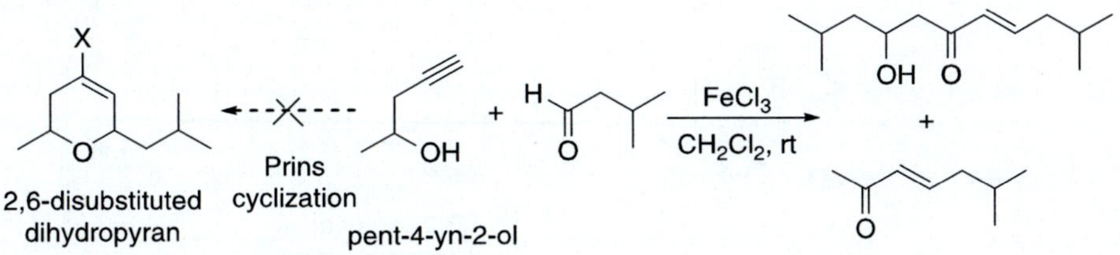

Scheme 22 Coupling of secondary homopropargylic alcohols and aldehydes promoted by iron (III) halides

As an extension of this work, the same authors explored such methodology for the synthesis of 2,6-disubstituted dihydropyrans using secondary homopropargylic alcohols (Scheme 10, route E). Surprisingly, the treatment of pent-4-yn-2-ol and 3-methylbutanal in the presence of $FeCl_3$ led to unsaturated (*E*)-β-hydroxyketone and (*E*)-α,β-unsaturated ketone in 2.5:1 ratio and 65% yield, without any trace of the expected Prins-type cyclic product (Scheme 22) [36]. To test the anion influence in this coupling, $FeCl_3$ and $FeBr_3$ were used in a comparative study for the reaction of pent-4-yn-2-ol ($R^1 = R^4 = H$, $R^2 = Me$) and several aldehydes. A range of aldehydes except for benzaldehyde was transformed into unsaturated β-hydroxy-ketones in moderate to good yields.

A probable mechanistic scenario involves the addition of the secondary homo-propargylic alcohols to the aldehydes promoted by ferric halide, thus generating an oxocarbenium ion that subsequently undergoes a 2-oxonium-[3,3]-sigmatropic rearrangement to an allenolate. Further intramolecular 1,3-oxygen transposition generates an unsaturated enolate. A subsequent coupling reaction with the suitable aldehyde or protonation leads to the unsaturated compounds (Scheme 23).

To support the 2,3-allenol as an intermediate in this reaction, 2,3-allenols were employed in test experiments as starting materials (Scheme 10, route F). The cross-over aldol product was obtained as the sole product, when the reaction was run in the presence of the suitable aldehyde. In the absence of an aldehyde the corresponding (*E*)-α,β-unsaturated ketone was obtained (Scheme 24) [37].

Several ways to suppress the 2-oxonium-[3,3]-rearrangements might be envi-sioned. Apart from the introduction of a bulky substituent R^4 at the aldehyde (Scheme 23) a similar steric repulsion between R^4 and R^3 might also be observed upon introduction of a bulky auxiliary at R^4. A proof-of-principle for this concept was observed upon by using of a trimethylsilyl group as substituent R^3 in the alkyne moiety (Scheme 25, R^3 = TMS). This improvement provided an efficient access to polysubstituted dihydropyrans via a silyl alkyne-Prins cyclization. Ab initio theoretical calculations support the proposed mechanism. Moreover, the use of enantiomerically enriched secondary homopropargylic alcohols yielded the cor-responding oxa-cycles with similar enantiomeric purity [38].

In furtherance of these studies, the reaction scope was broadened by employing homopropargylic amines to give the corresponding aza-cycles (Scheme 26) [39, 40]. Hence, the alkyne aza-Prins cyclization between homopropargyl tosyl amines

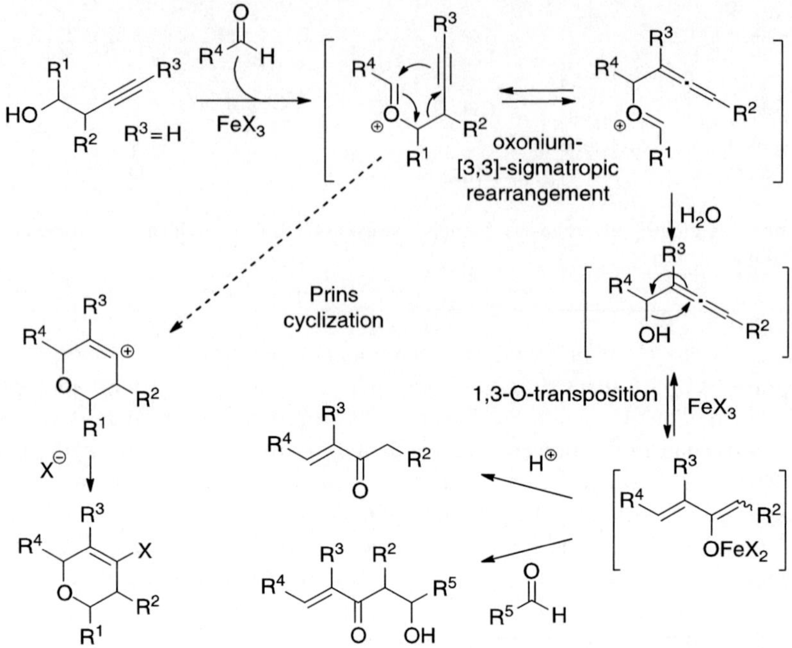

Scheme 23 Proposed mechanism for the addition of secondary homopropargylic alcohols to aldehydes promoted by iron(III) halides

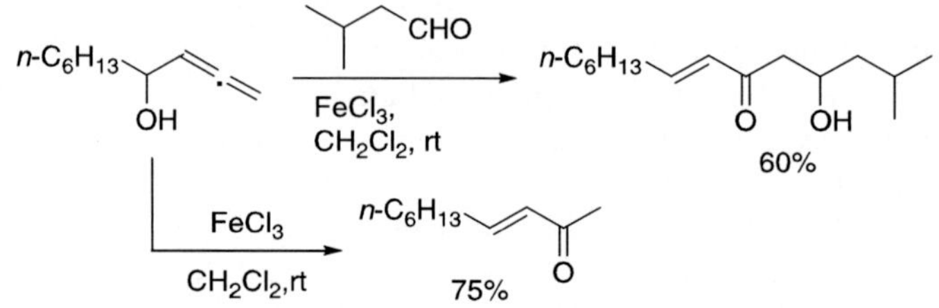

Scheme 24 Coupling of 2,3-allenols and aldehydes promoted by iron(III) halides

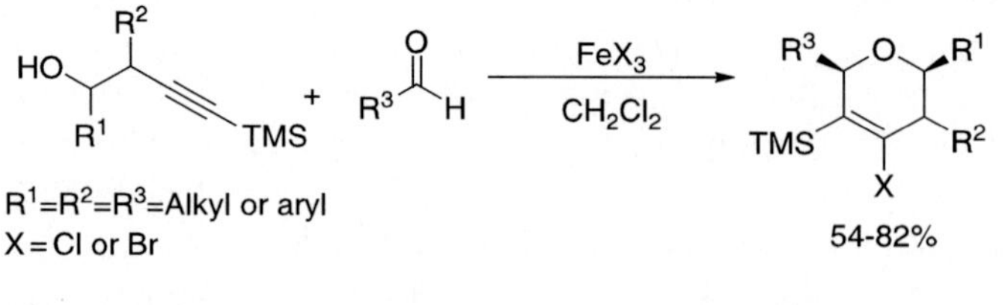

Scheme 25 Silyl alkyne-Prins cyclization of secondary homopropargylic alcohols and aldehydes using FeX_3 as a promoter

Scheme 26 Synthesis of 2-alkyl-4-halo-1-tosyl-1,2,5,6-tetrahydropyridines from *N*-tosyl homopropargyl amine and aldehydes using FeX_3 as promotor

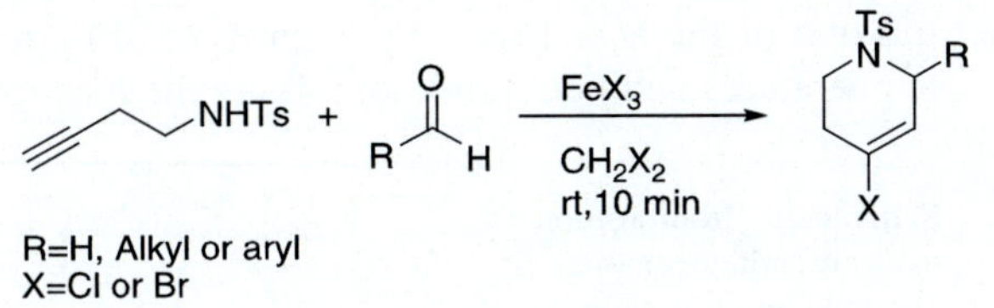

and aldehydes yielded 2-alkyl-4-halo-1-tosyl-1,2,5,6-tetrahydropyridines as sole products [41].

A plausible mechanism for this new alkyne aza-Prins cyclization is outlined in Scheme 27. Thus, reaction of the homopropargyl tosyl amine with an aldehyde promoted by ferric halide generates the *N*-sulfonyl iminium ion. This intermediate evolves to the corresponding piperidine, via the vinyl carbocation. Ab initio theoretical calculations support the proposed mechanism.

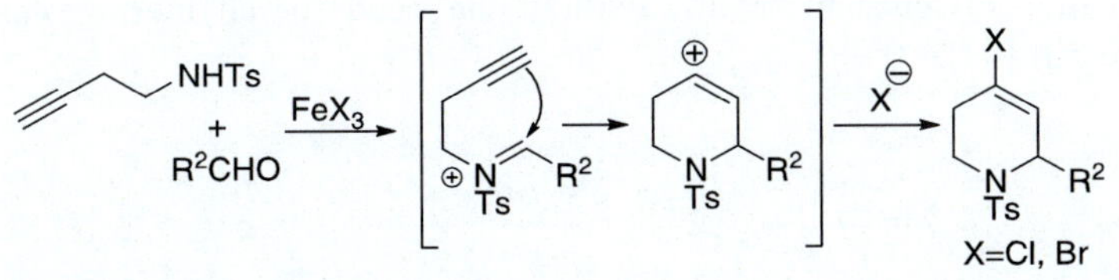

Scheme 27 Plausible mechanism of the alkyne aza-Prins cyclization promoted by iron(III)

In both oxa- and aza-alkyne Prins cyclization an unexpected halide exchange with halogenated solvents presumably caused by the vinyl cation intermediates was observed [37]. From a synthetic point of view, it is important to use the correct combination of FeX_3 and X-containing solvent in order to avoid the undesired halide scrambling (Scheme 28).

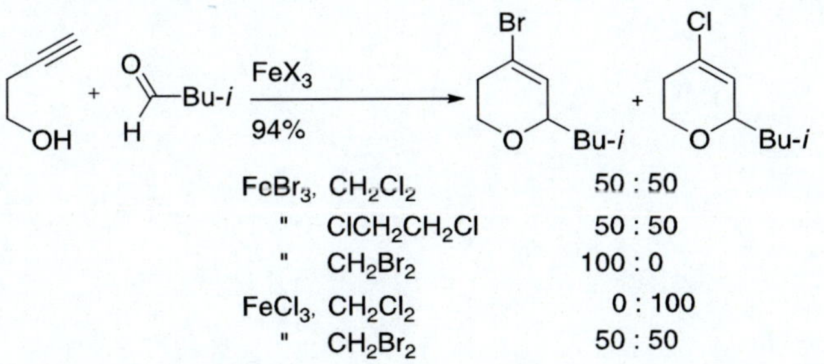

Scheme 28 Halide exchange with halogenated solvents during the acetylenic Prins-type cyclization

Wang et al. reported that $FeCl_3$- and $FeBr_3$-promoted cyclization/halogenation of alkynyl diethyl acetals formed (*E*)-2-(l-halobenzylidene or alkylidene)-substituted five-membered carbo- and heterocycles. It was found that the 1:1 molar ratio reaction of the acetal with $FeCl_3$ proceeded more efficiently at 0°C than at other temperatures and $FeCl_2$ did not initiate the reaction (Scheme 29) [42].

Scheme 29 Iron-promoted cyclization/halogenation of alkynyl diethyl acetals

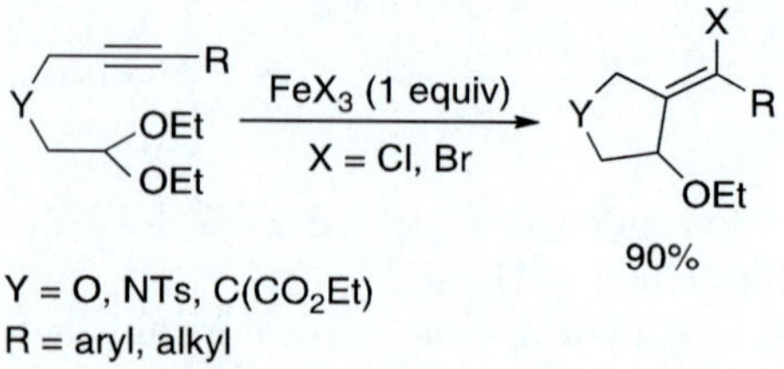

2.2.2 Aliphatic C–C Bond Formation Using Alkenes

Sakakura and coworkers reported the catalytic action of iron(III) triflate for the addition of acetylenes to olefins without the need for an inert gas atmosphere (Scheme 30) [43].

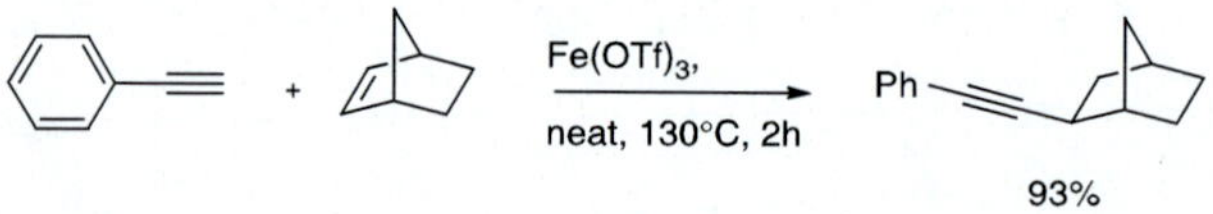

Scheme 30 $Fe(OTf)_3$-catalyzed addition of alkynes to alkenes

The reaction proceeds with isolated double bonds and electron-rich alkynes. Electron-withdrawing groups in the acetylene moiety decelerated the reaction. A plausible mechanism implies the activation of the olefin by coordination of the metal triflate followed by nucleophilic attack of the acetylene or acetylide (Scheme 31).

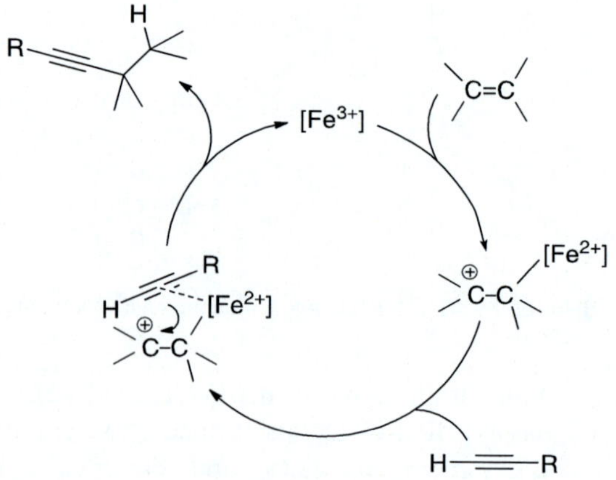

Scheme 31 Possible mechanism of iron(III)-catalyzed alkyne addition to alkenes

FeX_3 was also found to be an excellent promoter in the classical Prins cyclization (Scheme 10, route H), with the observation of a satisfactory reaction between 3-buten-1-ol and several aldehydes, affording the corresponding *cis*-4-halo-2-alkyl tetrahydropyrans in good yields [Eq. (1) in Scheme 32] [35]. In a similar manner, the methodology can be extended to the piperidine synthesis through an aza-Prins cyclization [Eq. (2), Scheme 32] [41].

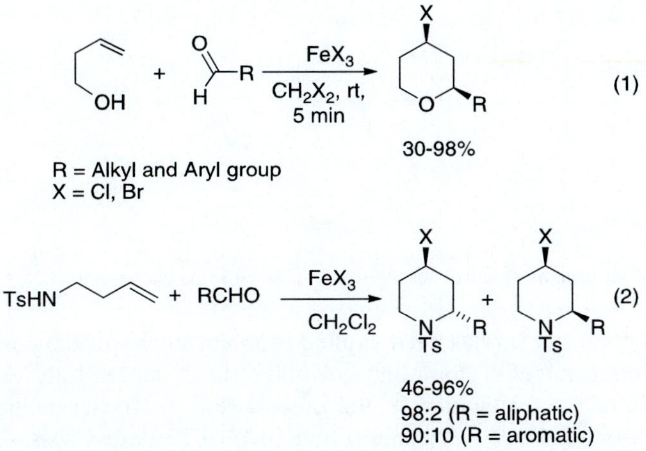

Scheme 32 Prins and aza-Prins cyclization using FeX_3 as a promoter

Whereas the Prins-type cyclizations reported in this and the preceeding chapter were performed using stoichiometric amounts of Fe salts as Lewis acids, a breakthrough in the field of catalysis was reported in 2009 when the first iron-catalyzed Prins- and aza-Prins cyclization was reported. The catalytic system, which is obtained by combining catalytic amounts of an iron salt with trimethylsilyl halides as a halide source, is widely applicable and promotes the construction of substituted six-membered oxa- and aza-cycles (Scheme 33) [44].

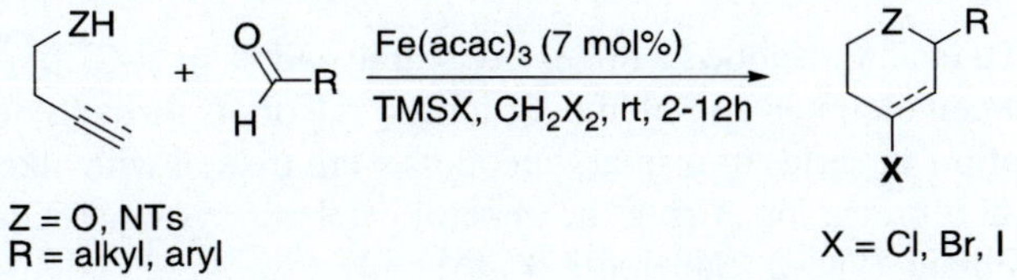

Scheme 33 The catalytic Prins cyclization leading to oxa- and aza-cycles

The catalytic cycle was devised relying on a ligand exchange between the iron complex and a chlorosilane, regenerating the iron(III) halide due to the more oxophilic character of the silicon (Scheme 34).

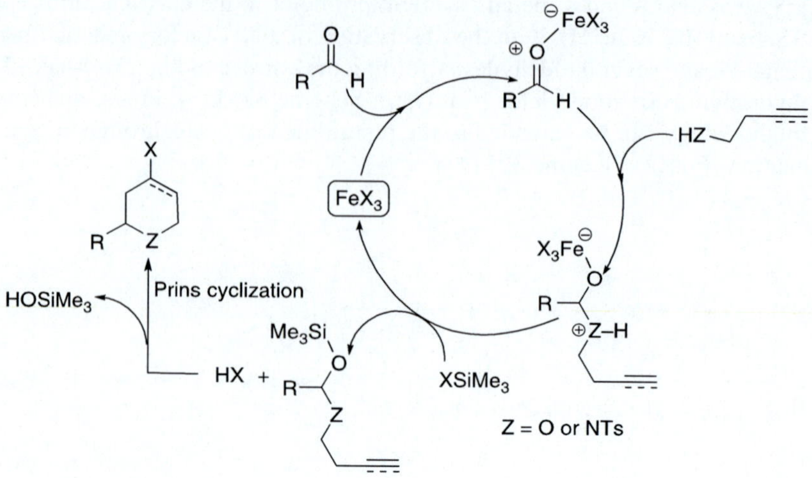

Scheme 34 Plausible mechanism for the iron-catalyzed Prins cyclization

This catalytic system has been applied to the synthesis of the core of complex antibiotic tetrapetalones. Hong and coworkers have successfully established the *N*-acyliminium ion cyclization in the preparation of 1-benzazepine derivatives. A combination of $FeCl_3$ (0.5 equiv) and TMSCl (2 equiv) was selected as an optimized condition for such stereoselective cyclization (Scheme 35) [45].

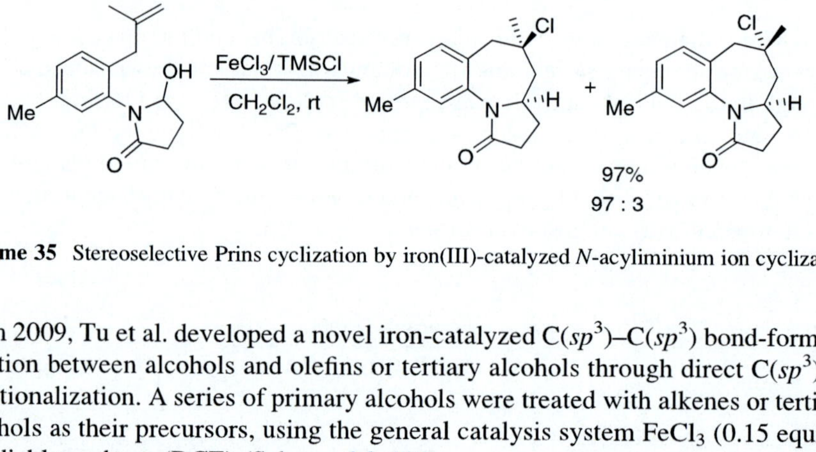

Scheme 35 Stereoselective Prins cyclization by iron(III)-catalyzed *N*-acyliminium ion cyclization

In 2009, Tu et al. developed a novel iron-catalyzed C(sp^3)–C(sp^3) bond-forming reaction between alcohols and olefins or tertiary alcohols through direct C(sp^3)–H functionalization. A series of primary alcohols were treated with alkenes or tertiary alcohols as their precursors, using the general catalysis system $FeCl_3$ (0.15 equiv)/ 1,2-dichloroethane (DCE) (Scheme 36) [46].

According to several deuterium-labeling experiments and previous results, a catalytic mechanism was proposed. Iron-initiated activation for cleavage of the C(sp^3)–H bond adjacent to oxygen of alcohol **a** is involved in the formation of **A**. The key step is the formation of a radical part **B** that evolves to free-radical species C. Subsequently, the metal hydride ($[Fe]^{IV}$–H) undergoes a hydrogen transfer to give the coupling product **c**, regenerating the iron catalyst $[Fe]^{III}$ (Scheme 37).

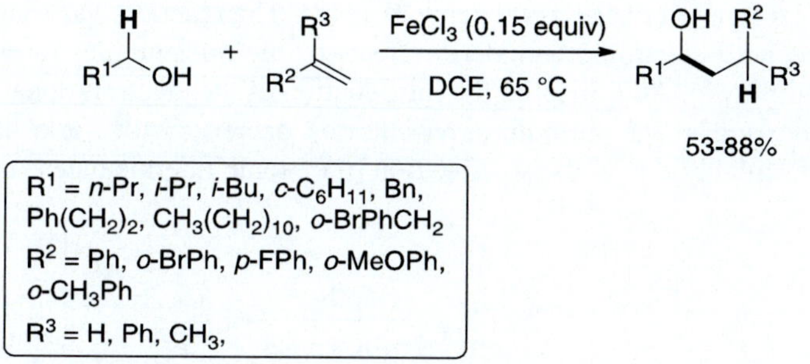

Scheme 36 Iron-catalyzed $C(sp^3)$–$C(sp^3)$ bond formation through $C(sp^3)$–H functionalization

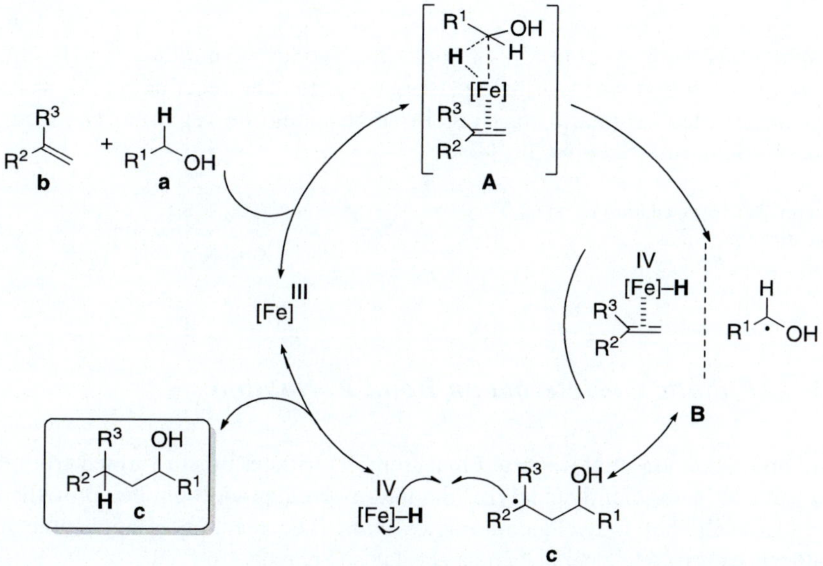

Scheme 37 Proposed mechanism of $C(sp^3)$–$C(sp^3)$ coupling catalyzed by iron(III)

3 C–Heteroatom Bond Formation

3.1 *Aromatic C–Heteroatom Bond Formation*

Iron(III) chloride is a common catalyst used in electrophilic aromatic substitutions. In addition to those applications outlined above for the construction of aromatic C–C bonds, such salts have also been used for the introduction of heteroatom-based functional groups at the aromatic ring [47].

In 2000, Takata et al. reported the synthesis of diarylsulfoximines by the Friedel–Craft reaction of sulfonimidoyl chloride with aromatic compounds in the presence of an equimolar amount of $FeCl_3$ (Scheme 38) [48]. The reaction was found to be very sensitive to both electronic and steric effects of the nucleophiles (aromatic hydrocarbons) employed. Additionally, the introduction of the electron-donating group on the benzene ring in the sulfonimidoyl chloride decreased the yield of the sulfoximines, presumably due to the stabilization of cationic intermediates.

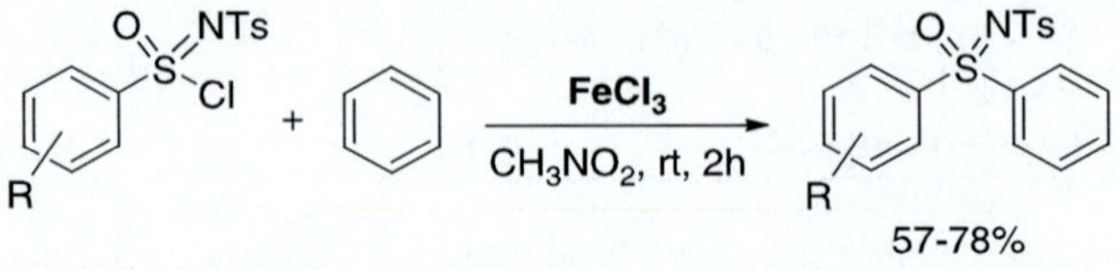

Scheme 38 Synthesis of diarylsulfoximines by Friedel–Crafts reactions

When used with *N*-chloro-, *N*-bromo-, and *N*-iodosuccinimide, iron(III) chloride catalyzes the introduction of halogens into arenes. The reaction works well even with deactivated aromatic rings but in some cases the regioselective course is difficult to control (Scheme 39) [49].

Scheme 39 Halogenation of aromatic compounds catalyzed by $FeCl_3$

3.2 *Aliphatic C–Heteroatom Bond Formation*

Choi and Sakakura et al. reported that iron(III) triflate, in situ formed from $FeCl_3$ and triflic acid, efficiently catalyzes the intermolecular addition of carboxylic acids to various alkenes to yield carboxylic esters. The reaction is applicable to the synthesis of unstable esters, such as acrylates (Scheme 40) [50].

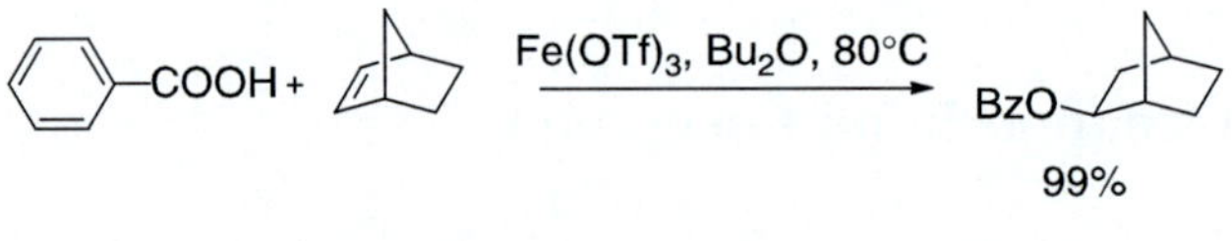

Scheme 40 Iron(III)-catalyzed intermolecular addition of carboxylic acids to alkenes

4 Rearrangements

Yadav's group reported that ketones undergo smooth rearrangement with $TMSN_3$ in the presence of catalytic amounts of $FeCl_3$ under mild conditions, to provide the corresponding amides, imides, and lactams in good yields with high selectivity. The azido-Schmidt reaction involves the addition of azide to the ketone followed by rearrangement and ring expansion. Similar to the classical Schmidt reaction, the mechanism most likely involves a Lewis-acid activation of the carbonyl group followed by the addition of azide. The resulting iminodiazonium ions, which undergo ring expansion by loss of nitrogen and subsequent migration of the anti-substituent to the electron-deficient nitrogen, give the corresponding amide (Scheme 41) [51].

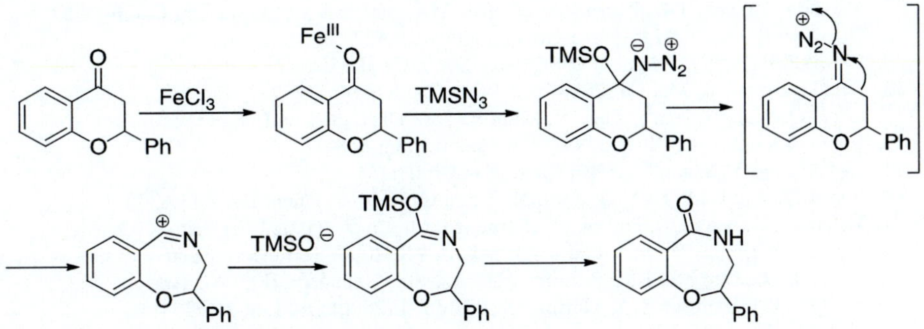

Scheme 41 Mechanism of the azido-Schmidt reaction catalyzed by $FeCl_3$

Acknowledgment We gratefully acknowledge financial support by the Spanish MICIIN (CTQ2008-06806-C02-01/BQU) and Canary Islands' ACIISI (PI 2007/022).

References

1. Pearson RG (1963) J Am Chem Soc 85:3533
2. Parr RG, Pearson RG (1983) J Am Chem Soc 105:7512
3. Corma A, García H (2003) Chem Rev 103:11
4. Alfarra A, Frackowiak E, Béguin F (2004) Appl Surf Sci 228:84
5. Pearson RG (2005) J Chem Sci 117:369
6. Aoshima S, Kanoka S (2009) Chem Rev 109:5245
7. Olah GA, Kobayashi S, Tashiro M (1972) J Am Chem Soc 94:7448
8. Kobayashi S, Busujima T, Nagayama S (2000) Chem Eur J 6:3491
9. Baaz M, Gutmann V (1963) In: Olah GA (ed) Friedel-Crafts and related reactions, chapter 5, vol 1. NewYork, Interscience
10. Zhang Y (1982) Inorg Chem 21:3889
11. Bolm C, Legros J, Le Paih J, Zani L (2004) Chem Rev 104:6217
12. Díaz DD, Miranda PO, Padrón JI, Martín VS (2006) Curr Org Chem 10:457
13. Sarhanw AAO, Bolm C (2009) Chem Soc Rev 38:2730
14. Watahiki T, Oriyama T (2002) Tetrahedron Lett 43:8959
15. Fürstner A, Leitner A (2002) Angew Chem Int Ed 41:609
16. Fürstner A, Leitner A, Méndez M, Krause H (2002) J Am Chem Soc 124:13856
17. Plietker B (ed) (2008) Iron catalysis in organic chemistry: reactions and applications. Wiley-VCH, Weinheim
18. Seddon KR (1997) J Chem Technol Biotechnol 68:351

19. Csihony S, Mehdi H, Horváth IT (2001) Green Chem 3:307
20. Jana U, Maiti S, Biswas S (2007) Tetrahedron Lett 47:7160
21. Wang Z, Sun X, Wu J (2008) Tetrahedron Lett 64:5013
22. Womack GB, Angeles JG, Fanelli VE, Heyer CA (2007) J Org Chem 72:7046
23. Womack GB, Angeles JG, Fanelli VE, Indradas B, Snowden RL, Sonnay P (2009) J Org Chem 74:5738
24. Huang W, Shen Q, Wang J, Zhou X (2008) J Org Chem 73:1586
25. Huang W, Hong L, Zheng P, Liu R, Zhou X (2009) Tetrahedron Lett 65:3603
26. Marcos R, Rodríguez-Escrich C, Herrerías CI, Pericàs MA (2008) J Am Chem Soc 130:16838
27. Miranda PO, Díaz DD, Padrón JI, Ramírez MA, Martín VS (2005) J Org Chem 70:57
28. Damiano JP, Postel M (1996) J Organomet Chem 522:303
29. Wu X-F, Bezier D, Darcel C (2009) Adv Synth Catal 351:367
30. Jana U, Biswas S, Maiti S (2008) Eur J Org Chem 5798
31. Liu Z-Q, Wang J, Han J, Zhao Y, Zhou B (2009) Tetrahedron Lett 50:1240
32. Xiang S-K, Zhang L-H, Jiao N (2009) ChemComm 6487
33. Ki P, Zhang Y, Wang L (2045) Chem Eur J 2009:15
34. Cao K, Zhang F-M, Tu Y-Q, Zhuo X-T, Fan C-A (2009) Chem Eur J 15:6332
35. Miranda PO, Díaz DD, Padrón JI, Bermejo J, Martín VS (1979) Org Lett 2003:5
36. Miranda PO, Ramírez MA, Padrón JI, Martín VS (2006) Tetrahedron Lett 47:283
37. Miranda PO, Carballo RM, Ramírez MA, Martín VS, Padrón JI (2007) Arkivoc iv:331
38. Miranda PO, Ramírez MA, Martín VS, Padrón JI (2006) Org Lett 8:1633
39. Mukai C, Sugimoto Y-I, Miyazawa K, Yamaguchi S, Hanaoka M (1998) J Org Chem 63:6281
40. Davis FA, Song M, Augustine AJ (2006) J Org Chem 71:2779
41. Carballo RM, Ramírez MA, Rodríguez ML, Martín VS, Padrón JI (2006) Org Lett 8:3837
42. Xu T, Yu Z, Wang L (2009) Org Lett 11:2113
43. Kohno K, Nakagawa K, Yahagi T, Choi J-C, Yasuda H, Sakakura T (2009) J Am Chem Soc 131:2784
44. Miranda PO, Carballo RM, Martín VS, Padrón JI (2009) Org Lett 11:357
45. Li C, Li X, Hong R (2009) Org Lett 11:4036
46. Zhang SY, Tu Y-Q, Fan C-A, Zhang F-M, Shi L (2009) Angew Chem Int Ed 48:8761
47. Bolm C, Legros J, Le Paih J, Le Zani L (2004) Chem Rev 104:6217
48. Furusho Y, Okada Y, Okada Y, Takata T (2000) Bull Chem Soc Jpn 73:2827
49. Tanemura K, Suzuki T, Nishida Y, Satsumabayashi K, Horaguchi T (2003) Chem Lett 32:932
50. Choi J-C, Kohno K, Masuda D, Yasuda H, Sakakura T (2008) Chem Commun 777
51. Yadav JS, Reddy BVS, Reddy UVS, Praneeth K (2008) Tetrahedron Lett 49:4742

Top Organomet Chem (2011) 33: 27–81
DOI: 10.1007/978-3-642-14670-1_2

Fe–H Complexes in Catalysis

Hiroshi Nakazawa and Masumi Itazaki

Abstract Organic syntheses catalyzed by iron complexes have attracted considerable attention because iron is an abundant, inexpensive, and environmentally benign metal. It has been documented that various iron hydride complexes play important roles in catalytic cycles such as hydrogenation, hydrosilylation, hydroboration, hydrogen generation, and element–element bond formation. This chapter summarizes the recent developments, mainly from 2000 to 2009, of iron catalysts involving hydride ligand(s) and the role of Fe–H species in catalytic cycles.

Keywords Catalysis · Electrochemical reduction · Hydroboration · Hydrogenation · Hydrosilylation · Iron hydride complex · Photochemical reduction

Contents

H. Nakazawa (✉) and M. Itazaki
Department of Chemistry, Graduate School of Science, Osaka City University, 3-3-138 Sugimoto, Sumiyoshi-ku, Osaka 558-8585, Japan
e-mail: nakazawa@sci.osaka-cu.ac.jp

1 Introduction

This chapter treats iron complexes with Fe–H bond(s). An H ligand on a transition metal is named in two ways, "hydride" and "hydrido." The term "hydrido" is recommended to be used for hydrogen coordinating to all elements by IUPAC recommendations 2005 [1]. However, in this chapter, the term "hydride" is used because it has been widely accepted and used in many scientific reports.

In 1931, Hieber and Leutert reported $Fe(CO)_4(H)_2$ not only as the first iron hydride complex but also as the first transition-metal hydride complex (FeH_2 was reported in 1929 from $FeCl_2$ and PhMgBr under a hydrogen atmosphere. However, it exists only in a gas phase) [2, 3]. The complex synthesized from $Fe(CO)_5$ and OH^- (Scheme 1) is isolable only at low temperature and decomposes at room temperature into $Fe(CO)_5$, $Fe(CO)_3$, and H_2.

Scheme 1 The first iron hydride complex

$$Fe(CO)_5 + 2OH^- \longrightarrow Fe(CO)_4(H)_2 + CO_3^{2-}$$

Since then, many iron hydride complexes have been prepared, isolated, characterized spectroscopically and, in some cases, by X-ray analyses. Hydrides show diagnostic signals in ^1H NMR spectra. They resonate to high field in a region (0 ~ –60 ppm) for diamagnetic iron hydride complexes. Paramagnetic hydride complexes, however, are very difficult to characterize. IR spectra are helpful because Fe–H stretching frequencies are observed in the range of 1,500–2,200 cm^{-1}. However, it should be noted that the intensities are often weak and hence the method is not entirely reliable. Crystallographic methods are generally powerful to obtain 3D arrangements and metrical data. However, hydrides, in some cases, are not detected because the hydride in the neighborhood of the metal is such a poor scatterer of X-rays. In addition, Fe–H bond distances should be carefully used because X-ray methods often underestimate the true Fe–H internuclear distance by approximately 0.1 Å. The X-ray structure of $[NEt_3H][HFe_3(CO)_{11}]$ having μ-H ligand was reported in 1965 [4]. This compound might be the first structurally characterized iron complex bearing an H ligand (Fig. 1). The first terminal hydride iron complex characterized by the X-ray analysis was $CpFeH(CO)(SiCl_3)_2$, in 1970 [5]. In 1972, the X-ray structure of *cis*-$[Fe(H)_2\{PPh(OEt)_2\}_4]$ was reported as the first dihydride iron complex [6] (Fig. 1).

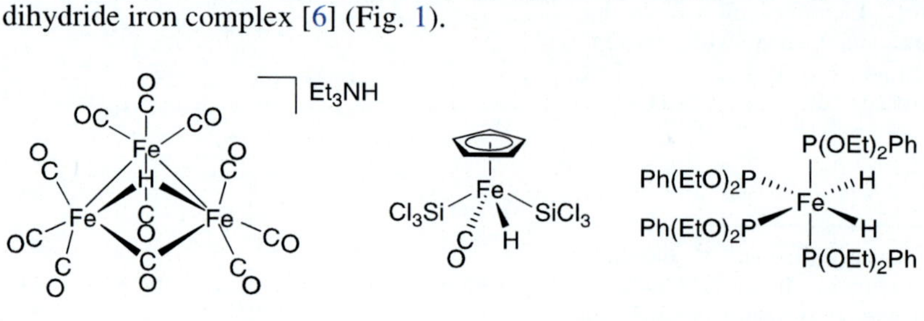

Fig. 1 The first X-ray crystal structures of three types of the iron hydride complexes

Iron hydride complexes can be synthesized by many routes. Some typical methods are listed in Scheme 2. Protonation of an anionic iron complex or substitution of hydride for one electron donor ligands, such as halides, affords hydride complexes. $NaBH_4$ and $LiAlH_4$ are generally used as the hydride source for the latter transformation. Oxidative addition of H_2 and E–H to a low valent and unsaturated iron complex gives a hydride complex. Furthermore, β-hydride abstraction from an alkyl iron complex affords a hydride complex with olefin coordination. The last two reactions are frequently involved in catalytic cycles.

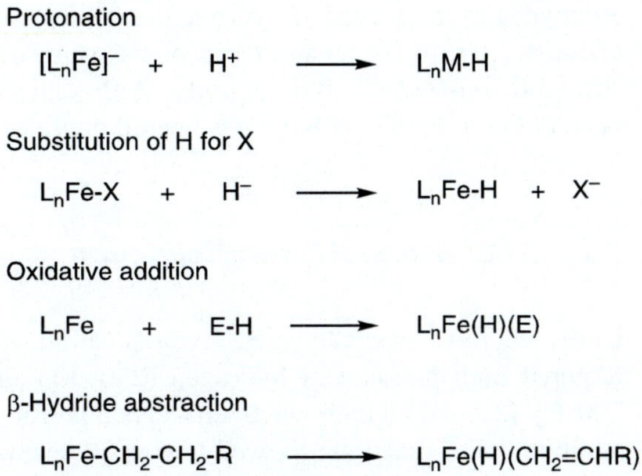

Scheme 2 Some typical synthetic routes of iron hydride complexes

A Fe–H bond is generally polarized as $Fe^{\delta+}$–$H^{\delta-}$ because H is more electronegative than Fe. However, iron hydride complexes impart much less negative charge to the hydride than early transition-metal hydride complexes.

Organic syntheses catalyzed by complexes of precious-metals (Pd, Pt, Rh, Ir, Ru, etc.) have attracted considerable attention because both catalytic activities and chemo- and region selectivities are superior to those of heterogeneous catalysts. In order to understand the underlying principles of homogenous catalysis, the reaction mechanisms are under intense investigation. For a large number of mechanisms, it has been well-established that various hydride complexes act as important intermediates in the catalytic cycle for hydorometalations such as, e.g., hydrosilylations or in hydrogenation reactions [7]. Furthermore, iron hydride complexes were shown to play pivotal roles in fundamental organometallic chemistry [8–11].

Precious metals have faced a significant price increase and the fear of depletion. By contrast, iron is a highly abundant metal in the crust of the earth (4.7 wt%) of low toxicity and price. Thus, it can be defined as an environmentally friendly material. Therefore, iron complexes have been studied intensively as an alternative for precious-metal catalysts within recent years (for reviews of iron-catalyzed organic reactions, see [12–20]). The chemistry of iron complexes continues to expand rapidly because these catalysts play indispensable roles in today's academic study as well as chemical industry.

This chapter summarizes the development of iron catalysts involving hydride ligand(s) and the role of the Fe–H species in the catalytic cycle in the past decade (2000–2009).

2 Hydrogenation

Reduction of unsaturated organic substrates such as alkenes, alkynes, ketones, and aldehydes by molecular dihydrogen or other H-sources is an important process in chemistry. In hydrogenation processes some iron complexes have been demonstrated to possess catalytic activity. Although catalytic intermediates have rarely been defined, the Fe–H bond has been thought to be involved in key intermediates.

2.1 *Alkene and Alkyne Reduction*

Until recently, iron-catalyzed hydrogenation reactions of alkenes and alkynes required high pressure of hydrogen (250–300 atm) and high temperature (around 200°C) [21–23], which were unacceptable for industrial processes [24, 25]. In addition, these reactions showed low or no chemoselectivity presumably due to the harsh reaction conditions. Therefore, modifications of the iron catalysts were desired.

In 2004, Peters and Daida reported the hydrogenation of olefins catalyzed by iron complexes **1–4** having the tris(diisopropylphosphino)borate ligand, $[PhBP^{iPr}{}_3]^-$, under mild conditions (low pressure and ambient temperature) (Table 1) [26]. Although these catalytic activities were not high, a plausible reaction mechanism has been proposed (Scheme 3). The iron(IV) trihydride phosphine complex **a** being spectroscopically well-defined undergoes hydrogen loss to give the coordinatively unsaturated mono hydride complex **d**. Two pathways from **a** to **d** are conceivable: (1) the trihydride complex **b**, which is a detectable species by NMR, and (2) the monohydride phosphine complex **c**, which can be structurally identified when $PR_3 = PMePh_2$. Olefin coordination to **d** and its insertion into the Fe–H bond affords the stable alkyl complex **e**. This iron alkyl complex reacts with hydrogen to release the alkane with regeneration of the active species **d**.

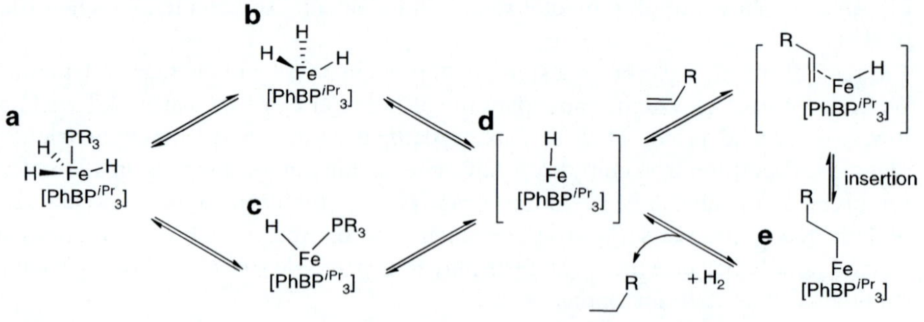

Scheme 3 Plausible pathway for hydrogenation of olefin catalyzed by the Fe–H complex

Table 1 Olefin hydrogenation reactions catalyzed by iron complexes

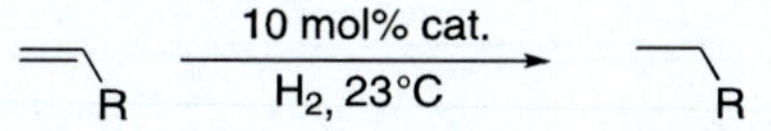

Catalyst	Substrate	H_2 (atm)	Time(min)	TOF(mol/h)
1	Styrene	4	78	7.7
2	Styrene	4	100	6.0
2	Styrene	1	410	1.5
2	Ethylene	4	25	24.0
2	1-Hexene	1	130	4.6
1	1-Hexene	1	115	5.2
2	Cyclooctene	1	55	10.9
2	2-Pentyne	1	370	1.6
3[a]	Styrene	1	1260	0.5
4	Styrene	1	1010	0.6

[a]Hydrogenation carried out at 50°C.

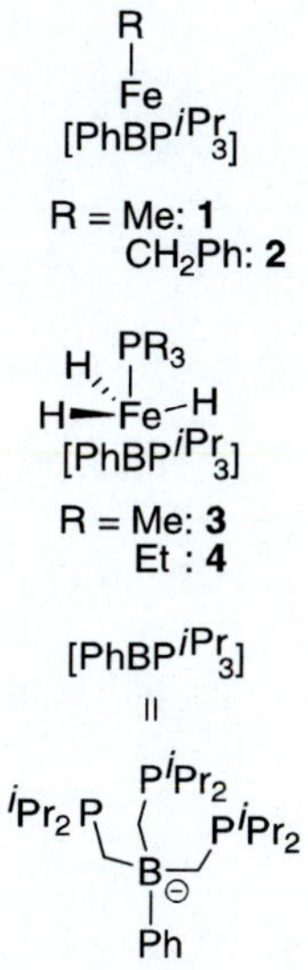

Bis(imino)pyridine iron complex **5** as a highly efficient catalyst for a hydrogenation reaction was synthesized by Chirik and coworkers in 2004 [27]. Complex **5** looks like a Fe(0) complex, but detailed investigations into the electronic structure of **5** by metrical data, Mössbauer parameters, infrared and NMR spectroscopy, and DFT calculations established the Fe(II) complex described as **5'** in Fig. 2 to be the higher populated species [28].

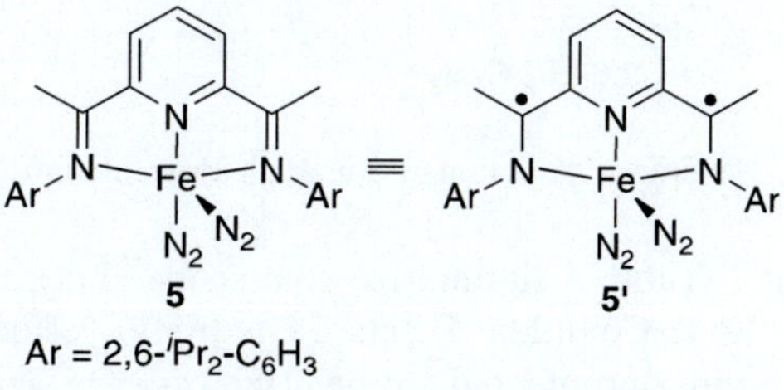

Fig. 2 Bis(imino)pyridine iron complex

Complex **5** was more active than the well-known precious-metal catalysts (palladium on activated carbon Pd/C, the Wilkinson catalyst $RhCl(PPh_3)_3$, and Crabtree's catalyst $[Ir(cod)(PCy_3)py]PF_6$) and the analogous *N*-coordinated Fe complexes **6–8** [29] for the hydrogenation of 1-hexene (Table 2). In mechanistic studies, the NMR data revealed that **5** was converted into the dihydrogen complex **9** via the monodinitrogen complex under hydrogen atmosphere (Scheme 4).

Table 2 Comparison of iron complexes with transition precious-metal catalysts for the hydrogenation of 1-hexene

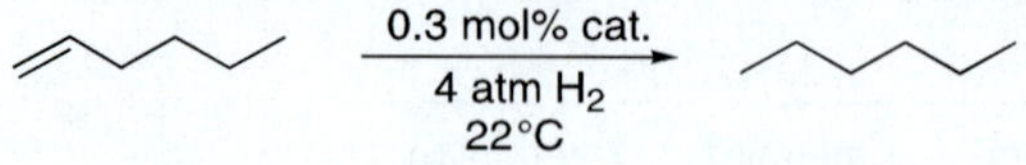

Catalyst	Time (min)	TOF (mol/h)	Ref.
5	12	1814	[27]
10% Pd/C	12	366	[27]
$RhCl(PPh_3)_3$	12	10	[27]
$[Ir(cod)(PCy_3)py]PF_6$	12	75	[27]
6[a]	1440	4	[29]
7	240	90	[29]
8	240	90	[29]

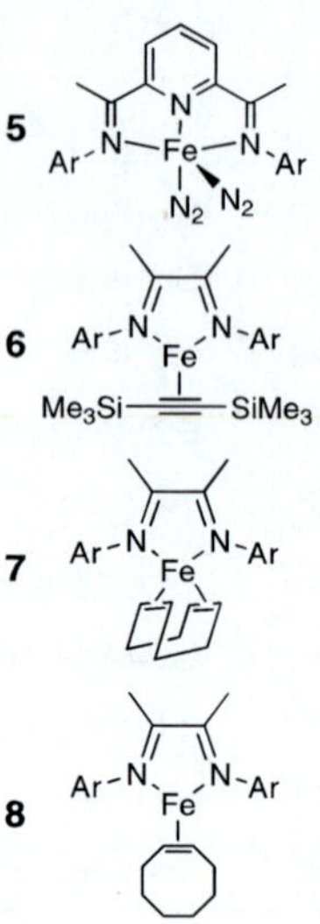

Ar = 2,6-^{i}Pr$_2$-C$_6$H$_3$

[a]Hydrogenation carried out at 60°C.

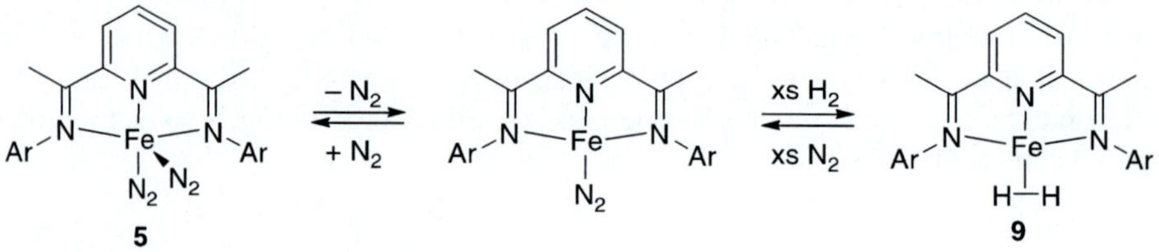

Ar = 2,6-^{i}Pr$_2$-C$_6$H$_3$

Scheme 4 Reaction of **5** under hydrogen atmosphere

Table 3 summarizes the scope and limitation of substrates for this hydrogenation. Complex **5** acts as a highly effective catalyst for functionalized olefins with unprotected amines (the order of activity: tertiary > secondary ≫ primary), ethers, esters, fluorinated aryl groups, and others [27, 30]. However, in contrast to the reduction of α,β-unsaturated esters decomposition of **5** was observed when α,β-unsaturated ketones (e.g., *trans*-chalcone, *trans*-4-hexen-3-one, *trans*-4-phenyl-3-buten-2-one, 2-cyclohexanone, carvone) were used (Fig. 3) [30].

With internal alkynes such as diphenylacetylene and 2-butyne, the perhydrogenated products were formed in the presence of complex **5** via the corresponding *cis*-2-alkene as an intermediate (Scheme 5). In case of terminal alkynes such as trimethylsilylacetylene, the desired product was not formed.

Table 3 Catalytic hydrogenation of various olefins with **5**

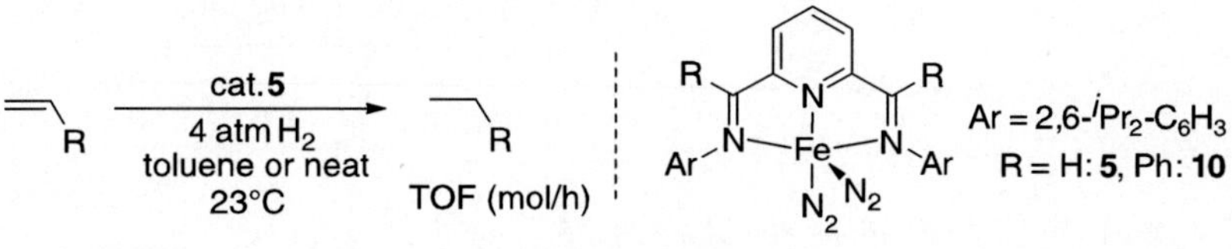

Substrate

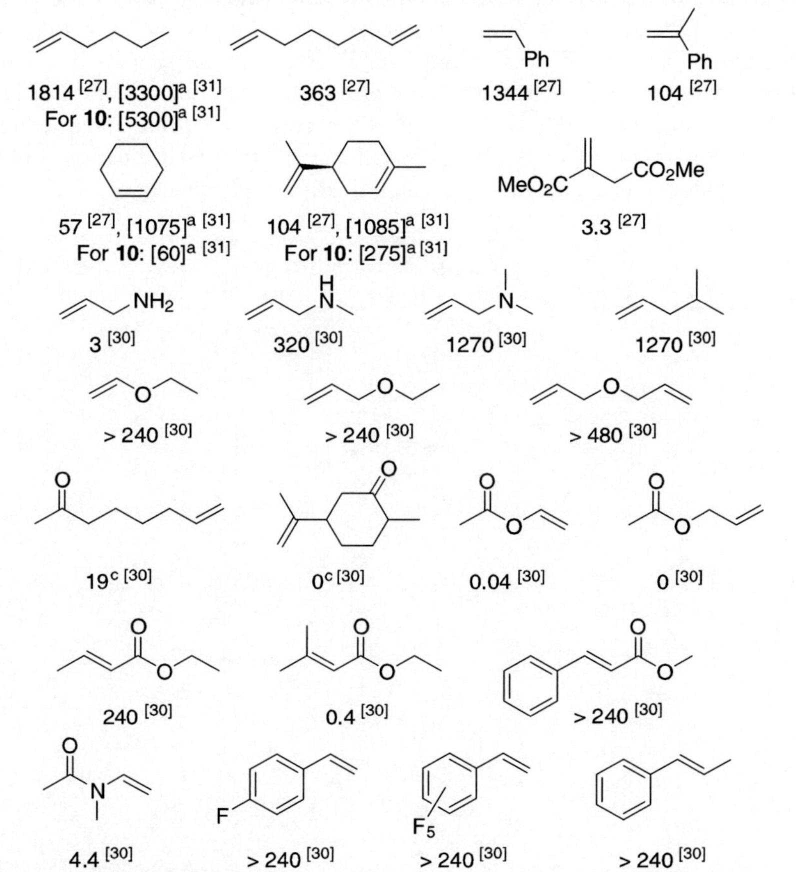

aIn brackets the results in pentane.
bThe product is (+)-*p*-meth-1-ene.
cReaction carried out at 65 °C

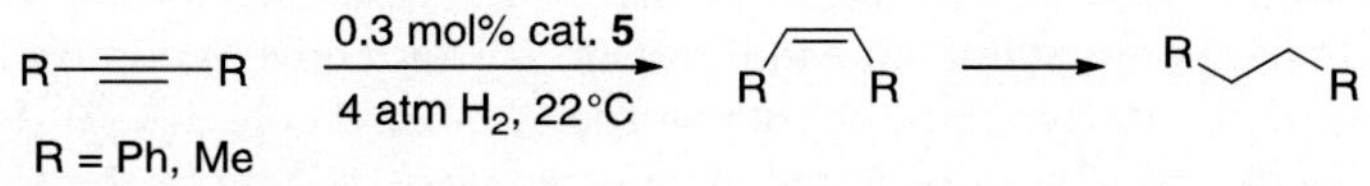

Scheme 5 Overreduction of alkynes

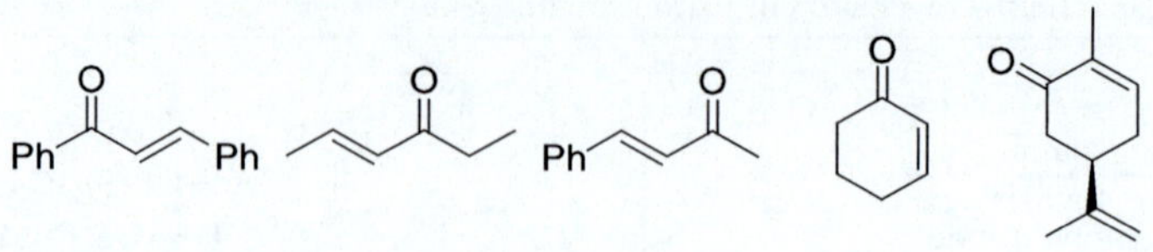

Fig. 3 Substrates leading to decomposition of catalyst **5**

The proposed catalytic cycle, which is based on experimental data, is shown in Scheme 6. Loss of 2 equiv. of N_2 from **5** (or alternatively 1 equiv. of N_2 or 1 equiv. of H_2 from complexes shown in Scheme 3) affords the active species **a**. Olefin coordination giving **b** is considered to be preferred over oxidative addition of H_2. Then, oxidative addition of H_2 to **b** provides the olefin dihydride intermediate **c**. Olefin insertion giving **d** and subsequent alkane reductive elimination yields the saturated product and regenerates the catalytically active species **a**.

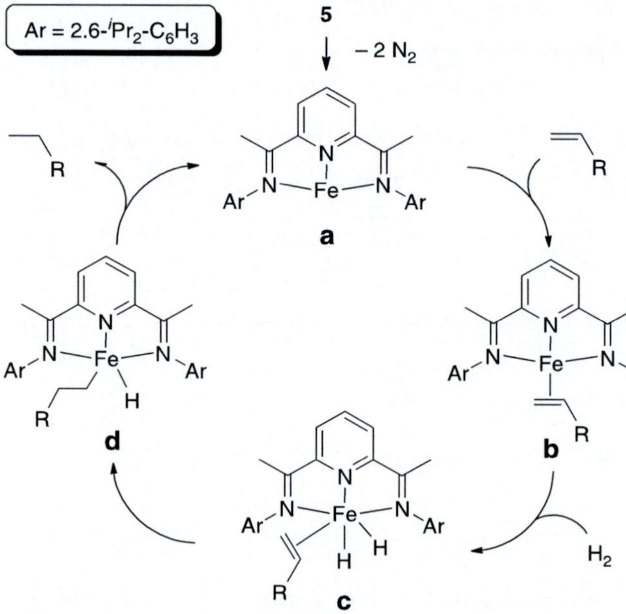

Scheme 6 Proposed mechanism for catalytic hydrogenation of olefin with **5**

Further investigations revealed that this hydrogenation is accelerated in pentane solution. These results are shown in brackets in Table 3 [31]. Under optimized reaction conditions high catalyst TOF up to 5,300 were achieved when **10** was used. In the absence of both hydrogen and nitrogen, **10** was converted into the η^6-arene complexes such as the bis(imino)pyridine iron η^6-phenyl complex, **10-Phenyl**, and the corresponding η^6-2,6-diisopropylphenyl complex, **10-Aryl**, in the 85:15 ratio in

C_6D_6 solution (Scheme 7). Both η^6-arene complexes were also determined by the X-ray diffraction and showed no reaction to hydrogen and olefins. Therefore, it was considered that the formation of the η^6-arene complexes was a deactivation pathway in the catalytic hydrogenation.

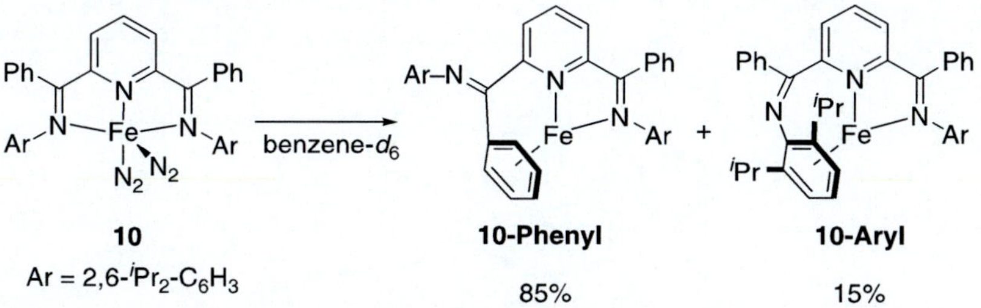

Scheme 7 Intramolecular arene coordination in bis(imino)pyridine iron complex **10**

2.2 *Ketone and Aldehyde Reduction*

Hydrogenation of substrates having a polar multiple C–heteroatom bond such as ketones or aldehydes has attracted significant attention because the alcohols obtained by this hydrogenation are important building blocks. Usually ruthenium, rhodium, and iridium catalysts are used in these reactions [32–36]. Nowadays, it is expected that an iron catalyst is becoming an alternative material to these precious-metal catalysts.

For transition-metal-catalyzed hydrogenation of ketones and aldehydes, H_2 or the combination of ^{i}PrOH with a base has been widely used as the hydrogen source (Scheme 8). In case of using H_2, the reaction is called "hydrogenation," whereas the reaction using the combination of ^{i}PrOH with a base is especially called "transfer hydrogenation."

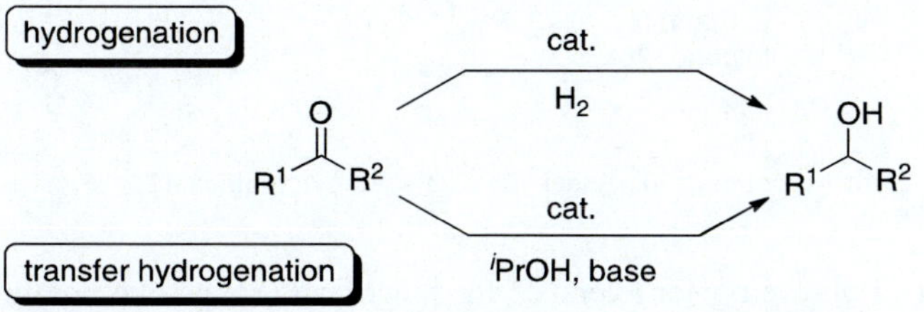

Scheme 8 Two pathways of hydrogenation reaction of ketones and aldehydes

With some transition-metal complexes, the ligand is not only an ancillary ligand. Similar to the transition-metal, it takes directly part in the hydrogen transfer process. Such ligand-metal bifunctional hydrogenation catalysis is dramatically changing the face of reduction chemistry (Scheme 9) (for reviews of ligand-metal bifunctional catalysis, see [32, 37–40]).

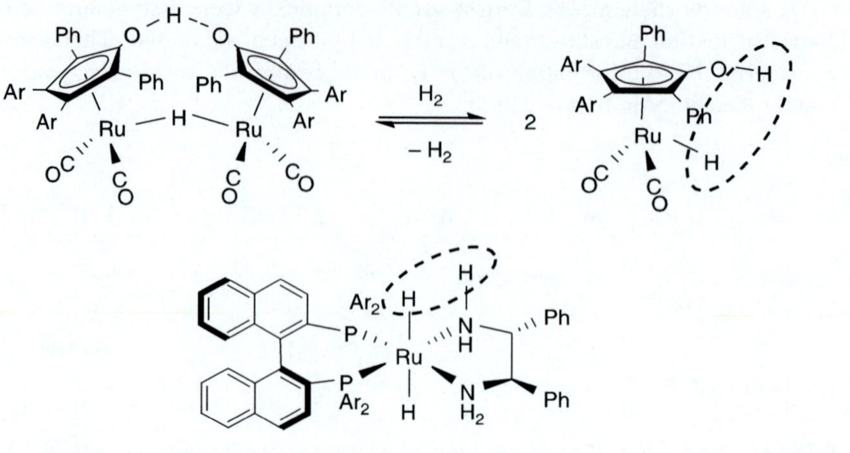

Scheme 9 Examples of ligand-metal bifunctional catalysts

These transition-metal catalysts contain electronically coupled hydridic and acidic hydrogen atoms that are transferred to a polar unsaturated species under mild conditions. The first such catalyst was Shvo's diruthenium hydride complex reported in the mid 1980s [41–44]. Noyori and Ikariya developed chiral ruthenium catalysts showing excellent enantioselectivity in the hydrogenation of ketones [45, 46].

In 2007, Casey showed that **11**, which corresponds to the Shvo's complex catalyzes hydrogenations of ketones and aldehydes [47]. Reaction of benzaldehyde in the presence of catalytic amount of **11** under H_2 (3 atm) afforded the corresponding benzylalcohol in 90% yield within 1 h (Scheme 10).

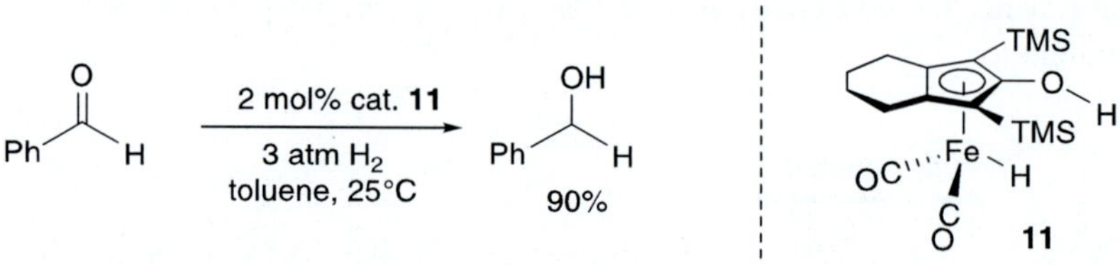

Scheme 10 Hydrogenation of benzaldehyde catalyzed by complex **11**

Catalyst **11** showed great tolerance for functional groups like unsaturated carbon–carbon bonds (alkenyl and alkynyl moieties), halides, electron-withdrawing and -donating groups, etc. (Table 4).

Although hydrogenation of *N*-benzylideneaniline in the presence of **11** afforded the corresponding product (eq. 1 in Scheme 11), the α,β-unsaturated ketone was converted into a mixture of unsaturated and saturated alcohols in the 42:56 ratio (eq. 2 in Scheme 11). Several substrates (nitrile derivatives, epoxides, esters, internal alkynes, and terminal alkenes), which are shown in Fig. 4, are not hydrogenated in this catalytic system.

Table 4 Iron complex-catalyzed hydrogenation of ketones

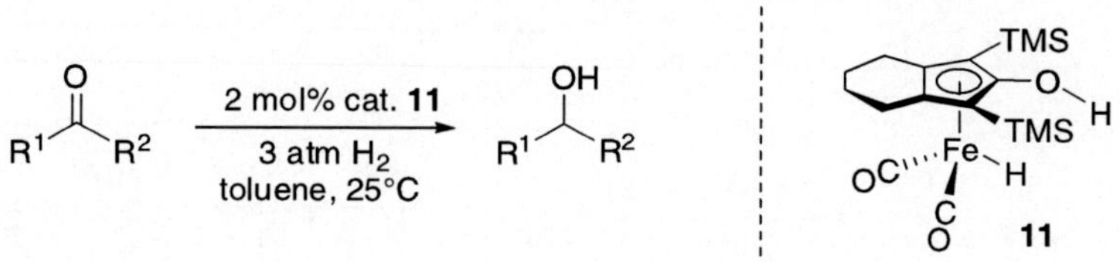

Substrate	Product	Yield[a]
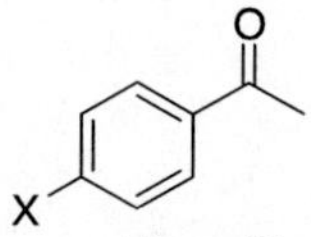	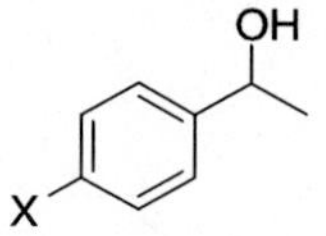	
X = *p*-H		83% (99%)
p-Br		91% (99%)
p-I		84% (99%)
p-NO_2		89% (99%)
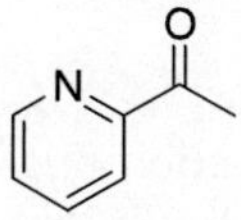	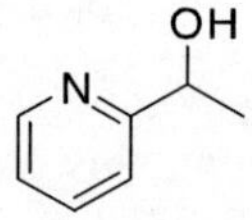	87% (100%)
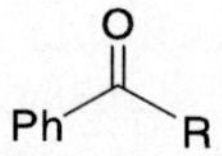	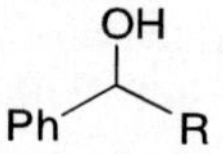	
R = Ph		55% (69%)
CF_3		91% (100%)
^{c}propyl		46% (50%)
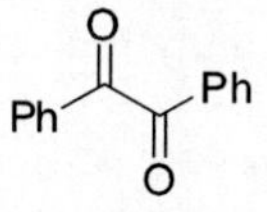	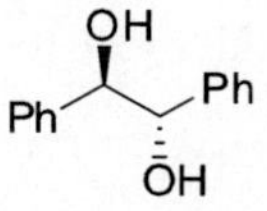	86% (100%) (*meso*/*dl* = 25)
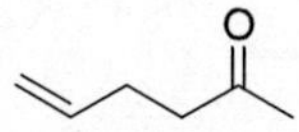	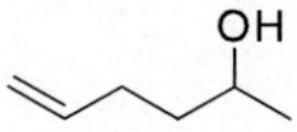	87% (100%)[b]
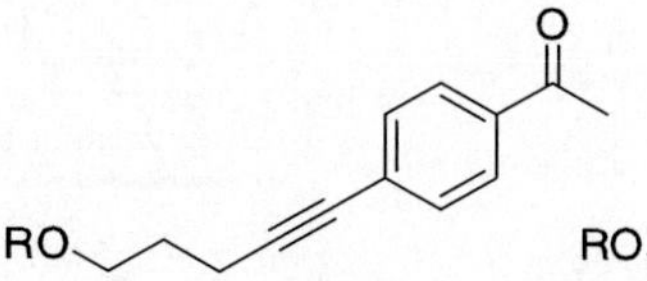	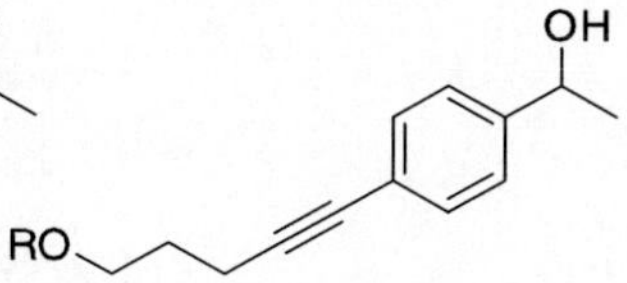	
R = H		57% (71%)
CH_2Ph		84% (87%)

[a]Isolated yield (NMR conversion in paretheses).
[b]Reaction was performed in diethylether.

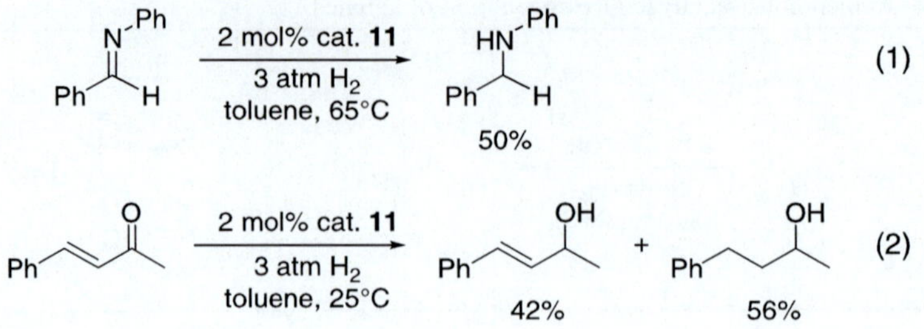

Scheme 11 Hydrogenation of *N*-benzylideneaniline and α,β-unsaturated ketone catalyzed by complex **11**

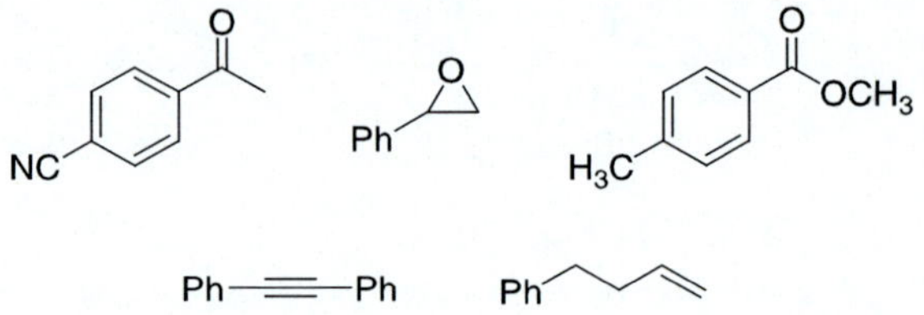

Fig. 4 Unreacted substrates for hydrogenation catalyzed by **11**

From a mechanistic point of view, this hydrogenation reaction might follow different mechanistic pathways (Scheme 12) [48]. One possibility is that intermediate **a**, which is generated by the transfer of acidic and hydridic hydrogen atoms of **11** to the C=O bond in aldehyde, reacts with H_2 to give dihydrogen complex **c** and then to give the corresponding alcohol and **11**. The second one is that alcohol complex **b** is formed as the initial product and is subsequently converted into **a** which reacts with H_2 to **c**. The subsequent pathway is similar. Finally, it is also possible that the full dissociation of the alcohol from **b** affords intermediate **a'** and then the addition of H_2 to **a'** reproduces **11** to complete the catalytic cycle.

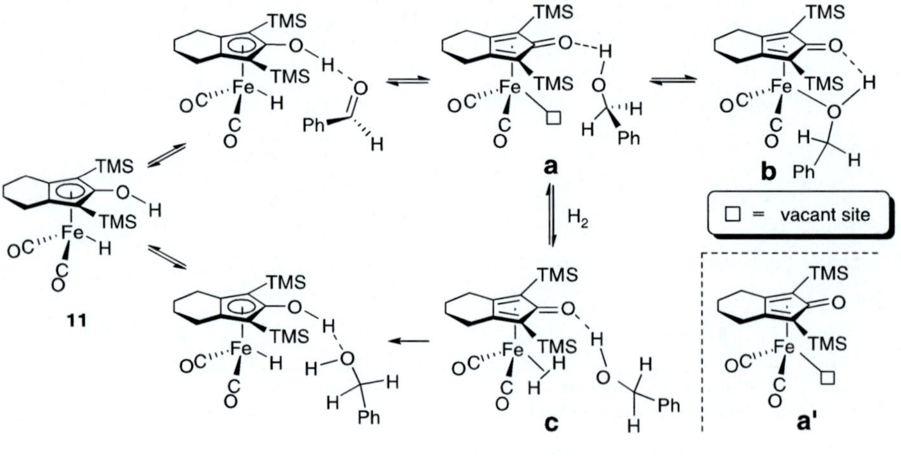

Scheme 12 Proposed catalytic cycle for Casey's hydrogenation

An iron complex-catalyzed enantioselective hydrogenation was achieved by Morris and coworkers in 2008 (Scheme 13) [49]. Reaction of acetophenone under moderate hydrogen pressure at 50°C catalyzed iron complex **12** containing a tetradentate diiminodiphosphine ligand in the presence of ^{t}BuOK afforded 1-phenylethanol with 40% conversion and 27% *ee*.

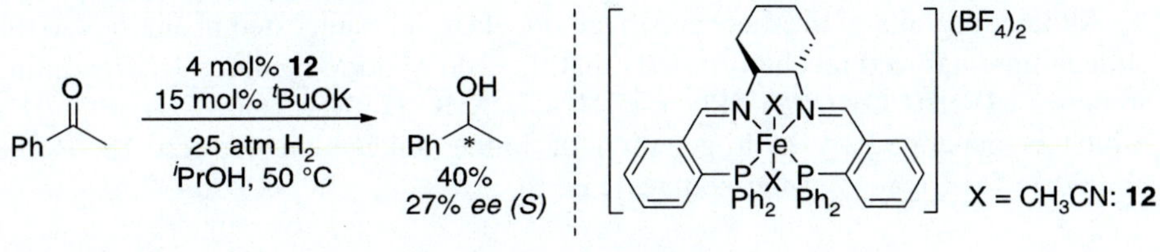

Scheme 13 Morris'asymmetric hydrogenation catalyzed by iron complex **12**

In addition, the related complexes **13** and **14** act as catalysts in enantioselective transfer hydrogenations (Table 5). The reactivity of acetophenone derivatives

Table 5 Enantioselective transfer hydrogenation catalyzed by **13** and **14**

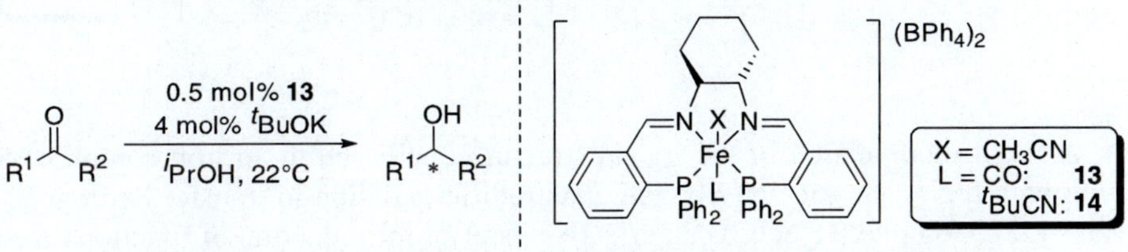

Substrate	Time (h)	Conv. (%)	*ee* (*S*) (%)	TOF (mol/h)
Ph-CO-Me[a]	0.4	95	29	907
Ph-CO-Me[b]	0.7	33	39	93
Ph-CO-Me	0.4	95	33	454
(2-Cl-C_6H_4)-CO-Ph	0.2	>99	18	995
(3-Cl-C_6H_4)-CO-Ph	0.4	99	24	495
(4-Cl-C_6H_4)-CO-Ph	0.2	94	26	938
(4-Br-C_6H_4)-CO-Ph	0.2	93	33	930
(4-Me-C_6H_4)-CO-Ph	0.6	86	33	279
(4-OMe-C_6H_4)-CO-Ph	0.5	69	23	260
Ph-CO-Et	3.6	95	61	26
$C_{10}H_7$-CO-Me	0.3	94	25	564
Ph$(CH_2)_2$-CO-Me	0.6	100	29	315
Ph-CO-Me[c]	2.6	34	76	28

[a] 0.25 mol% of **13** and 2 mol% of ^{t}BuOK were used
[b] 1 mol% of ^{t}BuOK was used
[c] Reaction catalyzed by **14**

having a ortho-halide-substituted aromatic ring is higher. Interestingly, this tendency is opposite to that for the ruthenium-catalyzed transfer hydrogenation reaction reported by Noyori [50]. Catalyst **13** is also efficient for transfer hydrogenations employing diphenylketone, benzaldehyde, or *N*-benzylideneaniline but the *ee* values were not observed. The best enantioselectivity (76% *ee*) was achieved by **14** with acetophenone.

Although details of the mechanism are not clear, it is suggested in analogy to the ruthenium-catalyzed mechanism [51] that the iron hydride complex having amine moieties $[FeH(CO)\{(R,R)\text{-}PPh_2(o\text{-}C_6H_4)CH_2NHC_6H_{10}NHCH_2(o\text{-}C_6H_4)PPh_2\}]^+$, which is generated by hydrogenation of imine linkage on catalyst **13**, is the candidate for a reactive intermediate (Fig. 5).

Fig. 5 Proposed catalytic intermediate

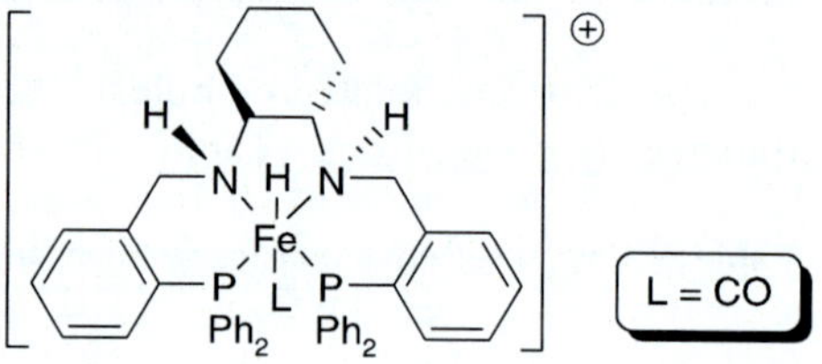

A slight modification of the ligand structure finally led to an iron complex **15** with both high catalytic activity and enantiodifferentiation in transfer hydrogenations [52]. Complex **15** acts as an effective catalyst for reduction of functionalized acetophenones containing alkyl groups, cyclic alkanes, chloro- or methoxy-groups on the aryl ring, and others (Table 6). Although the reaction rate is quite low up to 99%, the *ee* value is achieved with 2,2-dimethylpropiophenone. The highest TOF is observed by using 2-aceto-naphtone as substrate. In case of *N*-benzylideneaniline, the *ee* value is not observed.

In 2006, Beller and coworkers demonstrated two Fe-catalyzed transfer hydrogenations of aromatic and aliphatic ketones in the presence of ^{i}PrONa as a base and ^{i}PrOH as hydrogen source (Table 7). One is a biomimetic iron porphyrin system [53]. The highest TOF (642) was achieved by in situ generation of the catalyst from $Fe_3(CO)_{12}$ and *p*-Cl–C_6H_4-substituted porphyrin ligand **L3**. The other catalyst is an iron complex derived from the same iron-source upon combination of terpy and PPh_3 as ligands in a 1:1 ratio [54]. Other ratios of terpy and PPh_3, steric hindered phosphine ligands or bidentate phosphine ligands such as $Ph_2P(CH_2)_nPPh_2$ (n = 1, 2, 4, and 6), and analogous substituted-terpy were not effective. In both protocols, the base plays an important role for the catalytic activity. ^{i}PrONa and ^{t}BuONa were efficient, but other inorganic bases, for instance, LiOH and KOH as well as *N*-coordinative organic bases such as pyridine and NEt_3 showed low activities. In the absence of a base, a transfer hydrogenation was not observed.

Table 6 Fe-catalyzed transfer hydrogenation of acetophenone

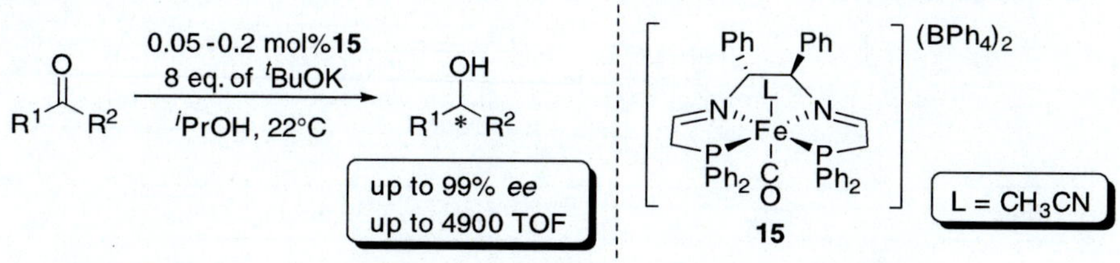

Substrate	Time (min)	Conv. (%)	*ee* (%)	TOF (mol/h)
Ph-CO-Me	30	90	82	3600
Ph-CO-Me[a]	30	75	84	3000
Ph-CO-Me[b]	30	80	83	3200
Ph-CO-Et	25	90	94	3375
Ph-CO-^{t}Bu	200	35	99	53
Ph-CO-(*cyclo*-C_4H_7)	40	95	94	1425
Ph-CO-(*cyclo*-C_6H_{11})	85	76	26	536
Ph$(CH_2)_2$-CO-Me	30	98	14	1960
(4-Cl-C_6H_4)-CO-Me	18	96	80	4800
(3-Cl-C_6H_4)-CO-Me	13	98	80	4523
(4-MeO-C_6H_4)-CO-Me	40	65	54	930
(3-MeO-C_6H_4)-CO-Me	30	80	85	2400
^{i}Pr-CO-Me	60	86	50	1280
1-Aceto-naphthone	60	93	92	1380
2-Aceto-naphthone	11	90	84	4900
Ph-CH=N-Ph	240	41	—	100

[a] ^{t}BuONa was used as a base.
[b] KOH was used as a base.

The scope and limitations for transfer hydrogenation employing either the iron porphyrin system or the combination of iron compound/terpy/PPh_3 are listed in Table 8. In most cases, the $FeCl_2$/terpy/PPh_3 system displays a higher activity. Except for chloromethyl- and cyclopropyl-acetophenone, the desired products were obtained in good to excellent yields. It should be noted that a ring opened product was not observed when cyclopropyl acetophenone was employed. Hence, a radical-type reduction pathway was excluded and a hydride mechanism appeared to be reasonable.

In mechanistic studies, monodeuterated alcohols were obtained by using ^{i}PrOD (Scheme 14). These results indicate that the intermediate for this transfer hydrogenation was not a dihydride complex but rather a monohydride complex, which was generally accepted by analogous transition-metal-catalyzed reactions [55–57].

Table 7 Fe-catalyzed transfer hydrogenation of acetophenone

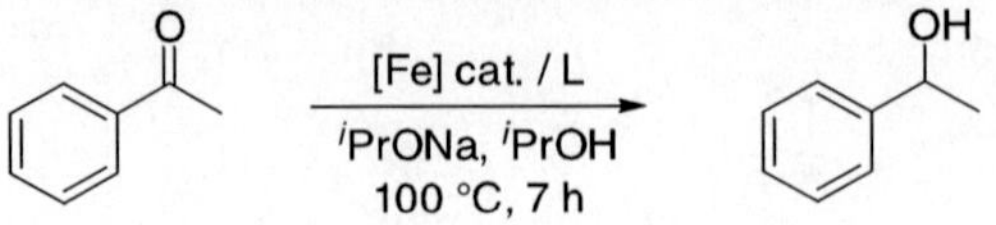

[Fe]	L		Yield(%)[a]	TOF (mol/h)[b]	Ref.
$Fe_3(CO)_{12}$	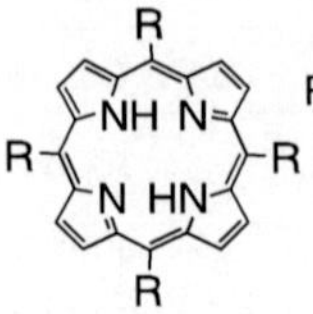 R = Ph:	**L1**	93	27	[53]
	4-pyridyl:	**L2**	94	27	[53]
	p-Cl-C_6H_4:	**L3**	90	26	[53]
		L3	45[c]	642[c]	[53]
$Fe_3(CO)_{12}$	terpy / PPh_3		95		[54]
$FeCl_2$	terpy / PPh_3		91		[54]

[a]Yield was determined by GC.
[b]Turnover frequencies were determined after 7 h.
[c]0.01 mol% of catalyst, which was the one-fiftieth of other iron porphyrin systems, was used.

Table 8 Fe-catalyzed transfer hydrogenation of various ketones[a]

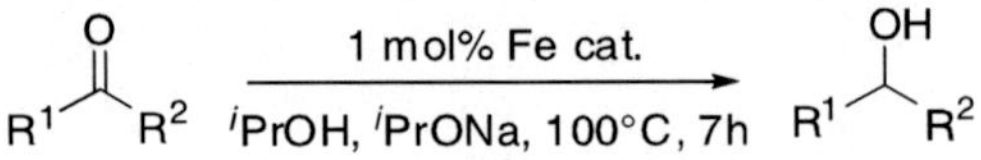

Product		$Fe_3(CO)_{12}$		$Fe_3(CO)_{12}$	$FeCl_2$
	L =	**L1**	**L2**	terpy / PPh_3	
		Yield (%)	Yield (%)	Yield (%)	Yield (%)
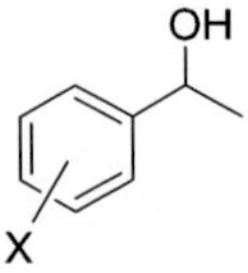	X = *p*-Cl	93	95	>99	97
	p-Me	50	68	84	83
	p-OMe	46	72	63	75
	o-OMe	>99	>99	>99	>99
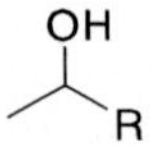	R = Cy	26 [71][b]	89	95	93
	^tBu	11 [55][b]	90	>99	>99
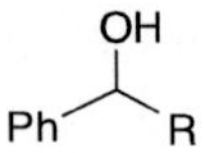	R = Et	92	87	81	92
	^{c}propyl	21	22	48	57
	CH_2Cl	<1	<1	5	8

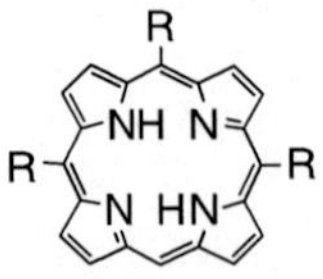

R = Ph: **L1**
4-pyridyl: **L2**

terpy =

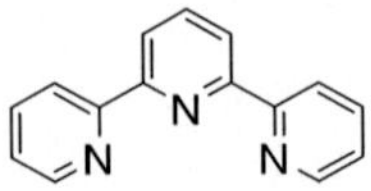

[a]Yield was determined by GC.
[b]In brackets the results after 24 h reaction time.

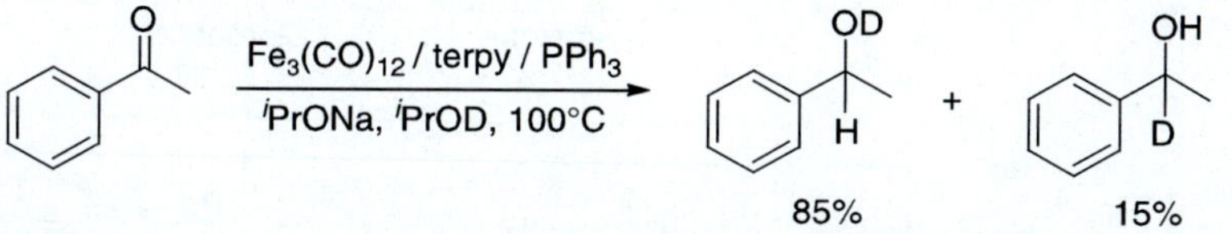

Scheme 14 Deuterium incorporation into acetophenone catalyzed by $[Fe_3(CO)_{12}]$/terpy/PPh_3/^iPrONa/^iPrOD system

The above-described reverse reaction (viz. the Fe-catalyzed dehydrogenation of alcohols to ketones/aldehydes) has been reported by Williams in 2009 (Table 9) [58]. In this reaction, the bicyclic complex **16** shows a sluggish activity, whereas the dehydrogenation of 1-(4-methoxyphenyl)ethanol catalyzed by the phenylated complex **17** affords the corresponding ketone in 79% yield when 1 equiv. (relative to **17**) of D_2O as an additive was used. For this oxidation reaction, 1-(4-methoxyphenyl) ethanol is more suitable than 1-phenylethanol and the reaction rate and the yield of product are higher.

Table 9 Fe-catalyzed dehydrogen conversion of alcohol to ketone[a]

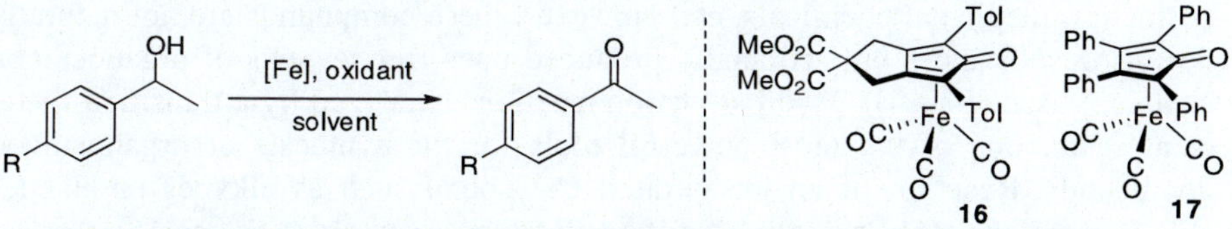

[Fe]	R	Oxidant	Solvent	T (°C)	Yield%,	Time
16 (0.1 equive.)	H	BQ	C_6D_6	65	1	16 h
17 (0.1 equive.)	H	BQ	C_6D_6	65	11	17 h
17 (0.1 equive.)	H	BQ	C_6D_6	65	24	4 d
17 (0.1 equive.)	OMe	–	$(CD_3)_2CO$	54	38	4 d
17 (0.2 equive.)[b]	OMe	–	$(CD_3)_2CO$	80	52	4 d
17 (0.5 equive.)[b]	OMe	–	$(CD_3)_2CO$	80	79	2 d

[a]Yield was determined by NMR spectroscopy. BQ = benzoquinone.
[b]Experiments were run with 1 equiv. of D_2O.

The proposed catalytic cycle for the dehydrogenation of alcohols to ketones is shown in Scheme 15. The initial reaction of **17** with H_2O affords the hydride complex **a** and CO_2. Dehydrogenation of **a** by acetone gives the active species **b** and 2-propanol. The subsequent reaction of **b** with the alcohol yields the corresponding ketone and regenerates **a** to complete the catalytic cycle.

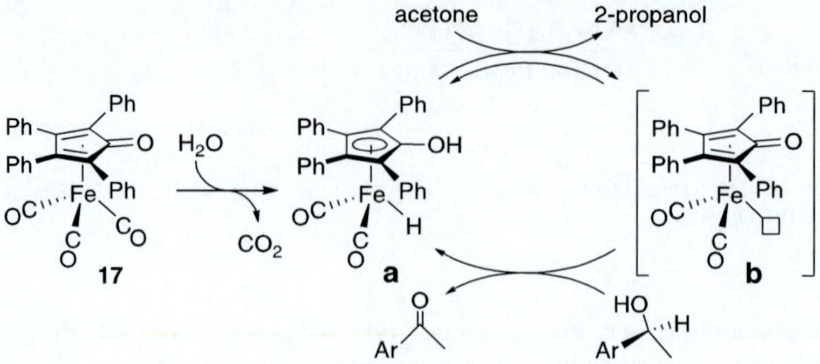

Scheme 15 Proposed catalytic cycle for dehydrogenation of alcohols to ketones

3 Hydrometalation

3.1 Hydrosilylation

Organosilicon compounds are widely used in our daily life as oil, grease, rubbers, cosmetics, medicinal chemicals, etc. However, these compounds are not naturally occurring substances but artificially produced ones (for reviews of organosilicon chemistry, see [59–64]). Hydrosilylation reactions catalyzed by a transition-metal catalyst are one of the most powerful tools for the synthesis of organosilicon compounds. Reaction of an unsaturated C–C bond such as alkynes or alkenes with hydrosilane affords a vinyl- or alkylsilane, respectively (Scheme 16).

Scheme 16 Hydrosilylation of C–C-multiple bonds

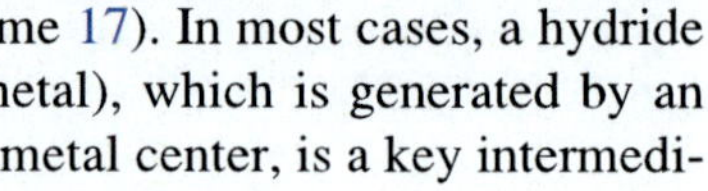

Employing ketones or aldehydes as starting materials, the corresponding silylethers are obtained. Thereafter, the oxidation or hydrolysis of the obtained silylethers gives the corresponding alcohols (Scheme 17). In most cases, a hydride (silyl) metal complex H–M–Si (M = transition-metal), which is generated by an oxidative addition of H–Si bond to the low-valent metal center, is a key intermediate in the hydrosilylation reaction.

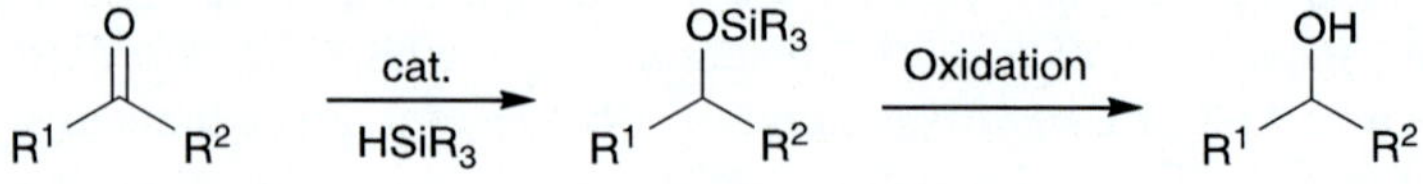

Scheme 17 Hydrosilylation of carbonyl groups

Bis(imino)pyridine iron complex **5** acts as a catalyst not only for hydrogenation (see **2.1**) but also for hydrosilylation of multiple bonds [27]. The results are summarized in Table 10. The reaction rate for hydrosilylations is slower than that for the corresponding hydrogenation; however, the trend of reaction rates is similar in each reaction. In case of *trans*-2-hexene, the terminal addition product hexyl (phenyl)silane was obtained predominantly. This result clearly shows that an isomerization reaction takes place and the subsequent hydrosilylation reaction delivers the corresponding product. Reaction of 1-hexene with H_2SiPh_2 also produced the hydrosilylated product in this system (eq. 1 in Scheme 18). However, the reaction rate for H_2SiPh_2 was slower than that for H_3SiPh. In addition, reaction of diphenylacetylene as an alkyne with phenylsilane afforded the monoaddition product due to steric repulsion (eq. 2 in Scheme 18).

Table 10 Olefine hydrosilylation catalyzed by **5**

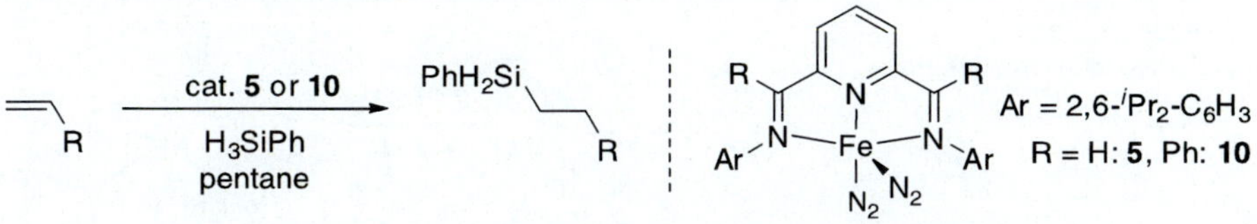

Substrate	Product	TOF(mol/h) 5	TOF(mol/h) 10
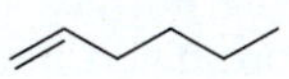	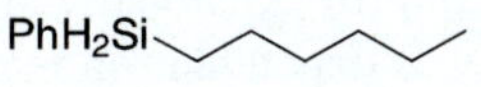	364 [27], 330 [31]	930 [31]
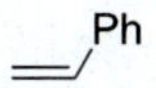	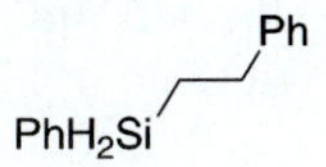	242 [27]	
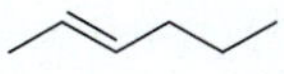	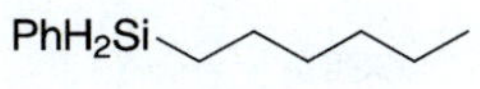	0.09[a] [27]	
	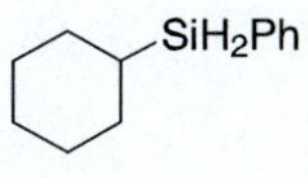	23 [27], 20 [31]	16 [31]
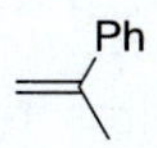	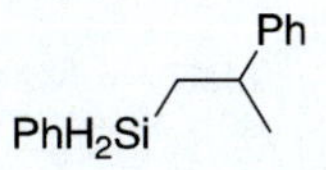	104 [27]	
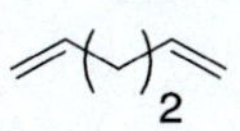	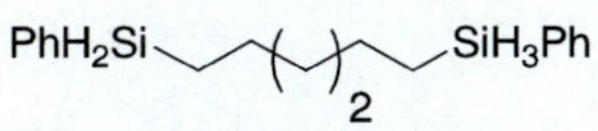	20 [27]	
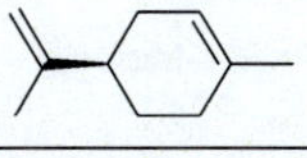	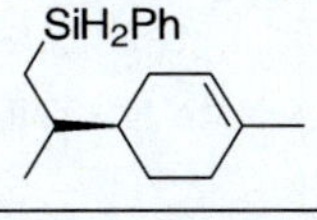	182 [27], 166 [31]	37 [31]

[a]25% of the internal hydrosilylation product was also obtained.

Scheme 18 Fe-catalyzed hydrosilylation of alkynes and alkenes

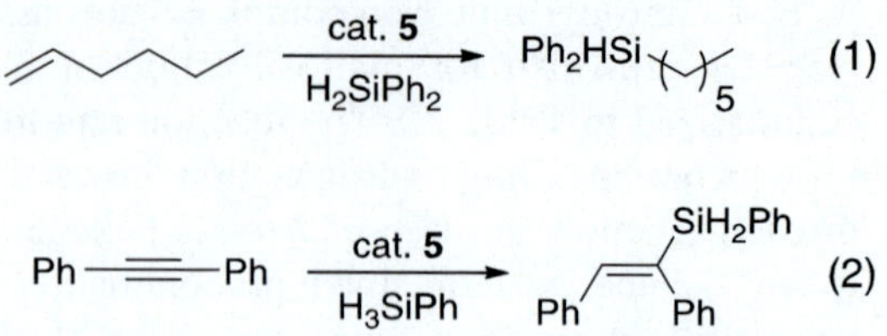

Stoichiometric reaction of **5** with phenylsilane produced the iron(0) bis(silane) σ-complex **18**, which was confirmed by the single-crystal X-ray analysis as well as SQUID (Superconducting QUantum Interference Device) magnetometry (Scheme 19). Complex **18** as a precatalyst showed high activity for the hydrosilylation of 1-hexene.

Scheme 19 Ligand exchange in hydrosilylations

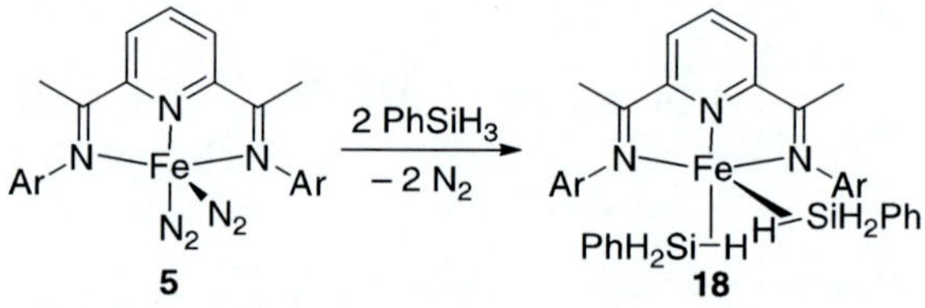

The hydrosilylation of various ketones catalyzed by a combination of $Fe(OAc)_2$ with a *N*-coordinated ligand was reported by Nishiyama and Furuta in 2007 [65]. In this reaction, the catalytic activity was dependant on the *N*-coordinated ligands employed (Scheme 20). The ligands on the left side of the dashed line were effective, whereas those on the right side led to no or low catalytic activity.

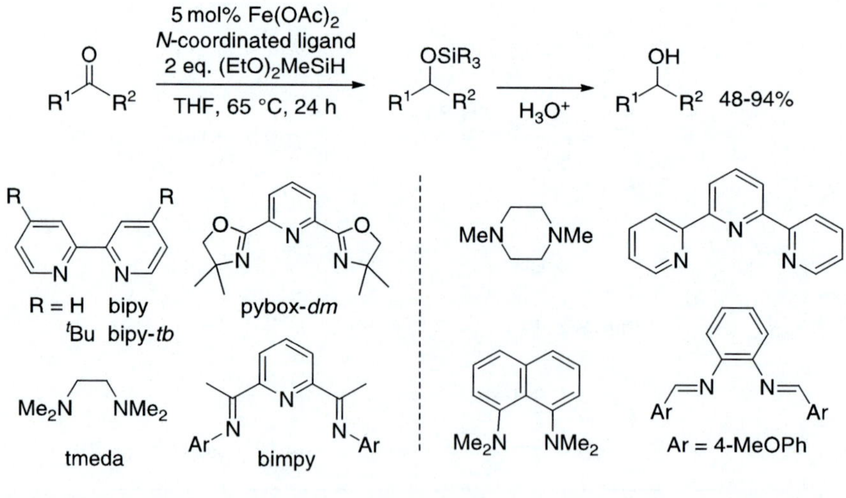

Scheme 20 Hydrosilylation of ketones catalyzed by $Fe(OAc)_2$ with the *N*-coordinated ligand

The $Fe(OAc)_2$/tmeda system showed good activity and chemoselectivity. The bromide and ester groups on the aryl ring of acetophenone were not reduced. Intriguingly, the catalytic activity was improved by using sodium thiophene-2-carboxylate (STC) instead of the *N*-ligand as a ligand for $Fe(OAc)_2$ [66]. With the $Fe(OAc)_2$/STC system, it was possible to use polymethylhydrosiloxane (PMHS) as well as $(EtO)_2MeSiH$ as a silyl source. However, in the presence of other iron sources such as $FeCl_3$ and $Fe(acac)_2$ with STC no hydrosilylation was not observed. In both systems, benzalacetone was reduced to the corresponding alcohol in good to excellent yields with a small amount of 1,4-reduction product. To check the scope of the catalytic reaction of the $Fe(OAc)_2$/STC system, both transfer hydrogenation and hydrogenation reactions (see **3.2.2**) were examined but the system was not very effective.

In 2007, the first asymmetric hydrosilylation was observed when a combination of $Fe(OAc)_2$ with the chiral ligand such as a pybox-*bn* (**L1**), bopa-*ip* (**L2**), and bopa-*tb* (**L3**) was used [65]. The results are summarized in Table 11. The best *ee* value (79% (*R*)) was attained by the combination of $Fe(OAc)_2$ with **L3**.

Table 11 The Fe-catalyzed enantioselective hydrosilylation of ketones

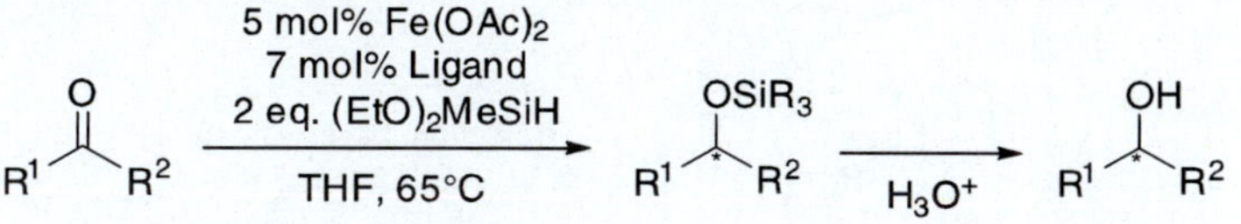

Product	Ligand	Yield(%)	ee(%)
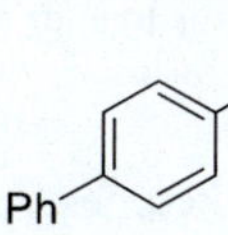	**L1**	93	37 (*R*)
	L2	82	57 (*R*)
	L3	82	79 (*R*)
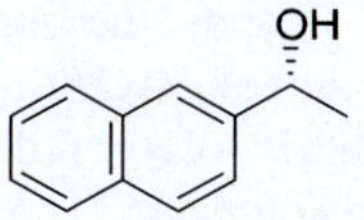	**L3**	59	65 (*R*)
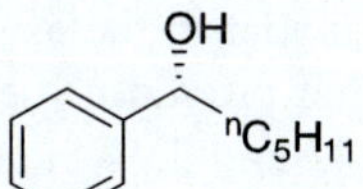	**L3**	39	59 (*R*)

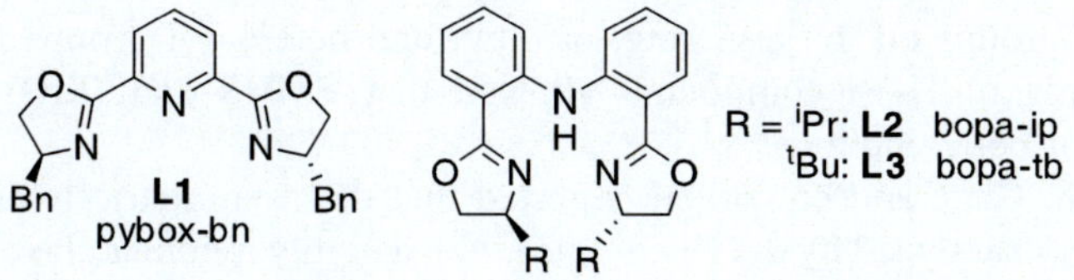

Beller and coworkers reported hydrosilylation reactions of organic carbonyl compounds such as ketones and aldehydes catalyzed by $Fe(OAc)_2$ with phosphorus ligands (Scheme 21). In case of aldehydes as starting materials, the $Fe(OAc)_2/PCy_3$ with polymethylhydrosiloxane (PMHS) as an H–Si compound produced the corresponding primary alcohols in good to excellent yields under mild conditions [67]. Use of other phosphorus ligands, for instance, PPh_3, bis(diphenylphosphino) methane (dppm), and bis(diphenylphosphino)ethane (dppe) decreased the catalytic activity. It should be noted that *trans*-cinnamaldehyde was converted into the desired alcohol exclusively and 1,4-reduction products were not observed.

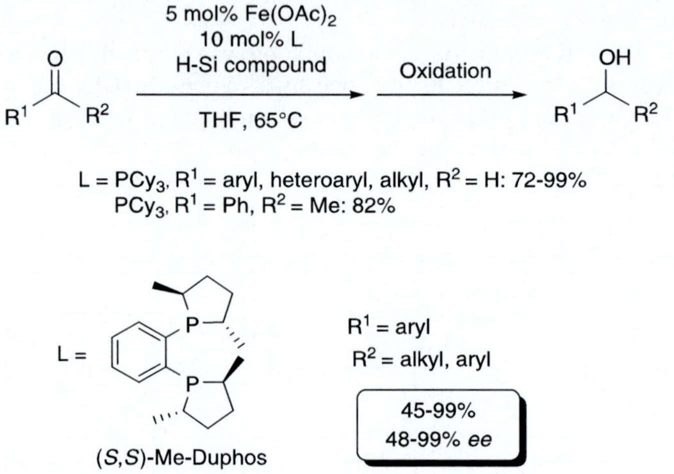

Scheme 21 Hydrosilylation catalyzed by an Fe complex with phosphorus ligands

An iron complex-catalyzed asymmetric hydrosilylation of ketones was achieved by using chiral phosphorus ligands [68]. Among various ligands, the best enantioselectivities (up to 99% *ee*) were obtained using a combination of $Fe(OAc)_2$/(*S*,*S*)-Me-Duphos in THF. This hydrosilylation works smoothly in other solvents (diethylether, *n*-hexane, dichloromethane, and toluene), but other iron sources are not effective. Surprisingly, this Fe catalyst (45% *ee*) was more efficient in the asymmetric hydrosilylation of cyclohexylmethylketone, a substrate that proved to be problematic in hydrosilylations using Ru [69] or Ti [70] catalysts (43 and 23% *ee*, respectively).

Both Fe-catalyzed systems are compatible with functionalized organic carbonyl compounds such as halides, sterically bulky groups and electron-withdrawing or -donating groups on the aryl ring, or alkyl and heteroaryl groups for aldehydes. A screening of H–Si compounds showed that PMHS and $(EtO)_2MeSiH$ were suitable for both reactions.

In 2008, Gade and coworkers reported that the asymmetric hydrosilylation of ketones was catalyzed by the Fe complex with a highly modular class of pincer-type ligand (Scheme 22) [71]. This Fe catalyst system showed both moderate to good enantioselectivities and yields at slightly elevated temperature. Although the yield of product increased when the reaction was carried out at 65°C, the enantiomeric excess of that decreased. By using sterically bulky dialkyl ketones such as adamantylmethylketon and *t*-butylmethylketone, both yield and enantiomeric excess of products decreased.

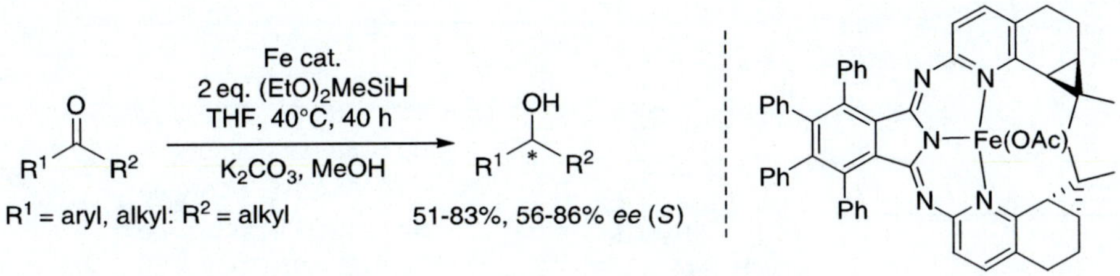

Scheme 22 Asymmetric hydrosilylation catalyzed by the Fe complex with the pincer-type ligand

Interestingly, the activity of the corresponding cobalt catalyst possessing a pincer-type ligand is higher than that of the iron complex. In addition, the cobalt complex also acts as a catalyst in asymmetric intermolecular cyclopropanations.

Hydrosilylations of aldehydes and ketones with Ph_2SiH_2 catalyzed by a bis(imino) pyridine iron complex were also achieved by Chirik and coworkers (Scheme 23) [72]. The catalytic activity of **19** is much higher than that of **20** except using 2-hexanone and 5-hexen-2-one as a ketone. The reaction rate increases with electron-withdrawing groups in the *para* position on the phenyl ring (the order of activity: CF_3 > H > OMe > ^{t}Bu > NMe_2). This tendency is similar to that observed for the corresponding hydrogenations, which was reported by Morris (see **2.2**) [49].

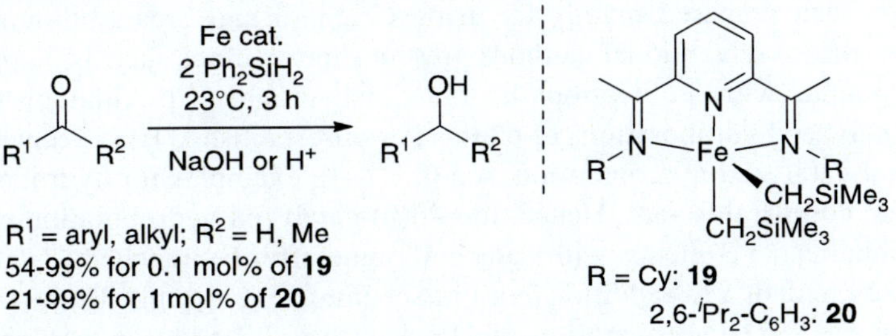

Scheme 23 Hydrosilylation catalyzed by a bis(imimo)pyridine iron complex

The comparison of a bis(imino)pyridine iron complex and a pyridine bis (oxazoline) iron complex in hydrosilylation reactions is shown in Scheme 24 [73]. Both iron complexes showed efficient activity at 23°C and low to modest enantioselectivites. However, the steric hindered acetophenone derivatives such as 2′,4′,6′-trimethylacetophenone and 4′-*tert*-butyl-2′,6′-dimethylacetophenone reacted sluggishly. The yields and enantioselectivities increased slightly when a combination of iron catalyst and $B(C_6F_5)_3$ as an additive was used.

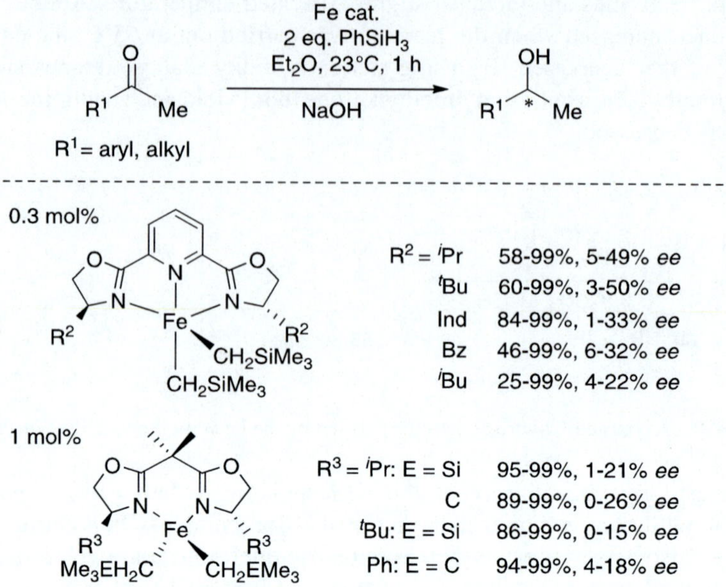

Scheme 24 The comparison of a bis(imino)pyridine iron complex and a pyridine bis(oxazoline) iron complex for hydrosilylation of ketones

3.2 *Hydroboration*

Allylboranes are widely used in organic synthesis (for recent examples, see [74–82]) and have been prepared mainly by using Grignard and organolithium reagents. However, these conventional methods present shortcomings such as inapplicability for compounds with an electrophilic functional group [75]. Although transition-metal-catalyzed hydroborations of olefins is well-established (for reviews of transition-metal-catalyzed hydroboration, see [83, 84]), examples for hydroborations of dienes are comparably rare. Hence, the Pd(0)-catalyzed hydroboration reaction of unfunctionalized 1,3-dienes with catecholborane affords branched (*Z*)-allylic boronates as a result of a 1,4-addition (eq. 1 in Scheme 25) [85], while Rh(I)- [86] and Ni (II)-catalyzed [87] hydroboration reactions afford 1,2-addition products mainly (eq. 2 in Scheme 25).

Scheme 25 Transition metal catalyzed hydroboration of 1,3-dienes

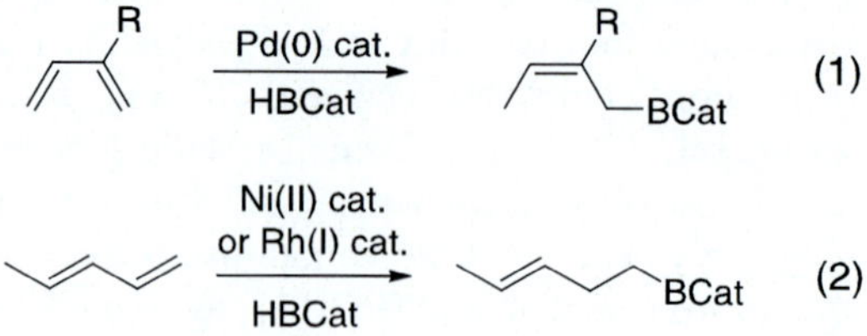

In 2009, Ritter and coworkers reported a selective Fe-catalyzed hydroboration of 1,3-dienes to produce linear (*E*)-γ-disubstituted allylboranes under mild conditions when a combination of **L1**·$FeCl_2$ and magnesium metal as a catalyst was used. The branched (*E*)-allylboranes were obtained by using **L2**·$FeCl_2$ instead of **L1**·$FeCl_2$ (Scheme 26) [88]. For the synthesis of 2-borylallylsilanes, this method was superior to the previously reported silaboration of allenes [89].

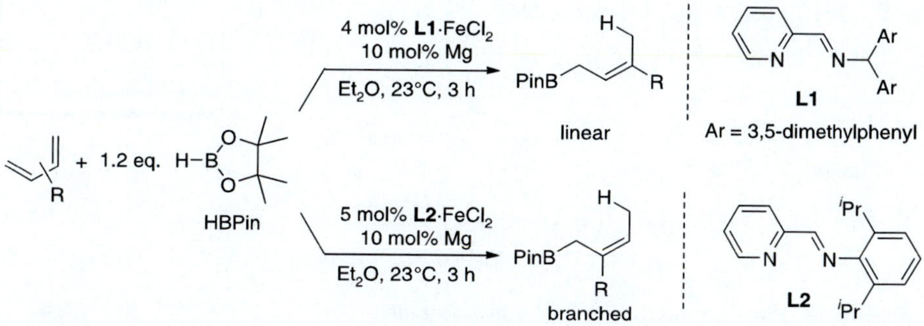

Scheme 26 1,4-Hydroboration of 1,3-diene derivatives with pinacolborane catalyzed by an iron complex

For a mechanistic investigation, hydroboration of myrcene with pinacolborane-d_1 was examined. A selective deuteration was observed at the methyl group of the hydroboration product (Scheme 27).

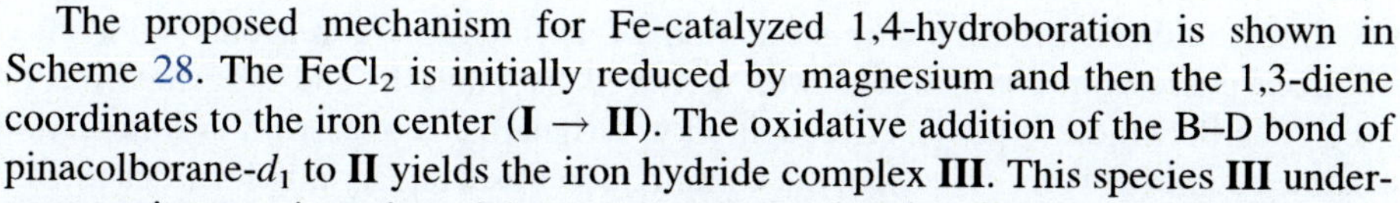

Scheme 27 Selected deuteration in hydroboration of myrcene catalyzed by $FeCl_2$ with **L1**

The proposed mechanism for Fe-catalyzed 1,4-hydroboration is shown in Scheme 28. The $FeCl_2$ is initially reduced by magnesium and then the 1,3-diene coordinates to the iron center (**I** → **II**). The oxidative addition of the B–D bond of pinacolborane-d_1 to **II** yields the iron hydride complex **III**. This species **III** undergoes a migratory insertion of the coordinated 1,3-diene into either the Fe–B bond to produce π-allyl hydride complex **IV** or the Fe–D bond to produce π-allyl boryl complex **V**. The π–σ rearrangement takes place (**IV** → **VI**, **V** → **VII**). Subsequently, reductive elimination to give the C–D bond from **VI** or to give the C–B bond from **VII** yields the deuterated hydroboration product and reinstalls an intermediate **II** to complete the catalytic cycle. However, up to date it has not been possible to confirm which pathway is correct.

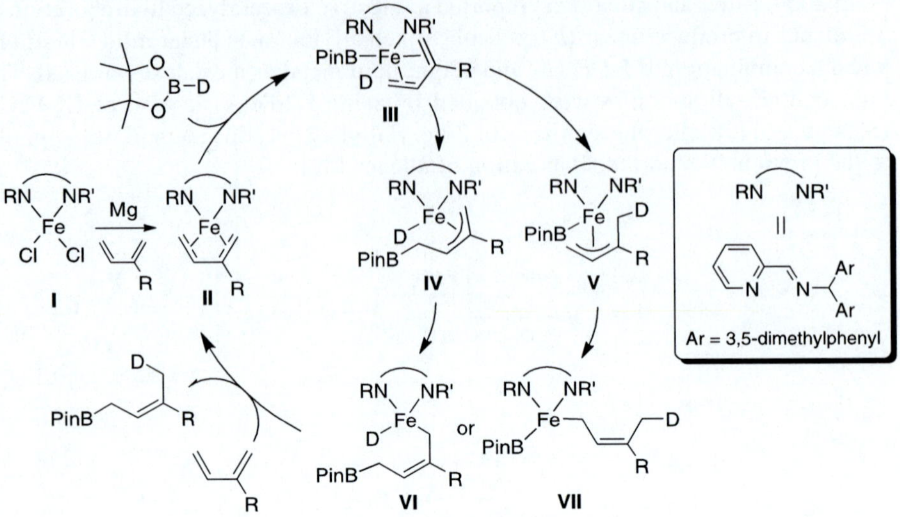

Scheme 28 Proposed mechanism for 1,4-hydroboration

4 Organic Synthesis

4.1 C–C and C–E Bond Formation

C–C and C–E (E = heteroatom) bond formations are valuable reactions in organic synthesis, thus these reactions have been achieved to date by considerable efforts of a large number of chemists using a precious-metal catalysts (e.g., Ru, Rh, and Pd). Recently, the application range of iron catalysts as an alternative for rare and expensive transition-metal catalysts has been rapidly expanded (for recent selected examples, see [12–20, 90–103]). In these reactions, a Fe–H species might act as a reactive key intermediate but also represent a deactivated species, which is prepared by β-H elimination.

Iron-catalyzed cross-coupling reaction of aryl Grignard reagents with alkyl halides possessing β-hydrogens was achieved by Hayashi and Nagano in 2004 (eq. 1 in Scheme 29) [104]. Although alkyl Grignard reagents with aryl and alkyl halides do not fit in this reaction, the combination of secondary alkyl halides as well as primary ones with aryl Grignard reagents are adaptable. In 2006, Ready and Zhang reported that the carbometalation-cross coupling of propargylic and homo-propargylic alcohols affords tri- and tetrasubstituted olefins with high regio- and stereo-selectivity catalyzed by Fe(III) complexes such as $Fe(acac)_3$ and $Fe(ehx)_3$ (ehx = 2-ethylhexanoate) (eq. 2 in Scheme 29) [105]. While Cu(I) salts show a low activity (<5% conversion) [106, 107], $Co(OAc)_2$ and $Ni(acac)_2$ catalysts afford the corresponding products in moderate yields.

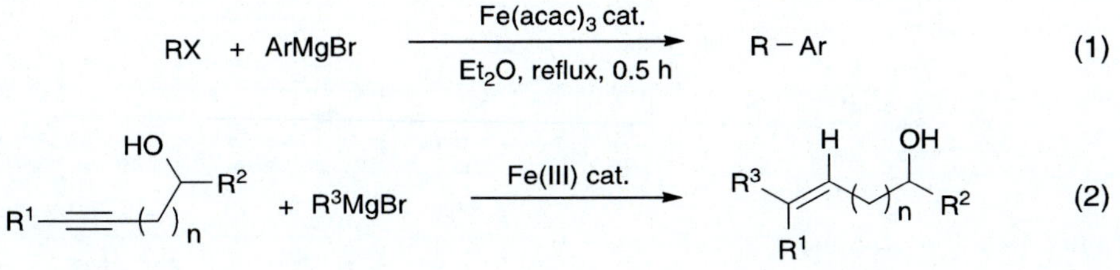

Scheme 29 Iron-catalyzed cross coupling reaction of aryl Grignard reagents with alkyl halides

The initially formed tetra-alkylferrate(II) represents the reactive intermediate in both reactions that undergoes a carboferration of the triple bond in eq. 2, Scheme 29. Transmetallation from Fe to Mg yields a vinyl-magnesium species, which liberates the desired olefin upon hydrolysis within the acidic work-up procedure. In the above two reactions, a competing β-hydride elimination from the ferrate yields the unreactive Fe–H species and hence is considered to be the deactivation step in the catalytic cycle.

In 2009, Ritter and coworkers reported a stereo- and regio-selective 1,4-addition of α-olefins to 1,3-dienes catalyzed by the mixture of $FeCl_2$, the iminopyridine ligand (**L1** or **L2**), in the presence of magnesium metal (Scheme 30) [108]. This combination is also adaptable to the 1,4-hydroboration of 1,3-dienes (see **3.2**).

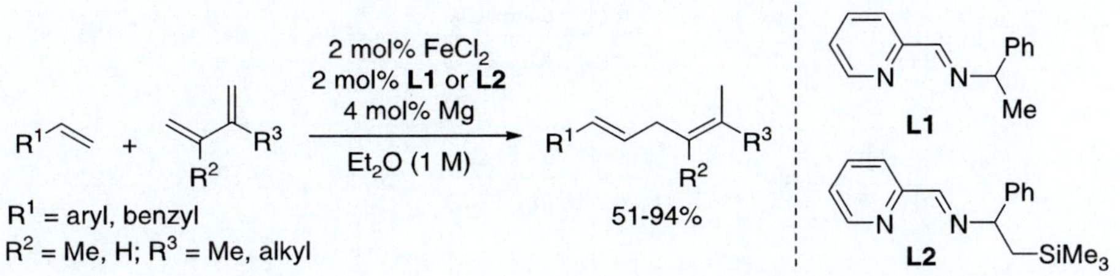

Scheme 30 Fe-catalyzed 1,4-addition of α-olefins to 1,3-dienes

In contrast to the hydroboration protocol, this reaction affords the linear 1,4-diene addition products in either case as the sole product using both **L1** or **L2** as a ligand. The system possesses a good degree of functional group tolerance for the functionalized styrenes with electron-withdrawing or -donating groups such as halides, ethers, and esters on the aryl ring.

The proposed catalytic cycle is shown in Scheme 31. Hence, $FeCl_2$ is reduced by magnesium and subsequently coordinates both to the 1,3-diene and α-olefin (**I** → **III**). The oxidative coupling of the coordinated 1,3-diene and α-olefin yields the allyl alkyl iron(II) complex **IV**. Subsequently, the π–σ rearrangement takes place (**IV** → **V**). The *syn*-β-hydride elimination (H_Z) gives the hydride complex **VI** from which the C–Hz bond in the 1,4-addition product is formed via reductive elimination with regeneration of the active species **II** to complete the catalytic cycle. Deuteration experiments support this mechanistic scenario (Scheme 32).

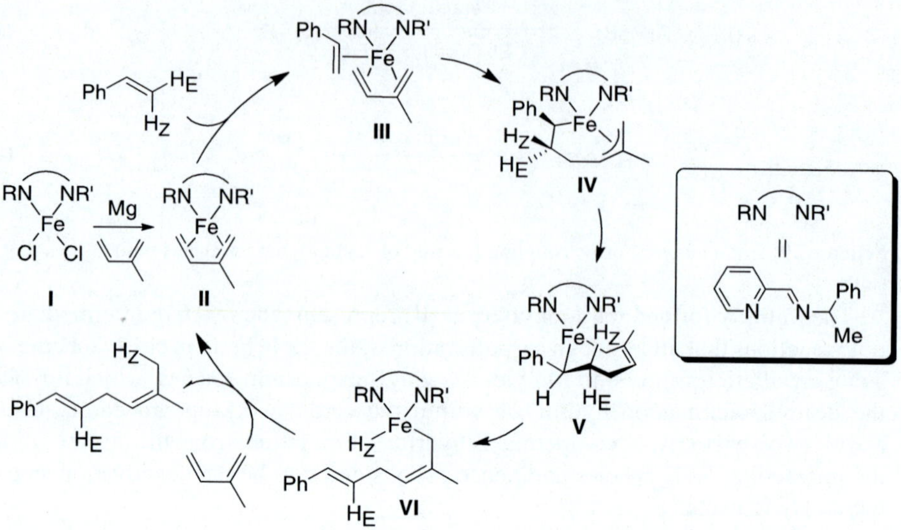

Scheme 31 Fe-catalyzed 1,4-addition of α-olefins to 1,3-dienes

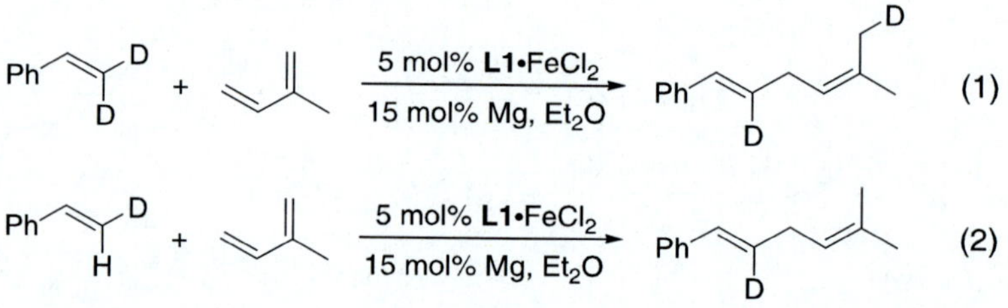

Scheme 32 Deuteration experiment for 1,4-addition of α-olefins to diene catalyzed by an iminopyridine-ferrous chloride complex

Iron-catalyzed C(sp^3)–C(sp^3) bond-forming cross-coupling reactions of alcohols with alkenes has been reported by Tu and coworkers in 2009 [109]. Reactions of primary alcohols with various alkenes in the presence of a catalytic amount of $FeCl_3$ in 1,2-dichloroethane afford the desired secondary alcohols as the cross-coupling products in moderate to good yields (Scheme 33). Iron sources such as

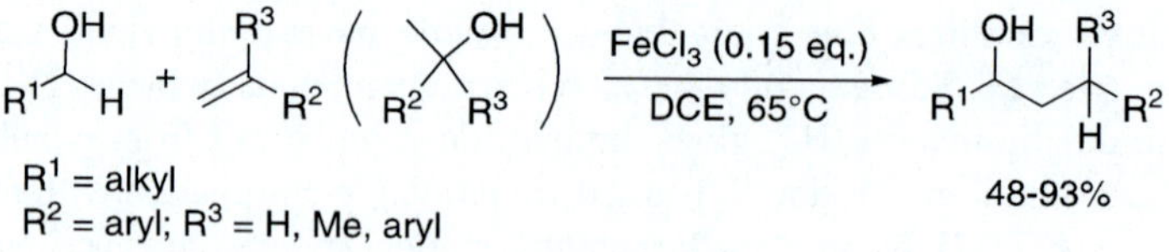

Scheme 33 Fe-catalyzed primary alcohols with various alkenes

$FeCl_2$, $Fe(acac)_3$, and $Fe(ClO_4)_3$ instead of $FeCl_3$, however, showed low activity. Tertiary alcohols, such as 3-phenylpropanol or cyclohexylmethanol instead of alkenes, are also adaptable to this reaction.

Based upon deuterium-labeling experiments, the following catalytic cycle was proposed (Scheme 34). The Fe(III) center coordinates to the olefin and interacts with the $C(sp^3)$–H bond adjacent to the oxygen atom of the alcohol to form intermediate **a**. Cleavage of the $C(sp^3)$–H bond gives radical pair **b**. Subsequently, both free-radical addition and dissociation afford a free-radical and a Fe(IV)–H species. The hydrogen transfer from the hydride complex to **c** gives the desired product and regenerates the $[Fe]^{III}$ catalyst for the next catalytic cycle.

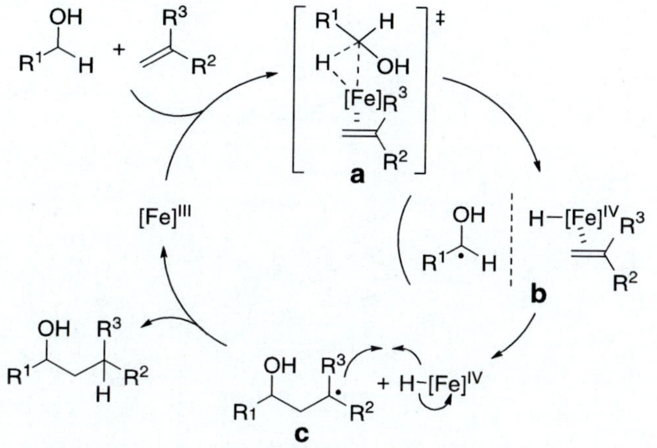

Scheme 34 Proposed catalytic cycle for the C–C bond formation

Beller and coworkers found in 2009 that alkynes react with amines under the CO pressure (20 bar) in the presence of catalytic amounts of $[Fe_3(CO)_{12}]$ to the corresponding succinimide in moderate to excellent yields (Scheme 35) [110]. Various terminal and internal alkynes and ammonia or primary amines are adaptable for this transformation. Furthermore, $[Fe(CO)_5]$ as an iron source showed high activity. The catalytic activity, however, decreased considerably when a phosphine ligand such as PPh_3 and $(^tBu)_2P(^nBu)$ was employed.

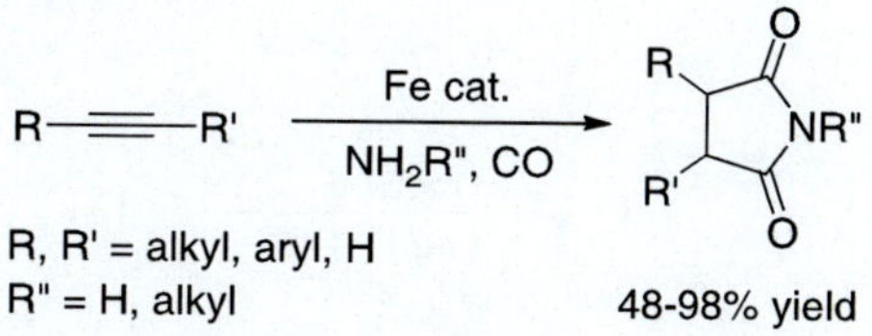

Scheme 35 Synthesis of succinimides catalyzed by an iron complex

The proposed reaction mechanism, which is assisted by Periasamy's stoichiometric reaction [111–118], is shown in Scheme 36. Initially, the reaction of $[Fe_3(CO)_{12}]$

with an amine gives both an "amine-$[Fe(CO)_4]$" and an $[Fe_2(CO)_8]$ species. Subsequently, $[Fe_2(CO)_8]$ reacts with alkynes under CO pressure to yield complexes **a** or **b**. The corresponding cyclic imides are obtained by the reaction of **a** or **b** with an excess amount of amine and CO via intermediates **c**–**d**. Although some mechanistic details are unapparent, hydride complex **d** and dihydrogen complex **e** are considered to be key intermediates.

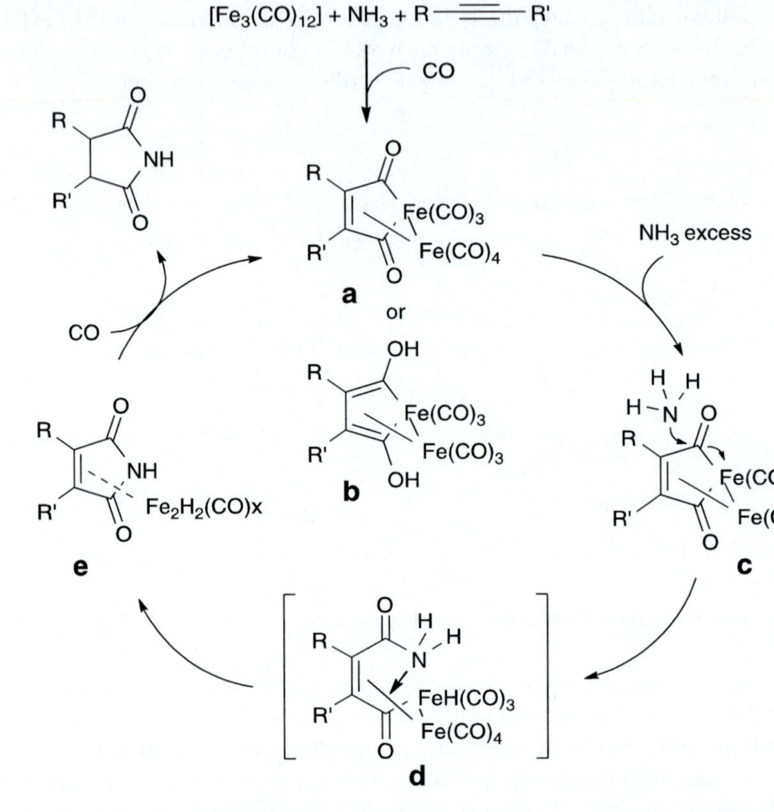

Scheme 36 Reaction mechanism of the formation of succinimides catalyzed by iron complexes

In 2009, Chirik reported a hydrogen-mediated reductive enyne cyclization catalyzed by the bis(imino)pyridine iron complex **5** (Scheme 37) [119]. In the

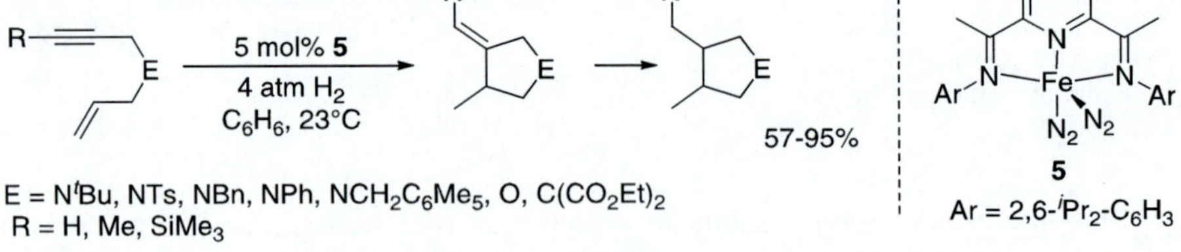

Scheme 37 Synthesis of succinimides catalyzed by an iron complex

presence of the Fe catalyst under hydrogen pressure, various 1,6-enynes are readily converted into the corresponding cyclic products. The unsaturated moiety on the product is subsequently hydrogenated to give the desired products such as pyrrolidine, tetrahydrofuran, and cyclopentane derivatives (see also **3.2.1**) [120].

The catalytic cycle, which is supported by stoichiometric and labeling experiments, is shown in Scheme 38. Loss of 2 equiv. of N_2 from **5** affords the active species **a**. Reaction of **a** with the 1,6-enyne gives the metallacycle complex **b**. Subsequently, **b** reacts with H_2 to give the alkenyl hydride complex **c** or the alkyl hydride complex **d**. Finally, reductive elimination constructs the C–H bond in the cyclization product and regenerates intermediate **a** to complete the catalytic cycle.

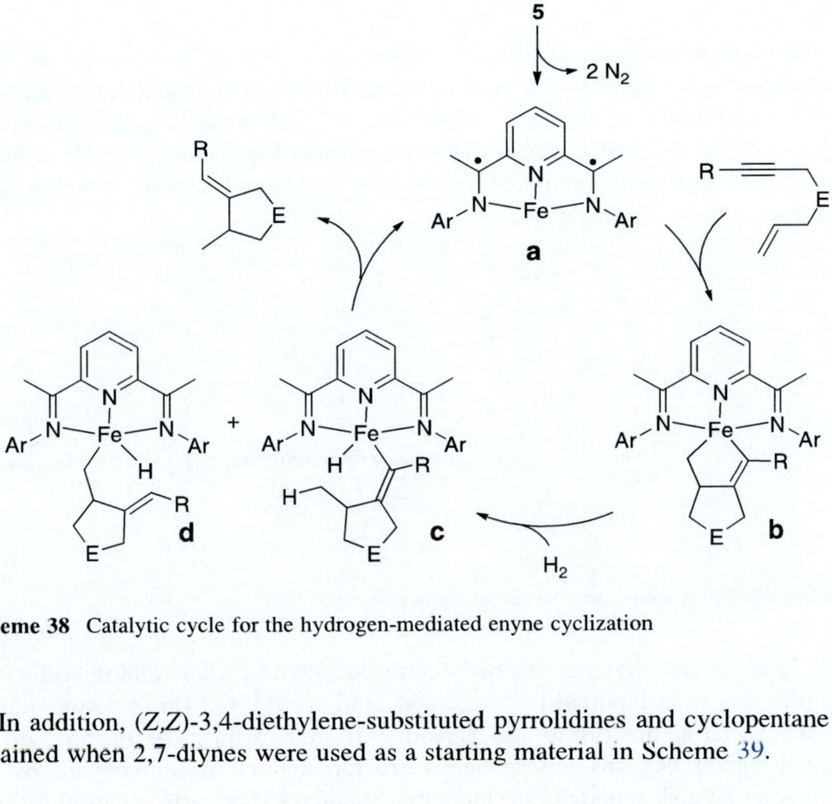

Scheme 38 Catalytic cycle for the hydrogen-mediated enyne cyclization

In addition, (*Z,Z*)-3,4-diethylene-substituted pyrrolidines and cyclopentane are obtained when 2,7-diynes were used as a starting material in Scheme 39.

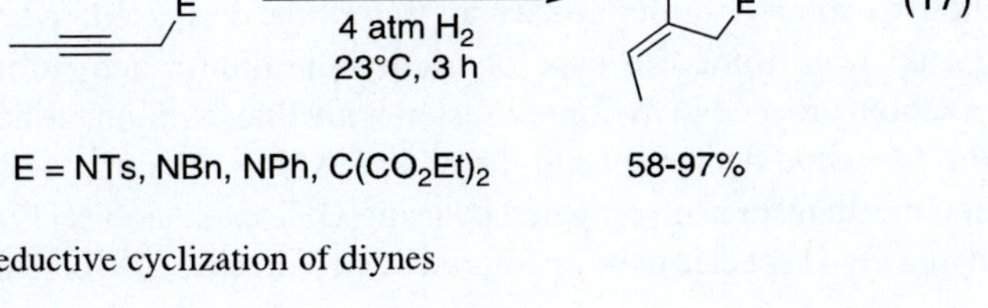

Scheme 39 Reductive cyclization of diynes

As an alternative method for the C–C bond formation, oligomerization and polymerization reactions of olefins catalyzed by a bis(imino)pyridine iron complex are also well known (Scheme 40) [121–124].

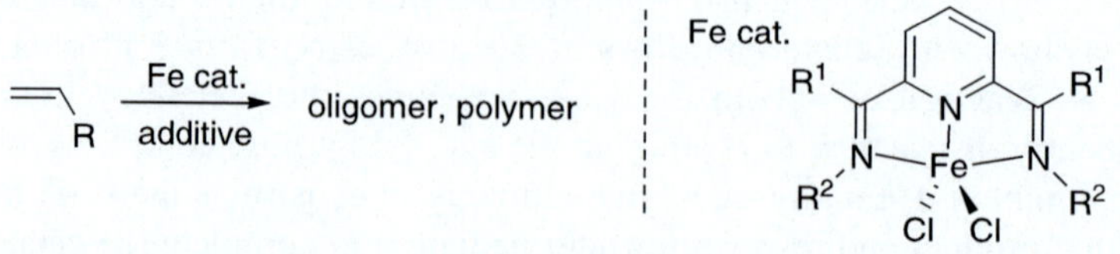

Scheme 40 Olefin polymerization catalyzed by a bis(imino)pyridine iron complex

The corresponding iron-catalyzed oligomerization of ethylene was developed by Gibson and coworkers [125]. A combination of an iron precatalyst with MAO (methyl aluminoxane) yields a catalyst that affords ethylene oligomers (>99% linear α-olefin mixtures). The activity of ketimine iron complexes (R^1 = Me) is higher than that of the aldimine analogs (R^1 = H) and also the α-value of the oligomer is better (Scheme 41).

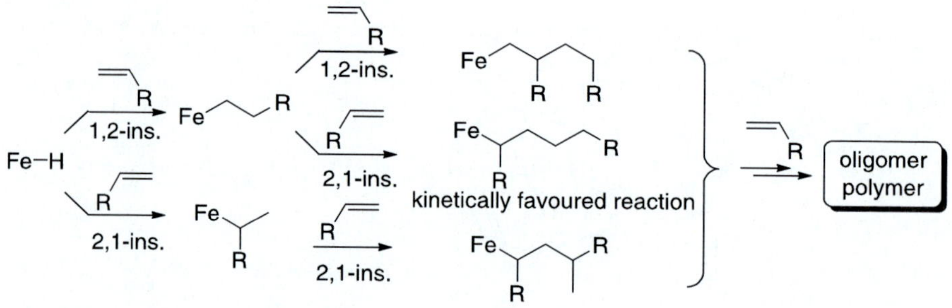

Scheme 41 The proposed mechanism of olefin polymerization

A head-to-head dimerization of α-olefin catalyzed by a bis(imino)pyridine iron complex has been reported by Small and Marcucci [126]. This reaction delivers linear internal olefins (up to 80% linearity) from α-olefins. The linearity of products, however, depends on the catalyst structure and the reaction conditions.

Fink and Babik reported that propylene polymerization was achieved by a bis (imino)pyridine iron complex with $Ph_3C[B(C_6F_5)_4]$ and trialkylaluminium as additives [127]. Both 3-methyl-^{n}butyl and ^{n}butyl endgroups were observed by ^{13}C NMR spectrum when triisobutylaluminium as an activator was used, whereas the only ^{n}propyl endgroup was formed in case of triethylaluminium activation. In addition, this polymerization proceeds two times faster with than without a hydrogen atmosphere, but the M_n value decreases and the M_w/M_n value rises up.

The reaction mechanism is investigated by using DFT calculations [128, 129] indicating formation of a Fe–H species to be an important step. In case of 1-butene, the synthesis of linear octenes (i.e., 1,2- and the successive 2,1-insertions of 1-butene) is kinetically favored rather than that of branched products (Scheme 41). Dimers are obtained via a competitive reaction of β-hydrogen elimination and β-hydrogen transfer termination. Further insertion reaction into the Fe–C bond affords higher oligomers and polymers.

The direct reductive amination (DRA) is a useful method for the synthesis of amino derivatives from carbonyl compounds, amines, and H_2. Precious-metal (Ru [130–132], Rh [133–137], Ir [138–142], Pd [143]) catalyzed reactions are well known to date. The first Fe-catalyzed DRA reaction was reported by Bhanage and coworkers in 2008 (Scheme 42) [144]. Although the reaction conditions are not mild (high temperature, moderate H_2 pressure), the hydrogenation of imines and/or enamines, which are generated by reaction of organic carbonyl compounds with amines, produces various substituted aryl and/or alkyl amines. A dihydrogen or dihydride iron complex was proposed as a reactive intermediate within the catalytic cycle.

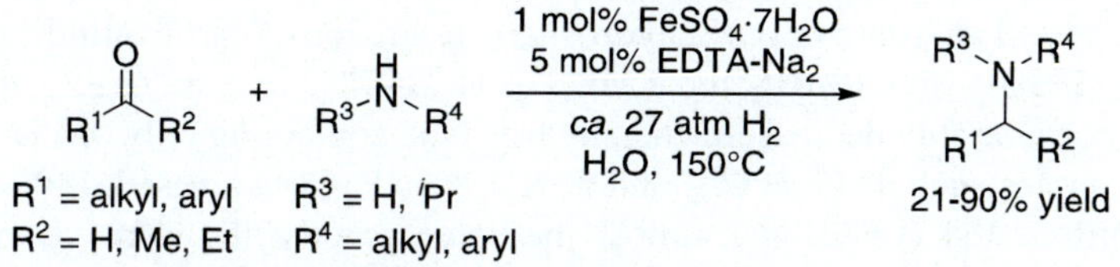

Scheme 42 Direct reductive amination of organic carbonyl compounds

4.2 *Others*

The dehydration of primary amides with hydrosilane catalyzed by iron carbonyl clusters, such as $[Et_3NH][HFe_3(CO)_{11}]$ and $Fe_2(CO)_9$, was achieved by Beller and coworkers in 2009 (Scheme 43) [145]. This reaction shows good functional group tolerance (e.g., such as aromatic, heteroaromatic, and aliphatic substrates).

A mechanistic proposal, which is based on the ruthenium-catalyzed dehydration reaction reported by Nagashima and coworkers [146], is shown in Scheme 44. Reaction of a primary amine with hydrosilane in the presence of the iron catalyst affords the bis(silyl)amine **a** and 2 equiv. of H_2. Subsequently, the isomerization of **a** gives the *N,O*-bis(silyl)imidate **b** and then elimination of the disiloxane from **b** produces the corresponding nitrile. Although the disiloxane and its monohydrolysis product were observed by ^{13}C and ^{29}Si NMR spectroscopy and by GC-Mass-analysis, intermediates **a** and **b** were not detected.

Scheme 43 Fe-catalyzed dehydration of amides to nitriles

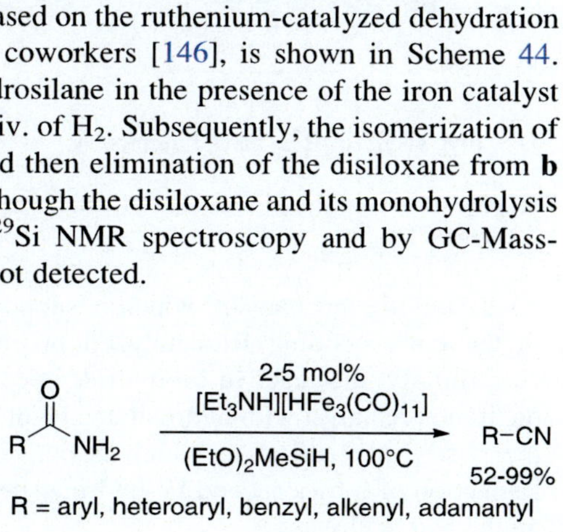

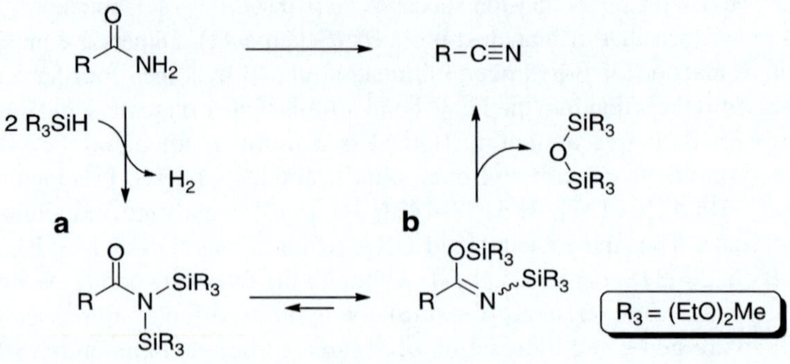

Scheme 44 Plausible pathway of Fe-catalyzed dehydration of primary amides to nitriles

In 2009, Beller (Scheme 45) [147] and Nagashima (Scheme 46) [148] independently reported an iron-catalyzed hydrosilane reduction of carboxamides to amines. Although inexpensive PMHS and TMDS as an H–Si source are usable, the yield of product considerably decreased when hydrosilane containing only one H–Si moiety or iron sources such as $Fe(acac)_2$ and FeX_2 ($X = F, Cl$) was used. In both thermal and photoassisted conditions, almost the same reactivities were observed upon using a combination of Fe catalyst with TMDS (Scheme 46).

Scheme 45 Fe-catalyzed deduction of amides to amines

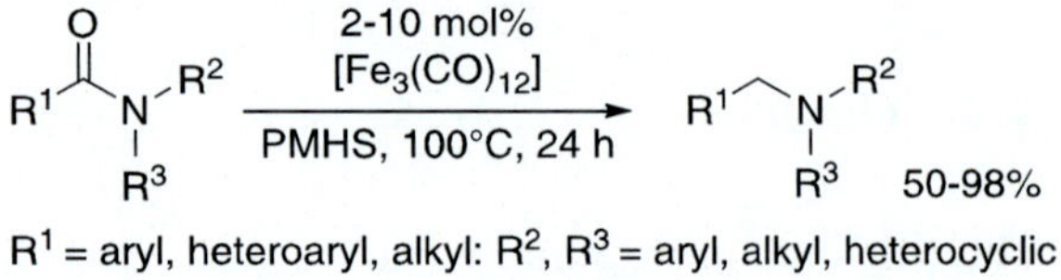

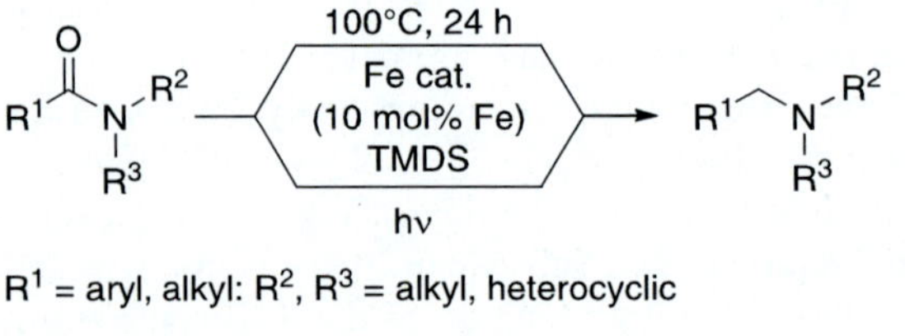

Fe cat.	Δ	hν
$Fe(CO)_5$	66-98%	69-94%
$Fe_3(CO)_{12}$	82-96%	68-95%

Scheme 46 Fe-catalyzed reduction of amides to amines under either thermal or photoassisted conditions

A catalytic mechanism, which is supported by deuterium-labeling experiments in the corresponding Ru-catalyzed procedure [146], is shown in Scheme 47. Accordingly, the reactive Fe-hydride species is formed in situ by the reaction of the iron precatalyst with hydrosilane. Hydrosilylation of the carboxyl group affords the *O*-silyl-*N*,*O*-acetal **a**, which is converted into the iminium intermediate **b**. Reduction of **b** by a second Fe-hydride species finally generates the corresponding amine and disiloxane.

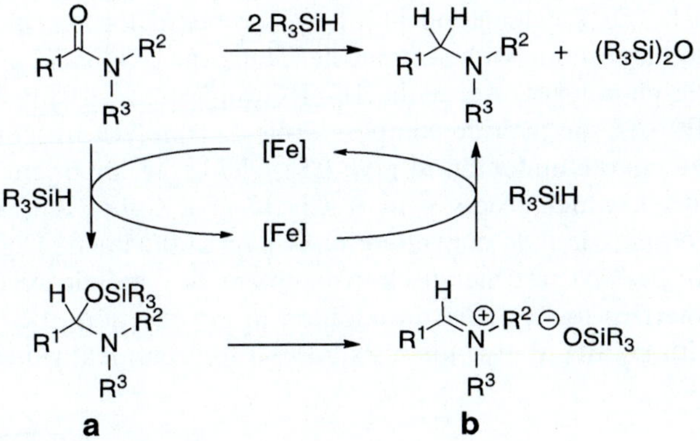

Scheme 47 Fe-catalyzed hydrosilane reduction of amides to amines

Oriyama and coworkers reported an iron-catalyzed reductive etherification of carbonyl compounds with triethylsilane and alkoxytrialkylsilane [149, 150] and alcohols (Scheme 48) [151].

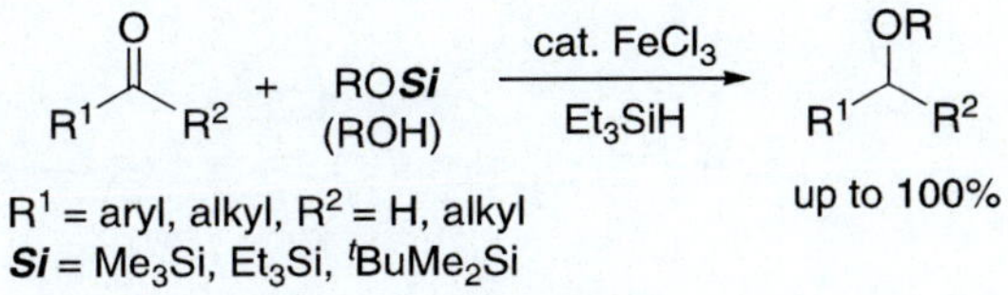

Scheme 48 Iron-catalyzed reductive etherification of carbonyl compounds

Although some methods for reductive etherifications of carbonyl compounds have been reported [152–162], the iron-catalyzed version possesses several advantages: (1) fairly short reaction times are needed, (2) not only trimethylsilyl ether but also triethylsilyl and ᵗbutyldimethylsilyl ethers and alcohols are adaptable, and (3) a broad substrate scope.

Our groups developed a catalytic C–CN bond cleavage of organonitriles catalyzed by the Fe complex (Scheme 49) [163, 164]. In this reaction, an organonitrile R–CN and Et_3SiH are converted into Et_3SiCN as a result of the C–CN bond cleavage and the Si–CN bond formation, and the R–H product. This is the first example of the catalytic C–CN bond cleavage of acetonitrile.

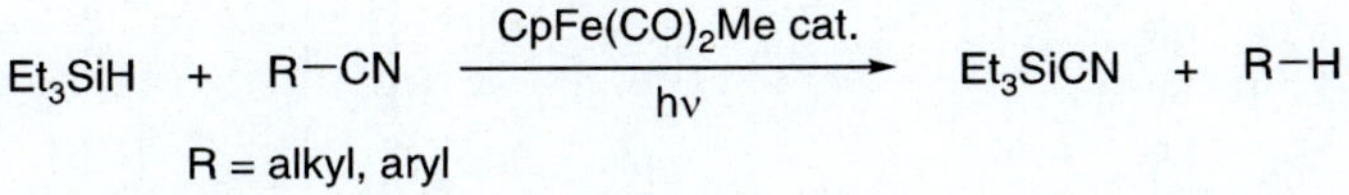

Scheme 49 Iron-catalyzed decyanation of nitriles

Supported by DFT calculation [165], we proposed the mechanism shown in Scheme 50 for the reaction with acetonitrile. Thus, one CO ligand in $CpFe(CO)_2Me$ is released via photolysis to give the 16-electron species **a** CpFe(CO)Me. In the presence of Et_3SiH, the hydride complex $CpFe(CO)Me(H)(SiEt_3)$ is formed. Subsequently, reductive elimination to give CH_4 yields the important intermediate **b** $CpFe(CO)(SiEt_3)$, which reacts with R–CN to give $CpFe(CO)(SiEt_3)(\eta^2\text{-NCR})$. This nitrile complex is then converted into $CpFe(CO)(Me)(\eta^1\text{-CNSiEt}_3)$. Finally, dissociation of Et_3SiNC regenerates **a** to complete the catalytic cycle. The released Et_3SiNC isomerizes to Et_3SiCN. In addition to the reaction of **b** with MeCN, **b** reacts also with Et_3SiH to give the bis(silyl)hydride complex, which is isolable.

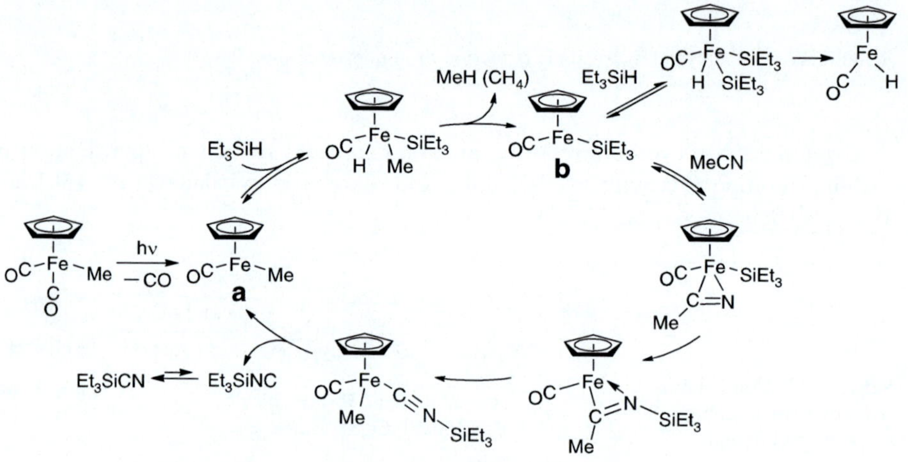

Scheme 50 Fe-catalyzed the C–CN bond cleavage of acetonitrile

We also reported that $CpFe(CO)_2Me$ acts as a precursor for the Si–O–Si bond formation reaction from hydrosilane and DMF (Scheme 51) [166, 167]. In this reaction, tertiary silanes and bis(silyl) compounds are converted into the corresponding disiloxanes and the polymers with $(\text{–R–Si–O–Si})_n$ backbone, respectively.

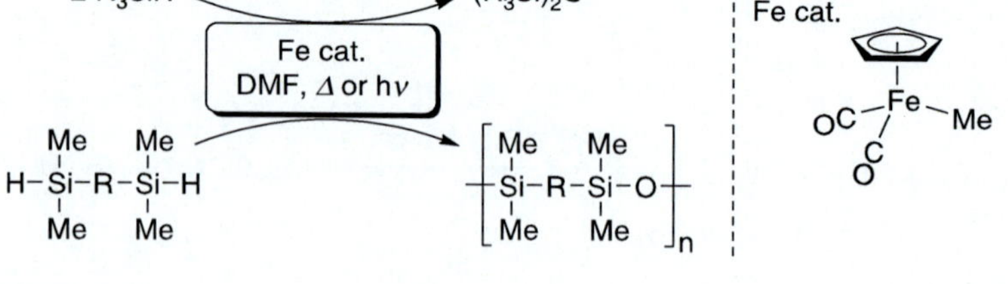

Scheme 51 The Fe-catalyzed Si–O–Si bond formation reaction from hydrosilanes and DMF

The reaction of $CpFe(CO)_2Me$ with R_3SiH gives the bis(silyl)hydride complex **21**. Photoreaction of **21** in DMF afforded the corresponding disiloxane (Scheme 52). We believe that the oxygen in the disiloxane is derived from DMF, because NMe_3 is concomitantly formed in this reaction. It is considered that the silyl species **a**, which is prepared via reductive elimination of R_3SiH from **21** in situ, is the active species within the catalytic cycle. Therefore, the generation of a bis(silyl)hydride species is the dormant step. We are currently studying the details of the reaction mechanism.

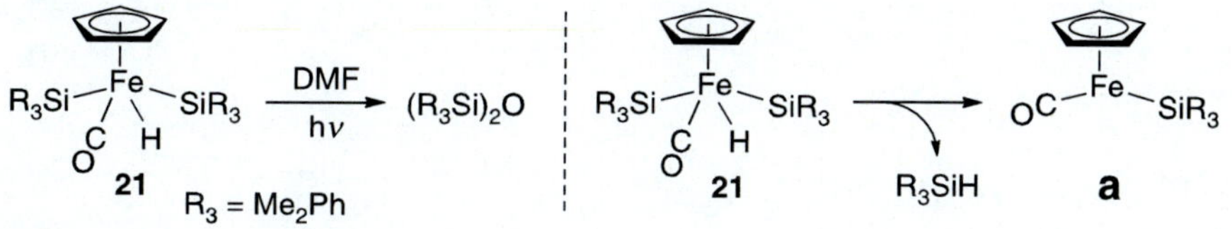

Scheme 52 Fe-catalyzed Si–O–Si bond formation reaction from hydrosilane and DMF

Isomerization reactions of allylic alcohols **a** to ketones **b** are catalyzed by various metal (e.g., Ru, Rh, Co, Ni, Mo, Ir, Pt) (Scheme 53) [168–172]. However, these metals are expensive and in some cases harsh reaction conditions are required.

Scheme 53 Transition metal catalyzed isomerization of allylic alcohols

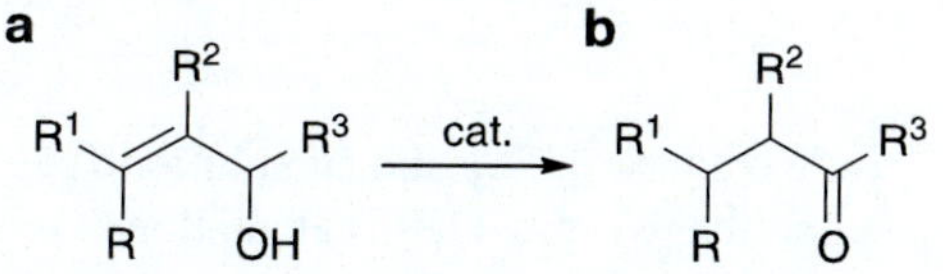

A mild, $Fe(CO)_5$-catalyzed isomerization of this type was reported by Grée and coworkers [173]. Allylic alcohols having mono-, di-, trisubstituted alkene are readily converted into their corresponding ketones, whereas polyunsaturated derivatives do not rearrange (Scheme 54).

The isomerization mechanism is clearly established by labeling experiments. The rearrangement of **a** to **c** via a π-allyl hydride complex **b** in the coordination sphere of the metal is a key step in this cayalytic cycle (Scheme 54) [174, 175]. In case of polyunsaturated derivatives, formation of a stable η^4 complexes (Scheme 55) is preferred over the rearrangement (**a** → **c**).

In addition, a 532 (visible) or 355 (UV region) nm laser-induced photoisomerization of allylic alcohols to aldehydes catalyzed by $[Fe_3(CO)_{12}]$ or $[Fe(CO)_4PPh_3]$ was developed by Fan [176]. In this reaction, key intermediates such as the π-allyl hydride species $[FeH(CO)_3(\eta^3\text{-}C_3H_3ROH)]$ (R = H, Me) were detected by pulsed laser FTIR absorption spectroscopy. These results strongly support the π-allyl mechanism of photoisomerization of allyl alcohols.

Furthermore, details of the isomerization of 1-alkenes into 2-alkenes were examined by deuteration experiments [177] and by using time-resolved IR spectroscopy in

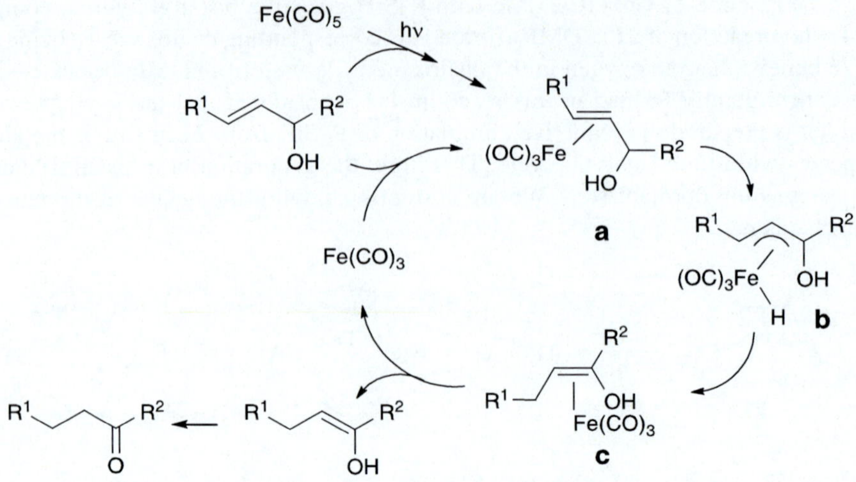

Scheme 54 Isomerization mechanism of allylic alcohols to ketones

Scheme 55 Photoreactions of $Fe(CO)_5$ with α,β-unsaturated dienol

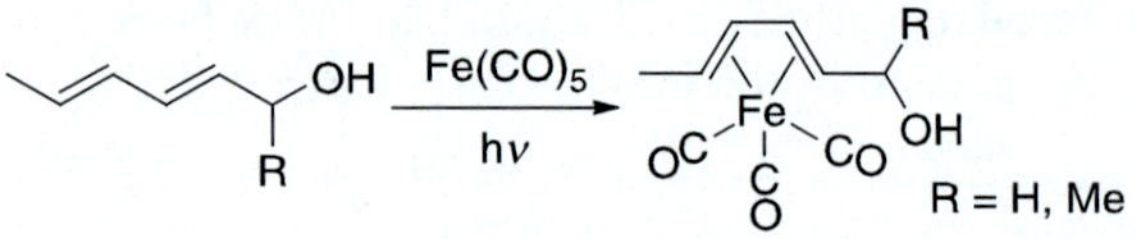

the gas phase [178] and in solution [179, 180]. The reaction mechanism, which is supported by the DFT calculations, is shown in Scheme 56. Thus, reaction of $Fe(CO)_5$ with a 1-alkene under photo-irradiation gives the η^2-alkene complex **a**.

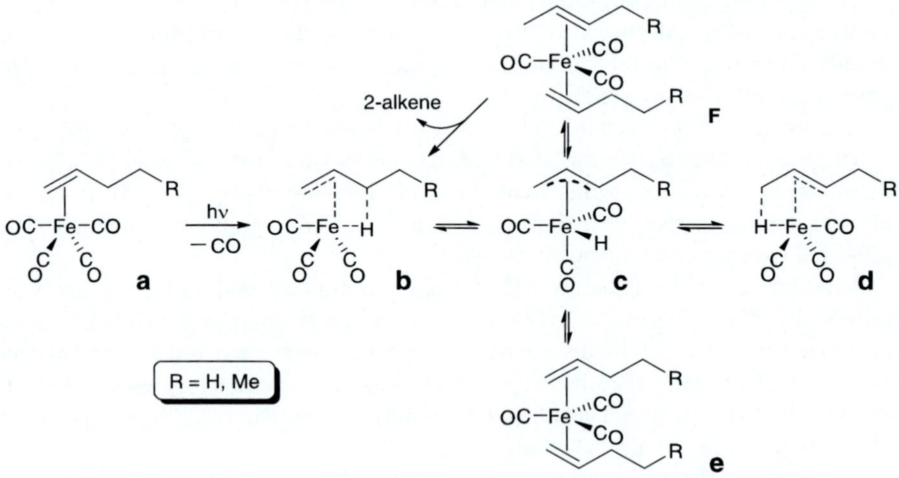

Scheme 56 Isomerization reaction of 1-alkene catalyzed by an iron complex

Complex **a** is readily converted into a Fe–γ-H agnostic complex **b** within an early picosecond timescale and then the π-allyl hydride complex **c** is generated by hydride abstraction. The energy level of the 2-alkene isomer **d**, which is calculated by DFT experiments, is similar to that of the 1-alkene complex **b**. In the next step, Fe $(CO)_3(\eta^2$-1-alkene$)(\eta^2$-2-alkene) **f**, which is generated via intramolecular isomerization of the coordinated 1-alkene to 2-alkene and the coordination of another 1-alkene, is a thermodynamically favored product rather than formation of a $Fe(CO)_3(\eta^2$-1-alkene$)_2$ **e**. Subsequently, release of the 2-alkene from **f** regenerates the active species **b** to complete the catalytic cycle.

5 Hydrogen Generation

H_2 serves as the alternative energy source relative to fossil fuels and biomass [181] because it is clean and environmentally friendly. Hence, catalytic hydrogen generation from water under mild conditions is one of the goals for the organometallic catalysis. One of the hopeful methods is the electrochemical reduction of protons by a hydrogenase mimic.

Based upon structural investigations by experts in the field of molecular biology, several Fe-containing complexes are found in the active side of natural proteins, e.g., a diiron complex in [FeFe]-hydrogenases (structure **a** in Fig. 6), a binuclear Fe–Ni complex in [NiFe]-hydrogenases (structure **b** in Fig. 6), or a monoiron complex in [Fe]-hydrogenases (structure **c** in Fig. 6). In [FeFe]-hydrogenases, two Fe centers are bridged by a CO ligand and a small organic moiety. However, it is not clear whether the moiety is a dithiolate ligand or not [182–187]. On the other hand, [NiFe]-hydrogenase is mainly constructed from both a large subunit containing the Ni–Fe cluster [188–190] and a small subunit, an iron–sulfur cluster ([4Fe–4S]). The waved

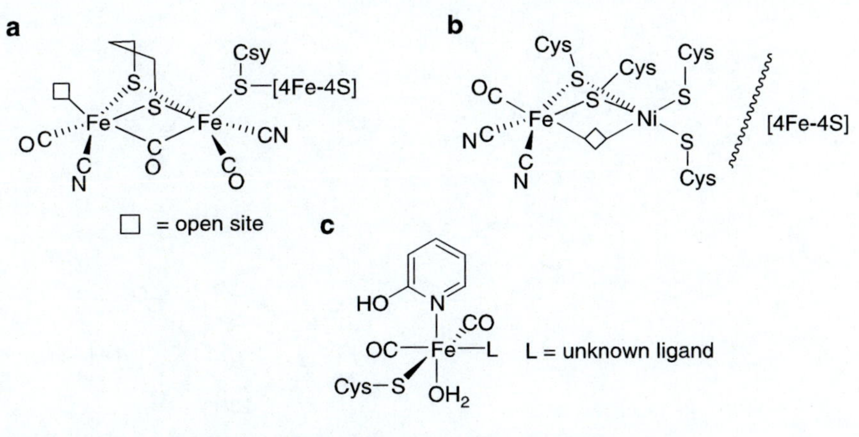

Fig. 6 The models for the active site of [FeFe]-, [NiFe]-, and [Fe]-hydrogenases

line stands for the interface between the hydrogenase large subunit and the small subunit [191–193]. The crystal structure of the model for the active site of the [Fe]-hydrogenases is not entirely clear [194, 195].

5.1 *Electrochemical Reduction*

Within the past 10 years, various biomimetic Fe model complexes were prepared and their catalytic activities in the electrochemical reduction of protons to H_2 were investigated (Scheme 57).

Scheme 57 Fe-catalyzed electrochemical reduction of protons to H_2

$$2H^+ + 2e^- \xrightarrow{\text{Fe cat.}} H_2$$

In 2001, Rauchfuss found that the dithiolate diiron complex $Et_4N[Fe_2\{\mu\text{-}S_2(CH_2)_3\}(CN)(CO)_4(PMe)_3]$ **22** acts as a catalyst in this reaction under strong acid conditions such as H_2SO_4, HCl, and HOTs [196]. In a preparative-scale reaction, a solution of 10^{-3} M of **22** with 50 equiv. H_2SO_4 in CH_3CN was electrolyzed at –1.2 V. Gas chromatographic analysis showed that the yield of produced H_2 was 100 ($\pm$10)%. The proposed catalytic cycle is shown in Scheme 58. Initially, **22** is converted into the μ-hydride complex **23** by metal protonation. A subsequent second protonation at the N atom of the CN^- ligand affords **23H**$^+$. Two-electron reduction of **23H**$^+$ yields H_2 and regenerates **22**. In this cycle, the μ-hydride complex **23** was isolated by a stoichiometric reaction of **22** with H_2SO_4 [196] and its structure was confirmed by the X-ray analysis [197].

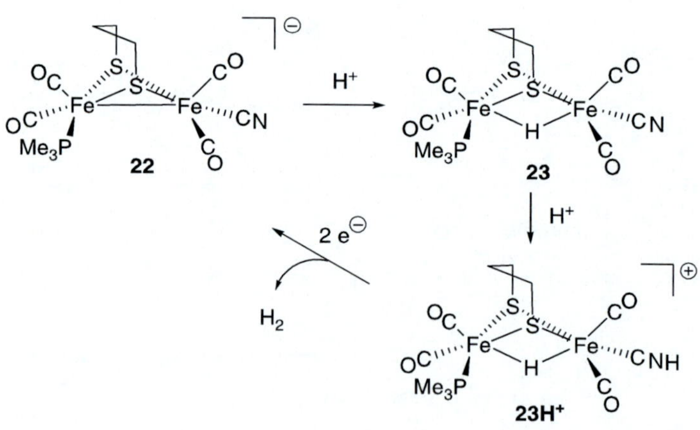

Scheme 58 Proposed mechanism of the reduction of protons to H_2 catalyzed by **22**

Further investigations of the related dithiolate diiron complexes [$Fe_2\{\mu\text{-}S_2(CH_2)_3\}(CO)_4(L)_2$] (L = CO or phosphine) were conducted by Rauchfuss's group and Darensbourg's group [197–199] and revealed the catalytic activities to be strongly dependant on the nature of the ligand L. Two proposed mechanisms for these complexes are shown in Fig. 7. EECC mechanism (Fig. 7a) is adopted when L is a CO ligand, whereas ECCE mechanism (Fig. 7b) is likely for the phosphine complex.

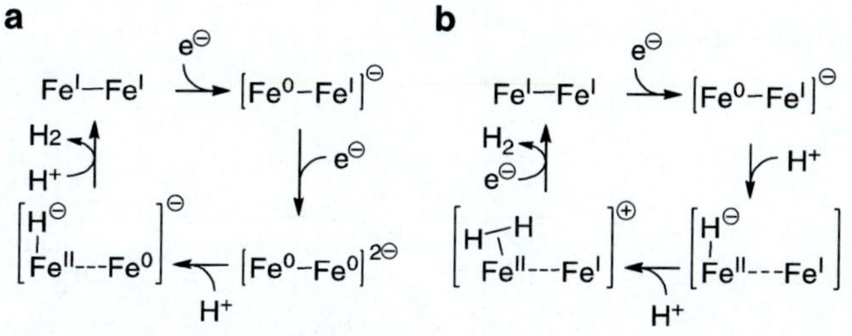

Fig. 7 Two proposed mechanisms for H_2 production from [$Fe_2\{\mu\text{-}S_2(CH_2)_3\}(CO)_4(L)_2$]

The μ-hydride diiron complex [$(\mu\text{-H})Fe_2(BC)(CO)_4(PMe_3)_2$]$PF_6$ **24** (BC = benzo-[*c*]-cinnoline), which was prepared by protonation of [$Fe_2(BC)(CO)_4(PMe_3)_2$] as a model for the active site of the [FeFe]-hydrogenase, was reported by Rauchfuss in 2007 (Scheme 59) [200]. The structure of **24** was confirmed by the X-ray analysis. The electrochemical reduction of protons was attained by using **24** as a catalyst in the presence of *p*-toluenesulfonic acid (pK_a = 8.73).

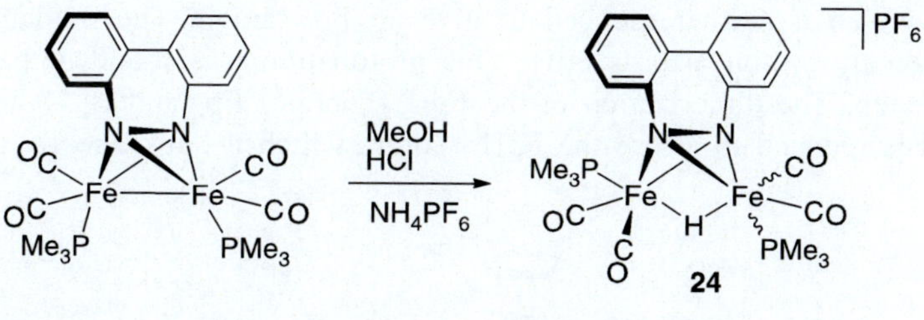

Scheme 59 Protonation reaction of [$Fe_2(BC)(CO)_4(PMe_3)_2$] to form **24**

In the same year, Evans and coworkers reported the electrochemical reduction of protons to H_2 catalyzed by the sulfur-bridged dinuclear iron complex **25** as a hydrogenase mimic in which acetic acid was used as a proton source [201]. The proposed mechanism for this reaction is shown in Scheme 60. The reduction of **25** readily affords **25**$^{2-}$ via a one electron reduction product **25**$^{-}$. Protonation of **25**$^{2-}$ by acetic acid produces the μ-hydride complex **25H**$^{-}$ which is reduced to **25H**$^{2-}$. The obtained **25H**$^{2-}$ reacts with acetic acid to generate molecular hydrogen

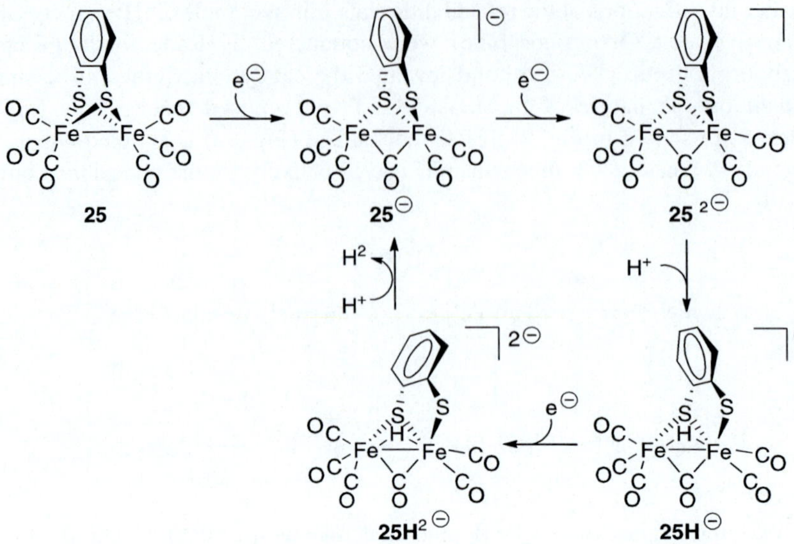

Scheme 60 Proposed mechanism of the reduction of protons to H_2 catalyzed by **25**

and **25**$^-$ to complete the catalytic cycle. This ECEC mechanism was confirmed by calculations and electrochemical experiments.

Evans found that molecular hydrogen was efficiently generated by the reaction of a simple diiron complex [CpFe(CO)$_2$]$_2$ (Fp$_2$) with acetic acid (pK_a = 22.3) in acetonitrile [202]. Electrochemical simulations revealed that Fp$_2$, [CpFe(CO)$_2$]$^-$ (Fp$^-$), and [CpFe(CO)$_2$H] (FpH) were key intermediates in this catalytic mechanism (Scheme 61). Reduction of Fp$_2$ produces both an Fp$^-$ anion and an Fp• radical, which is further reduced to give an Fp$^-$ anion. The oxidation of the Fp$^-$ anion by proton affords FpH. This protonation was found to be the rate-limiting step. The dimerization of the FpH generates Fp$_2$ and H_2. Alternatively, the FpH is reduced to afford the FpH$^-$ anion, which is subsequently protonated

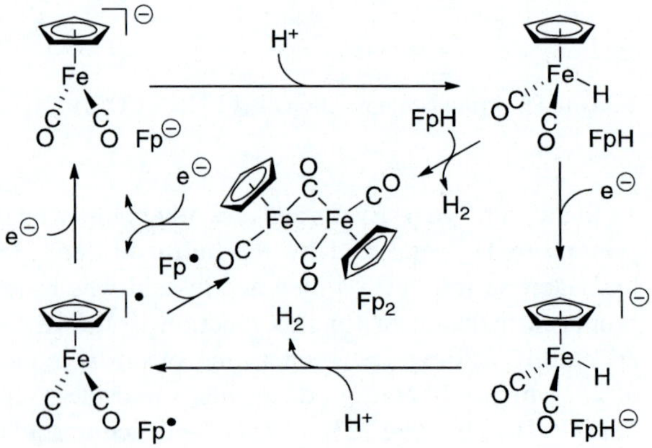

Scheme 61 [CpFe(CO)$_2$]$_2$ reduction followed by catalytic reduction of protons to H_2

with release of H_2. This system was adaptable to weaker acids such as 2-BrC_6H_4OH ($pK_a = 23.9$), 4-BrC_6H_4OH ($pK_a = 25.5$), and 4-$^tBuC_6H_4OH$ ($pK_a = 27.5$). In contrast, **25** showed no activity when 4-$^tBuC_6H_4OH$ was used.

Chiang and coworkers synthesized a dimer of compound **26** in which two diiron subunits are linked by two azadithiolate ligands as a model of the active site for the [FeFe]-hydrogenase [203]. Protonation of **26** afforded the μ-hydride complex **[26-2H^{N}2H^{Fe}]$^{4+}$** via the initially protonated spieces **[26-2H^N]$^{2+}$** (Scheme 62). These three complexes were also characterized by the X-ray diffraction analyses. H_2-generation was observed by electrochemical reduction of protons catalyzed by **26** in the presence of HBF_4 as a proton source. It was experimentally ascertained that **[26-2H^{N}2H^{Fe}]$^{4+}$** was converted into **26** by four irreversible reduction steps in the absence of HBF_4.

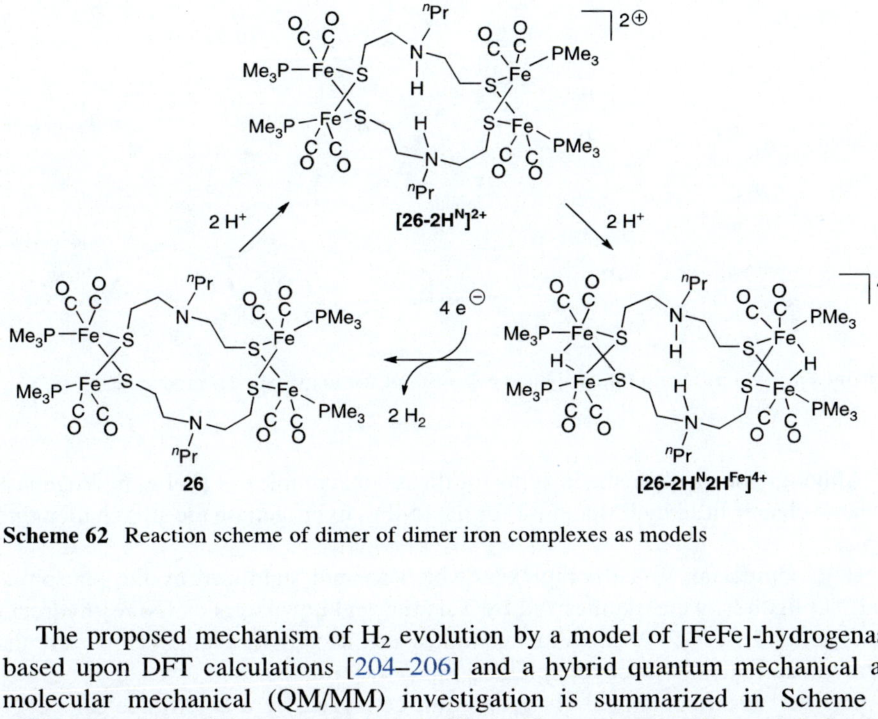

Scheme 62 Reaction scheme of dimer of dimer iron complexes as models

The proposed mechanism of H_2 evolution by a model of [FeFe]-hydrogenases based upon DFT calculations [204–206] and a hybrid quantum mechanical and molecular mechanical (QM/MM) investigation is summarized in Scheme 63 [207]. Complex **I** is converted into **II** by both protonation and reduction. Migration of the proton on the N atom to the Fe center in **II** produces the hydride complex **III**, and then protonation affords **IV**. In the next step, two pathways are conceivable. One is that the molecular hydrogen complex **VI** is synthesized by proton transfer and subsequent reduction (Path **a**). The other proposed by De Gioia, Ryde, and coworkers [207] is that the reduction of **IV** affords **VI** via the terminal hydride complex **V** (Path **b**). Dehydrogenation from **VI** regenerates **I**.

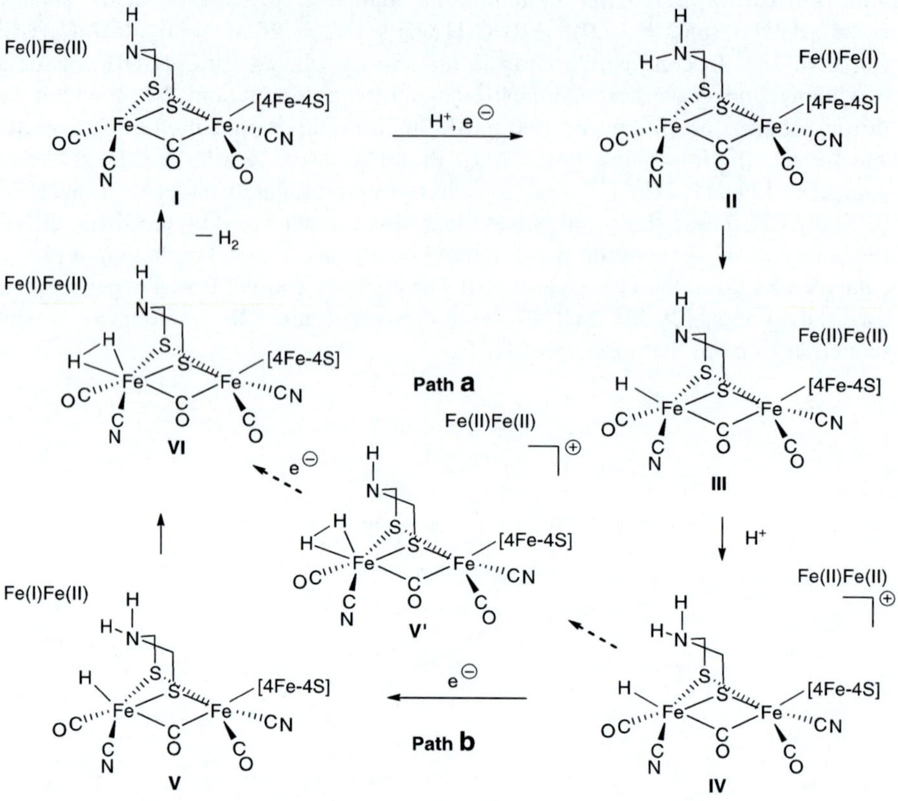

Scheme 63 Two pathways for catalytic mechanism of the Enzymatic H_2 production

Although the mechanistic insights for the model complex of [FeFe] hydrogenase became clearer little by little, those for the [NiFe] hydrogenase are still challenging topics.

Heterodinuclear Ni–Fe complexes, which are not stabilized by the phosphine and NO ligands, were synthesized by Tatsumi and coworkers as [NiFe] hydrogenase mimics [208–210]. Several examples are shown in Fig. 8. However, the catalytic activities of these complexes are not ascertained.

The first electrochemical H_2 generation catalyzed by a hetero-nuclear Fe–Ni complex $[Ni(L)Fe_2(CO)_6]$ (**27**) [$L^{2-} = (CH_3C_6H_3S_2)_2(CH_2)_3{}^{2-}$] (Fig. 9) with trifluoroacetic acid was reported by Schöder and coworkers in 2006 [211]. Based on their electrochemical behavior, spectroscopic data, and DFT calculations of **27**, an EECC mechanism was ruled out and therefore an ECCE or ECEC mechanism involving the formation of Fe^{II}–H^- and Ni^{III}–H^- intermediates is likely. In this cycle, six catalytic turnovers were achieved. This value is comparable to those for $[Fe_2(X)(CO)_4(PMe_3)_2]$ [X = 2(EtS$^-$), pdt^{2-}, edt^{2-}, xyldt^{2-}] (ca. 6–30 TON) [198, 199].

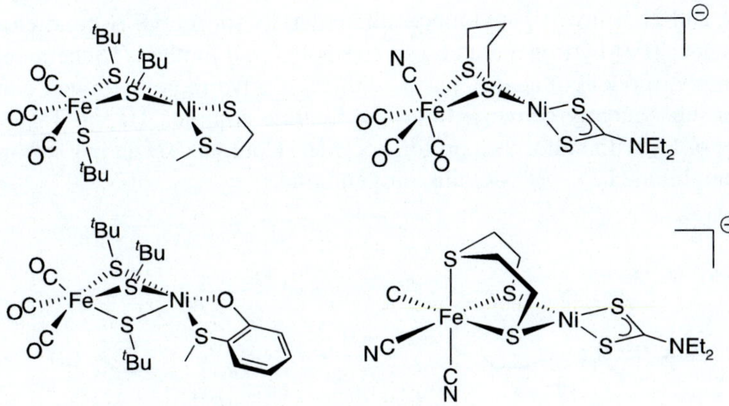

Fig. 8 The model complexes for [NiFe]-hydrogenases

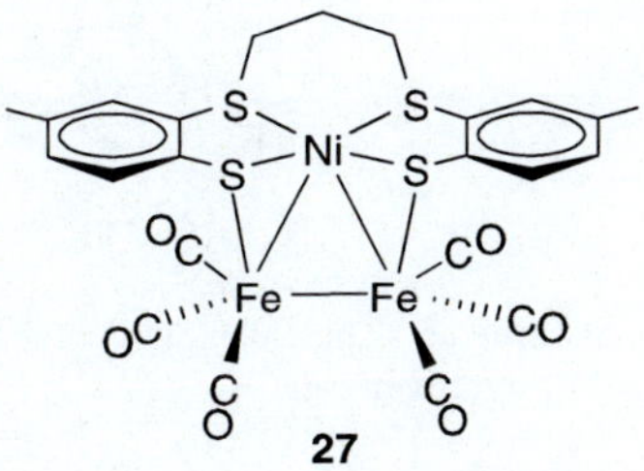

27

Fig. 9 Electrochemical H_2 generation catalyzed by complex **27** as a model for the [NiFe] hydrogenase

In 2009, Rauchfuss and coworkers succeeded in the synthesis of the Fe–μ-H–Ni complex $[(CO)_3Fe(pdt)(\mu\text{-}H)Ni(dppe)]BF_4$ **28** (pdt = 1,3-propanedithiolate, dppe = 1,2-$C_2H_4(PPh_2)_2$) as a model for [NiFe]-hydrogenases (Scheme 64) [212]. The structure of **28** was characterized by X-ray crystallographic analysis. This is the first example of an Fe–Ni thiolato hydride complex. Evolution of H_2 by electrochemical reduction of CF_3CO_2H ($pK_a = 12.65$) was observed in the presence of the catalytic amounts of **28**.

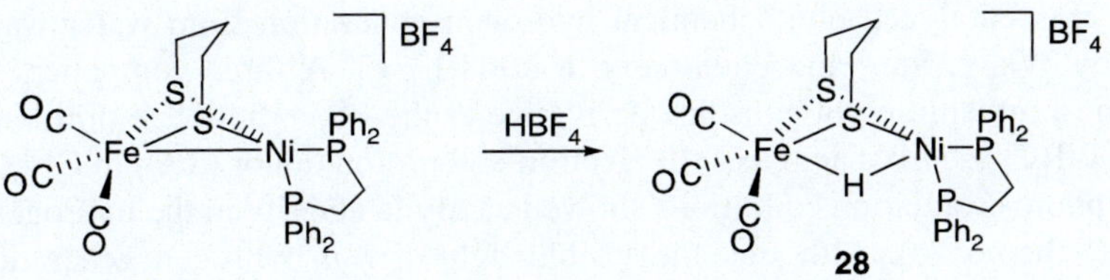

28

Scheme 64 Protonation reaction of $(CO)_3Fe(pdt)Ni(dppe)$ to form **28**

Yang and Hall investigated mechanistic details for the H–H bond cleavage by mononuclear [Fe]-hydrogenases using computational methods (Scheme 65) [213]. Accordingly, model complex **I** reacts with H_2 to give the dihydrogen complex **II**, which is subsequently converted into the hydride complex **III** via $\mathbf{TS_{II,III}}$ for the cleavage of H_2 with sulfur as a proton accepter. Complex **III** shows strong interaction through Fe–$H^{\delta-} \cdots H^{\delta+}$–O dihydrogen bond.

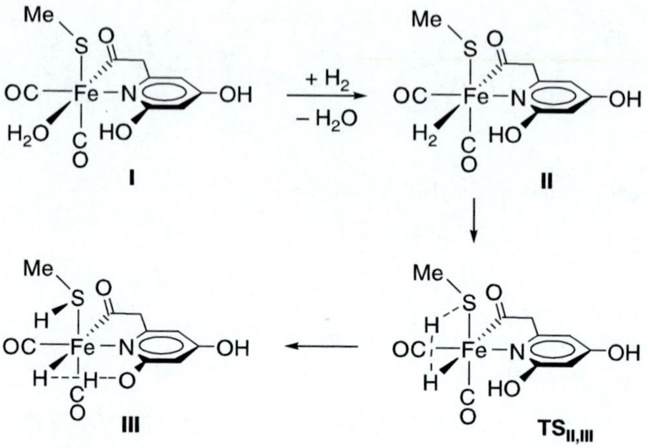

Scheme 65 The calculated mechanism of H–H bond cleavage reaction of the model complex for [Fe]-hydrogenases

5.2 *Photochemical Reduction*

Transition-metal catalyzed photochemical reactions for hydrogen generation from water have recently been investigated in detail. The reaction system is composed of three major components such as a photosensitizer (PS), a water reduction catalyst (WRC), and a sacrificial reagent (SR). Although noble-metal complexes as WRC have been used [214–230], examples for iron complexes are quite rare. It is well known that a hydride as well as a dihydrogen (or dihydride) complex plays important roles in this reaction.

An iron-catalyzed photochemical hydrogen generation from water was developed by Wang, Sun, and coworkers in 2008 [231]. A three component catalyst system (a ruthenium polypyridine derivative as the PS, a dithiolate diiron complex as the WRC, ascorbic acid as both electron and proton donor (SR)) in CH_3CN/H_2O under photo-irradiation conditions showed catalytic activity in the hydrogen generation (Scheme 66). Although the visible light-driven hydrogen generation from water proved successful, the catalytic activity is not high (up to 4.3 TON based on the iron complex).

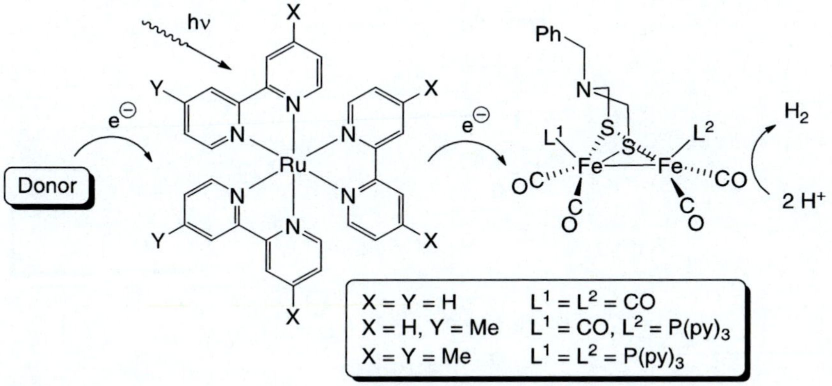

Scheme 66 Photochemical hydrogen generation catalyzed by the model complex of [FeFe]-hydrogenases

Very recently, Hammarström, Ott, and coworkers found that the catalytic activity was significantly increased (up to 200 TON based on the iron complex) when a diiron complex having a 3,6-dichlorobenzene-1,2-dithiolate ligand (Fig. 10) instead of a benzylazadithiolate ligand was used as a WRC (see ref. [64] in [232]).

Fig. 10 Diiron complex having a 3,6-dichlorobenzene-1,2-dithiolate ligand as a catalyst for the hydrogen generation

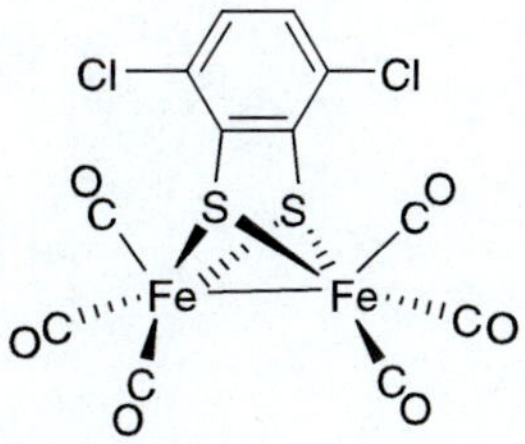

Recently, Beller and coworkers reported another efficient Fe-catalyzed light-driven hydrogen generation system [233]. The highest activity for the iron complex was achieved by using the combination of [Ir(bpy)(ppy)$_2$][PF$_6$]as the PS, an iron(0) carbonyl complex as the WRC, and triethylamine (TEA) as the SR under photolysis conditions. The maximum TON was 322 based on the iron complex. The mechanism for the photochemical reduction of water, which is similar to the one proposed by Wang, Sun, and coworkers [231], is shown in Scheme 67. The iron carbonyl species **a** is generated by reduction of [Fe(CO)$_5$] by the PS$^-$, which is generated by photoreaction of PS and TEA as the SR. Protonation and reduction of **a** affords the hydride species **b**, which is subsequently protonated to give dihydrogen with regeneration of **a**.

Two different types of zinc-porphyrins coordinated diiron complex act as catalysts for the photochemical reduction hydrogen evolution from water. In this system

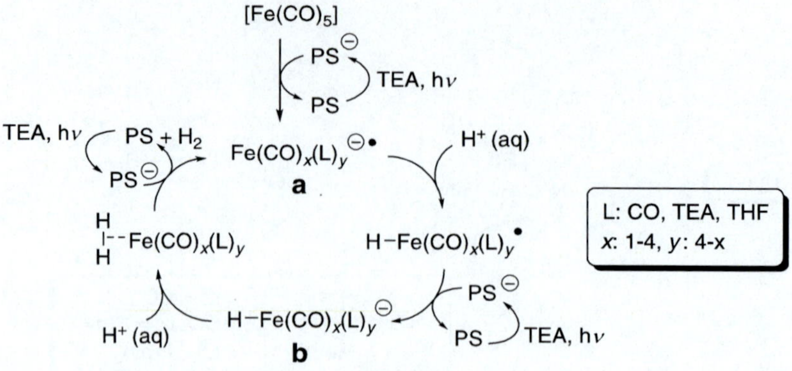

Scheme 67 The proposed mechanism for iron-catalyzed light-driven hydrogen generation from water

achieved by Hartl and coworkers (Scheme 68) [234], two different chromophores (***Zn*^1^** and ***Zn*^2^**) and a tertiary amine N^iPr_2Et as a SR for the electron donor are required.

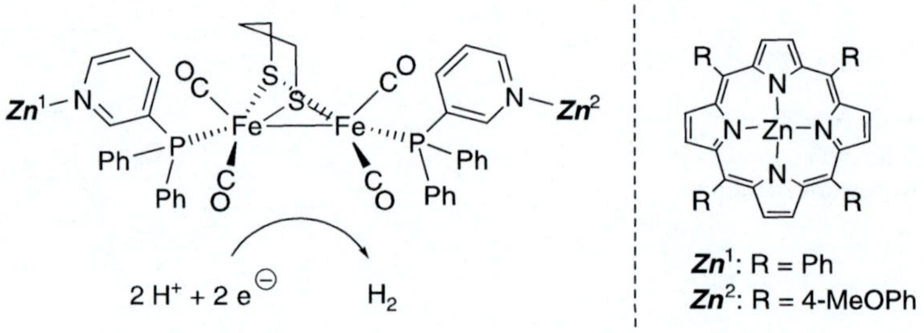

Scheme 68 The molecular hydrogen evolution catalyzed by self-assembled Zn–Fe–Zn complex

Acknowledgment We would like to show our respect for the great efforts of all authors, whose names were listed in the references. We wish to thank Dr. Yuji Suzaki for the reference collection.

Appendix

The isolated iron hydride complexes introduced in this chapter are listed in Table 12, where the hydride chemical shifts in the ^{1}HNMR spectra and the Fe–H bond distances are summarized.

Table 12

Complex		^{1}H NMR hydride signal (δ, ppm)	X-ray analysis Fe-H bond distance (Å)	Ref.
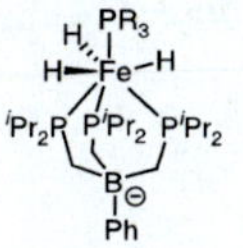	R_3 = Me_3: **3**	δ 13.72 (s)[a]	1.44(2), 1.46(3), 1.51(2)	[26]
	Et_3: **4**	δ 13.16 (s)[a]		[26]
	$MePh_2$	δ 13.16 (s)[a]		[26]
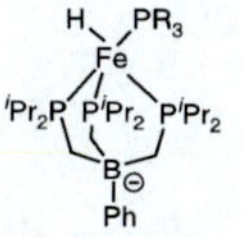	R_3 = Me_3	δ 2.34 (br,s)	1.511(14)	[26]
	$MePh_2$	N.D.	1.363(13)	[26]
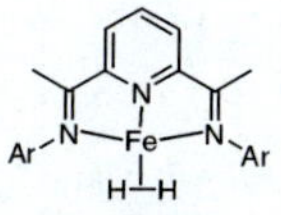 **9**		N.D.		[27]
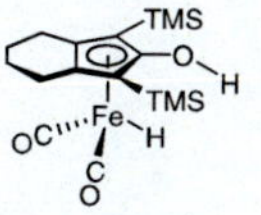 **11**		δ 11.62 (s)	1.38(2)	[235] [47] (for catalytic reaction)
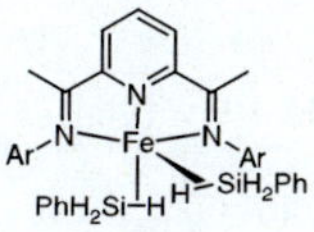 **18**		δ 6.69 (s)	1.45(3), 1.51(3)	[27]
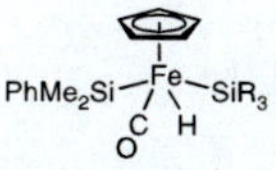	R_3 = Me_2Ph: **21**	δ 13.22 (s)	1.44(3)	[166,167]
	Et	δ 13.71 (s)	1.46(3)	
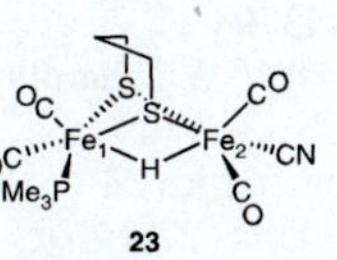 **23**		δ 17.08 (d, J_{PH} = 24 Hz)	1.70(2) for Fe_1 1.63(2) for Fe_2	[197]
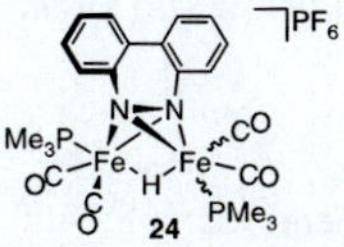 **24**		δ 15.479 (dd, J_{PH} = 8, 26.5 Hz, *unsym*)		[200]
		δ15.621 (dd, J_{PH} = 9 Hz, *sym*)	1.62(3) for *sym*	[200]
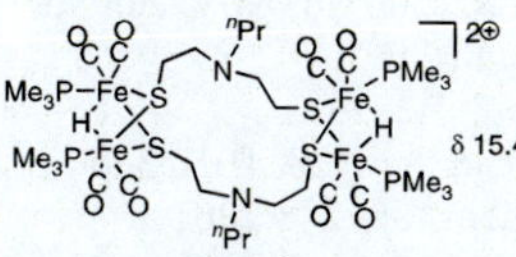 **[26-2HFe]$^{2+}$**		δ 15.46 (d, J_{PH} = 21 .5 Hz)	1.620[b], 1.668[b]	[203]
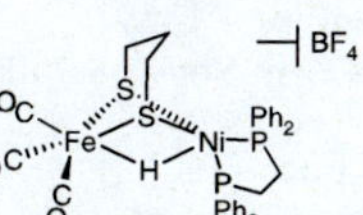 **28**		δ 3.53 (tt, J_{PH} = 6 Hz, J_{HH} = 0.6 Hz)	1.46(6)	[212]

[a] $^1H\{^{31}P\}$ NMR measurement.

[b] The bond distances were calculated by Mercury 2.2 for Windows [236] using CIF data.

References

1. Connelly NG, Damhus T, Hartshorn RM, Hutton AT (eds) (2005) Nomenclature of Inorganic Chemistry IUPAC Recommendations 2005
2. Hieber W, Leutert F (1931) Naturwissenschaften 19:360
3. Weichselfelder T, Thiede B (1926) Justus Liebigs Ann Chem 447:64
4. Dahl LF, Blount JF (1965) Inorg Chem 4:1373
5. Manojlović-Muir L, Muir KW, Ibers JA (1970) Inorg Chem 9:447
6. Guggenberger LJ, Titus DD, Flood MT, Marsh RE, Orio AA, Gray HB (1972) J Am Chem Soc 94:1135
7. Esteruelas MA, Oro LA (1998) Chem Rev 98:577
8. Morris RH, Sawyer JF, Shiralian M, Zubkowski JD (1985) J Am Chem Soc 107:5581
9. Bianchini C, Peruzzini M, Zanobini F (1988) J Organomet Chem 354:C19
10. Field LD, Messerle BA, Smernik RJ (1997) Inorg Chem 36:5984
11. Stoppioni P, Mani F, Sacconi L (1974) Inorg Chim Acta 11:227
12. Plietker B (ed) (2008) Iron catalysis in organic chemistry, Wiley-VCH, Weinheim
13. Bolm C, Legros J, Le Paih J, Zani L (2004) Chem Rev 104:6217
14. Fürstner A, Martin R (2005) Chem Lett 34:624
15. Correa A, Mancheño OG, Bolm C (2008) Chem Soc Rev 37:1108
16. Sherry BD, Fürstner A (2008) Acc Chem Res 41:1500
17. Gaillard S, Renaud J-L (2008) ChemSusChem 1:505
18. Enthaler S, Junge K, Beller M (2008) Angew Chem 120:3363, Angew Chem Int Ed 47:3317
19. Sarhan AAO, Bolm C (2009) Chem Soc Rev 38:2730
20. Chen MS, White MC (2007) Science 318:783
21. Frankel EN, Emken EA, Peters HM, Davison VL, Butterfield RO (1964) J Org Chem 29:3292
22. Frankel EN, Emken EA, Davison VL (1965) J Org Chem 30:2739
23. Harmon RE, Gupta SK, Brown DJ (1973) Chem Rev 73:21
24. Schroder MA, Wrighton MS (1976) J Am Chem Soc 98:551
25. Kismartoni LC, Weitz E, Cedeno DL (2005) Organometallics 24:4714
26. Daida EJ, Peters JC (2004) Inorg Chem 43:7474
27. Bart SC, Lobkovsky E, Chirik PJ (2004) J Am Chem Soc 126:13794
28. Bart SC, Chlopek K, Bill E, Bouwkamp MW, Lobkovsky E, Neese F, Wieghardt K, Chirik PJ (2006) J Am Chem Soc 128:13901
29. Bart SC, Hawrelak EJ, Lobkovsky E, Chirik PJ (2005) Organometallics 24:5518
30. Trovitch RJ, Lobkovsky E, Bill E, Chirik PJ (2008) Organometallics 27:1470
31. Archer AM, Bouwkamp MW, Cortez M-P, Lobkovsky E, Chirik PJ (2006) Organometallics 25:4269
32. Noyori R, Ohkuma T (2001) Angew Chem Int Ed 40:40
33. Ikariya T, Murata K, Noyori R (2006) Org Biomol Chem 4:393
34. Shvo Y, Czarkie D, Rahamim Y, Chodosh DF (1986) J Am Chem Soc 108:7400
35. Casey CP, Singer SW, Powell DR, Hayashi RK, Kavana M (2001) J Am Chem Soc 123:1090
36. Santosh Laxmi YR, Bäckvall J-E (2000) Chem Commun 611
37. Lau CP, Ng SM, Jia G, Lin Z (2007) Coord Chem Rev 251:2223
38. Noyori R, Kitamura M, Ohkuma T (2004) Proc Natl Acad Sci USA 101:5356
39. Clapham SE, Hadzovic A, Morris RH (2004) Coord Chem Rev 248:2201
40. Conley BL, Pennington-Boggio MK, Boz E, Williams TJ (2010) Chem Rev 110:2294
41. Blum Y, Czarkie D, Rahamim Y, Shvo Y (1985) Organometallics 4:1459
42. Shvo Y, Czarkie D, Rahamim Y, Chodosh DF (1986) J Am Chem Soc 108:7400
43. Menashe N, Shvo Y (1991) Organometallics 10:3885
44. Menashe N, Salant E, Shvo Y (1996) J Organomet Chem 514:97
45. Noyori R, Ohkuma T (1999) Pure Appl Chem 71:1493

46. Doucet H, Ohkuma T, Murata K, Yokozama T, Kozawa M, Katayama E, England AF, Ikariya T, Noyori R (1998) Angew Chem Int Ed 37:1703
47. Casey CP, Guan H (2007) J Am Chem Soc 129:5816
48. Casey CP, Guan H (2009) J Am Chem Soc 131:2499
49. Sui-Seng C, Freutel F, Lough AJ, Morris RH (2008) Angew Chem Int Ed 47:940
50. Hashiguchi S, Fujii A, Takehara J, Ikariya T, Noyori R (1995) J Am Chem Soc 117:7562
51. Rautenstrauch V, Hoang-Cong X, Churlaud R, Abdur-Rashid K, Morris RH (2003) Chem Eur J 9:4954
52. Mikhailine A, Lough AJ, Morris RH (2009) J Am Chem Soc 131:1394
53. Enthaler S, Erre G, Tse MK, Junge K, Beller M (2006) Tetrahedron Lett 47:8095
54. Enthaler S, Hagemann B, Erre G, Junge K, Beller M (2006) Chem Asian J 1:598
55. Zassinovich G, Mestroni G, Gladiali S (1992) Chem Rev 92:1051
56. Gladiali S, Mestroni G (2004) In: Beller M, Bolm C (eds) Transition metals for organic synthesis, 2nd edn. Wiley-VCH, Weinheim, p 145
57. Gladiali S, Alberico E (2006) Chem Soc Rev 35:226
58. Thorson MK, Klinkel KL, Wang J, Williams TJ (2009) Eur J Inorg Chem 295
59. Ojima I (1989) In: Patai S, Rapport Z (eds) The chemistry of organosilicon compounds, Wiley, New York
60. Brook MA (2000) Silicon in organic, organometallic, and polymer chemistry. Wiley, New York
61. Marciniec B (ed) (1992) Comprehensive handbook on hydrosilylation. Pergamon, Oxford
62. Marciniec B (1996) In: Cornils B, Herrmann WA (eds) Applied homogeneous catalysis with organometallic compounds, vol 1. Weinheim, Wiley-VCH, Chap 2
63. Speier JL (1979) Adv Organomet Chem 17:407
64. Marciniec B, Guliñski J (1993) J Organomet Chem 446:15
65. Nishiyama H, Furuta A (2007) Chem Commun 760
66. Furuta A, Nishiyama H (2008) Tetrahedron Lett 49:110
67. Shaikh NS, Junge K, Beller M (2007) Org Lett 9:5429
68. Shaikh NS, Enthaler S, Junge K, Beller M (2008) Angew Chem Int Ed 47:2497
69. Nishibayashi Y, Takei I, Uemura S, Hidai M (1998) Organometallics 17:3420
70. Yun J, Buchwald SL (1999) J Am Chem Soc 121:5640
71. Langlotz BK, Wadepohl H, Gade LH (2008) Angew Chem Int Ed 47:4670
72. Tondreau AM, Lobkovsky E, Chirik PJ (2008) Org Lett 10:2789
73. Tondreau AM, Darmon JM, Wile BM, Floyd SK, Lobkovsky E, Chirik PJ (2009) Organometallics 28:3928
74. Brown HC, Zaidlewicz M (2001) In: Organic Syntheses via Boranes, Vol 2. Aldrich Chemical Company, Milwaukee, p 118
75. Yamamoto Y, Asao N (1993) Chem Rev 93:2207
76. Bubnov YN (1987) Pure Appl Chem 59:895
77. Brown HC, Zaidlewicz M (2001) Organic syntheses via boranes, Vol 2. Aldrich Chemical Company, Milwaukee, WI, pp 189 and 207
78. Kennedy JWJ, Hall DG (2003) Angew Chem Int Ed 42:4732
79. Sugiura M, Hirano K, Kobayashi S (2004) J Am Chem Soc 126:7182
80. Solin N, Wallner OA, Szabó KJ (2005) Org Lett 7:689
81. Lou S, Moquist PN, Schaus SE (2007) J Am Chem Soc 129:15398
82. Sebelius S, Olsson VJ, Wallner OA, Szabó KJ (2006) J Am Chem Soc 128:8150
83. Burgess K, Ohlmeyer MJ (1991) Chem Rev 91:1179
84. Beletskaya I, Pelter A (1997) Tetrahedron 53:4957
85. Satoh M, Nomoto Y, Miyaura N, Suzuki A (1989) Tetrahedron Lett 30:3789
86. Zaidlewicz M, Meller J (1997) Tetrahedron Lett 38:7279
87. Matsumoto Y, Hayashi T (1991) Tetrahedron Lett 32:3387
88. Wu JY, Moreau B, Ritter T (2009) J Am Chem Soc 131:12915
89. Suginome M, Ohmori Y, Ito Y (2000) J Organomet Chem 611:403
90. Nakamura M, Matsuo K, Ito S, Nakamura E (2004) J Am Chem Soc 126:3686

91. Sapountzis I, Lin W, Kofink CC, Despotopoulou C, Knochel P (2005) Angew Chem 117:1682, Angew Chem Int Ed 44:1654
92. Plietker B (2006) Angew Chem 118:6200, Angew Chem Int Ed 45:6053
93. Cahiez G, Habiak V, Duplais C, Moyeux A (2007) Angew Chem 119:4442, Angew Chem Int Ed 46:4364
94. Guérinot A, Reymond S, Cossy J (2007) Angew Chem 119:6641, Angew Chem Int Ed 46:6521
95. Fürstner A, Martin R, Krause H, Seidel G, Goddard R, Lehmann CW (2008) J Am Chem Soc 130:8773
96. Carril M, Correa A, Bolm C (2008) Angew Chem 120:4940, Angew Chem Int Ed 47:4862
97. Czaplik WM, Mayer M, Wangelin AJ (2009) Angew Chem 121:616, Angew Chem Int Ed 48:607
98. Yoshikai N, Matsumoto A, Norinder J, Nakamura E (2009) Angew Chem 121:2969, Angew Chem Int Ed 48:2925
99. Cahiez G, Foulgoc L, Moyeux A (2009) Angew Chem 121:3013, Angew Chem Int Ed 48:2969
100. Li Y-Z, Li B-J, Lu X-Y, Lin S, Shi Z-J (2009) Angew Chem 121:3875, Angew Chem Int Ed 48:3817
101. Li Z, Cao L, Li C-J (2007) Angew Chem 119:6625, Angew Chem Int Ed 46:6505
102. Li Z, Yu R, Li H (2008) Angew Chem 120:7607, Angew Chem Int Ed 47:7497
103. Zhang Y, Li C-J (2007) Eur J Org Chem 4654
104. Nagano T, Hayashi T (2004) Org Lett 6:1297
105. Zhang D, Ready JM (2006) J Am Chem Soc 128:15050
106. Duboudin JG, Jousseaume B, Bonakdar A, Saux A (1979) J Organomet Chem 168:227
107. Lu Z, Ma S (2006) J Org Chem 71:2655
108. Moreau B, Wu JY, Ritter T (2009) Org Lett 11:337
109. Zhang S-Y, Tu Y-Q, Fan C-A, Zhang F-M, Shi L (2009) Angew Chem Int Ed 48:8761
110. Driller KM, Klein H, Jackstell R, Beller M (2009) Angew Chem Int Ed 48:6041
111. Periasamy M, Rameshkumar C, Radhakrishnan U (1997) Tetrahedron Lett 38:7229
112. Periasamy M, Rameshkumar C, Radhakrishnan U, Brunet J-J (1998) J Org Chem 63:4930
113. Rameshkumar C, Periasamy M (2000) Organometallics 19:2400
114. Rameshkumar C, Periasamy M (2000) Tetrahedron Lett 41:2719
115. Periasamy M, Mukkanti A, Raj DS (2004) Organometallics 23:6323
116. Periasamy M, Mukkanti A, Raj DS (2004) Organometallics 23:619
117. Rameshkumar C, Periasamy M (2000) Synlett 1619
118. Periasamy M, Rameshkumar C, Mukkati A (2002) J Organomet Chem 649:209
119. Sylvester KT, Chirik PJ (2009) J Am Chem Soc 131:8772
120. Jang H-Y, Krische MJ (2004) Acc Chem Res 37:653
121. Small BL, Brookhart M, Bennett AMA (1998) J Am Chem Soc 120:4049
122. Small BL, Brookhart M (1998) J Am Chem Soc 120:7143
123. Britovsek GJP, Gibson VC, Kimberley BS, Maddox PJ, McTavish SJ, Solan GA, Whitea AJP, Williams DJ (1998) Chem Commun 849
124. Britovsek GJP, Bruce M, Gibson VC, Kimberley BS, Maddox PJ, Mastroianni S, McTavish SJ, Redshaw C, Solan GA, Strömberg S, White AJP, Williams DJ (1999) J Am Chem Soc 121:8728
125. Britovsek GJP, Mastroianni S, Solan GA, Baugh SPD, Redshaw C, Gibson VC, White AJP, Williams DJ, Elsegood MRJ (2000) Chem Eur J 6:2221
126. Small BL, Marcucci AJ (2001) Organometallics 20:5738
127. Babik ST, Fink G (2002) J Mol Cat A: Chem 188:245
128. Khoroshun DV, Musaev DG, Vreven T, Morokuma K (2001) Organometallics 20:2007
129. Raucoules R, de Bruin T, Raybaud P, Adamo C (2009) Organometallics 28:5358
130. Kadyrov R, Riermeier TH (2003) Angew Chem 115:5630, Angew Chem Int Ed 42:5472
131. Bunlaksananusorn T, Rampf F (2005) Synlett 2682

132. Hamid MHSA, Williams JMJ (2007) Chem Commun 725
133. Tararov VI, Börner A (2005) Synlett 203
134. Tararov VI, Kadyrov R, Riermeier TH, Fischer C, Börner A (2004) Adv Synth Catal 346:561
135. Tararov VI, Kadyrov R, Riermeier TH, Börner A (2002) Adv Synth Catal 344:200
136. Tararov VI, Kadyrov R, Riermeier TH, Börner A (2000) Chem Commun 1867
137. Kadyrov R, Riermeier TH, Dingerdissen U, Tararov V, Börner A (2003) J Org Chem 68:4067
138. Kitamura M, Lee D, Hayashi S, Tanaka S, Yoshimura M (2002) J Org Chem 67:8685
139. Imao D, Fujihara S, Yamamoto T, Ohta T, Ito Y (2005) Tetrahedron 61:6988
140. Fujita K-I, Yamaguchi R (2005) Synlett 560
141. Chi Y, Zhou Y-G, Zhang X (2003) J Org Chem 68:4120
142. Blank B, Madalska M, Kempe R (2008) Adv Synth Catal 350:749
143. Robichaud A, Nait Ajjou A (2006) Tetrahedron Lett 47:3633
144. Bhor MD, Bhanushali MJ, Nandurkar NS, Bhanage BM (2008) Tetrahedron Lett 49:965
145. Zhou S, Addis D, Das S, Junge K, Beller M (2009) Chem Commun 4883
146. Hanada S, Motoyama Y, Nagashima H (2008) Eur J Org Chem 4097
147. Zhou S, Junge K, Addis D, Das S, Beller M (2009) Angew Chem Int Ed 48:9507
148. Sunada Y, Kawakami H, Imaoka T, Motoyama Y, Nagashima H (2009) Angew Chem Int Ed 48:9511
149. Iwanami K, Seo H, Tobita Y, Oriyama T (2005) Synthesis 183
150. Iwanami K, Yano K, Oriyama T (2005) Synthesis 2669
151. Iwanami K, Yano K, Oriyama T (2007) Chem Lett 36:38
152. Doyle MP, DeBruyn DJ, Kooistra DA (1972) J Am Chem Soc 94:3659
153. Nicolaou KC, Hwang C-K, Nugiel DA (1989) J Am Chem Soc 111:4136
154. Izumi M, Fukase K (2005) Chem Lett 34:594
155. Kato J-I, Iwasawa N, Mukaiyama T (1985) Chem Lett 743
156. Sassaman MB, Kotian KD, Prakash GKS, Olah GA (1987) J Org Chem 52:4314
157. Hatakeyama S, Mori H, Kitano K, Yamada H, Nishizawa M (1994) Tetrahedron Lett 35:4367
158. Komatsu N, Ishida J, Suzuki H (1997) Tetrahedron Lett 38:7219
159. Bajwa JS, Jiang X, Slade J, Prasad K, Repič O, Blacklock TJ (2002) Tetrahedron Lett 43:6709
160. Yang W-C, Lu X-A, Kulkarni SS, Hung S-C (2003) Tetrahedron Lett 44:7837
161. Chandrasekhar S, Chandrashekar G, Babu BN, Vijeender K, Venkatram Reddy K (2004) Tetrahedron Lett 45:5497
162. Wada M, Nagayama S, Mizutani K, Hiroi R, Miyoshi N (2002) Chem Lett 248
163. Nakazawa H, Kamata K, Itazaki M (2005) Chem Commun 4004
164. Nakazawa H, Itazaki M, Kamata K, Ueda K (2007) Chem Asian J 2:882
165. Nakazawa H, Kawasaki T, Miyoshi K, Suresh CH, Koga N (2004) Organometallics 23:117
166. Itazaki M, Ueda K, Nakazawa H (2009) Angew Chem Int Ed 48:3313; ibid 48: 6938
167. Sharma HK, Pannell KH (2009) Angew Chem Int Ed 48:7052
168. Trost BM, Kulawiec RJ (1993) J Am Chem Soc 115:2027 (and references cited therein)
169. Bäckvall JE, Andreasson U (1993) Tetrahedron Lett 34:5459
170. Marko IE, Gautier A, Tsukazaki M, Llobet A, Plantalech-Mir E, Urch CJ, Brown SM (1999) Angew Chem Int Ed 38:1960
171. Slugovc C, Rüba E, Schmid R, Kirchner K (1999) Organometallics 18:4230
172. Ni Bricourt H, Monflier E, Carpentier J-F, Mortreux A (1998) Eur J Inorg Chem 1739
173. Cherkaoui H, Soufiaoui M, Grée R (2001) Tetrahedron 57:2379
174. Hendrix WT, Cowherd FG, von Rosenberg JL (1968) Chem Commun 97
175. Strauss JU, Ford PW (1975) Tetrahedron Lett 2917
176. Chong TS, Tan ST, Fan WY (2006) Chem A Eur J 12:5128
177. Shi B, O'Brien RJ, Bao S, Davis BH (2001) J Catal 199:202
178. Taber DF, Kanai K, Jiang Q, Bui G (2000) J Am Chem Soc 122:6807

179. Glascoe EA, Sawyer KR, Shanoski JE, Harris CB (2007) J Phys Chem C 111:8789
180. Sawyer KR, Glascoe EA, Cahoon JF, Schlegel JP, Harris CB (2008) Organometallics 27:4370
181. Navarro RM, Pena MA, Fierro JLG (2007) Chem Rev 107:3952
182. Peters JW, Lanzilotta WN, Lemon BJ, Seefeldt LC (1998) Science 282:1853
183. Nicolet Y, de Lacey AL, Vernede X, Fernandez VM, Hatchikian EC, Fontecilla-Camps JC (2001) J Am Chem Soc 123:1596
184. Nicolet Y, Piras C, Legrand P, Hatchikian CE, Fontecilla-Camps JC (1999) Structure 7:13
185. Peters JW, Lanzilotta WN, Lemon BJ, Seefeldt LC (1998) Science 282:1853
186. Nicolet Y, de Lacey AL, Vernede X, Fernandez VM, Hatchikian EC, Fontecilla-Camps JC (2001) J Am Chem Soc 123:1596
187. Frey M (2002) Chembiochem 3:153
188. Volbeda A, Garcin E, Piras C, de Lacey AL, Fernandez VM, Hatchikian EC, Frey M, Fontecilla-Camps JC (1996) J Am Chem Soc 118:12989
189. Volbeda A, Charon M-H, Piras C, Hatchikian EC, Frey M, Fontecilla-Camps JC (1995) Nature 373:580
190. Volbeda A, Martin L, Cavazza C, Matho M, Faber BW, Roseboom W, Albracht SPJ, Garcin E, Rousset M, Fontecilla-Camps JC (2005) J Biol Inorg Chem 10:239
191. Ogata H, Hirota S, Nakahara A, Komori H, Shibata N, Kato T, Kano K, Higuchi Y (2005) Structure 13:1635
192. de Lacey AL, Fernandez VM, Rousset M (2005) Coord Chem Rev 249:1596
193. Vincent KA, Belsey NA, Lubitz W, Armstrong FA (2006) J Am Chem Soc 128:7448
194. Huhmann-Vincent J, Scott BL, Kubas GJ (1999) Inorg Chim Acta 294:240
195. Shima S, Pilak O, Vogt S, Schick M, Stagni MS, Meyer-Klaucke W, Warkentin E, Thauer RK, Ermler U (2008) Science 321:572
196. Gloaguen F, Lawrence JD, Rauchfuss TB (2001) J Am Chem Soc 123:9476
197. Gloaguen F, Lawrence JD, Rauchfuss TB, Bénard M, Rohmer M-M (2002) Inorg Chem 41:6573
198. Mejia-Rodriguez R, Chong D, Reibenspies JH, Soriaga MP, Darensbourg MY (2004) J Am Chem Soc 126:12004
199. Chong D, Georgakaki IP, Mejia-Rodriguez R, Sanabria-Chinchilla J, Soriaga MP, Darensbourg MY (2003) Dalton Trans 3:4158
200. Volkers PI, Rauchfuss TB (2007) J Inorg Biochem 101:1748
201. Felton GAN, Vannucci AK, Chen J, Lockett LT, Okumura N, Petro BJ, Zakai UI, Evans DH, Glass RS, Lichtenberger DL (2007) J Am Chem Soc 129:12521
202. Felton GAN, Vannucci AK, Okumura N, Lockett LT, Evans DH, Glass RS, Lichtenberger DL (2008) Organometallics 27:4671
203. Chiang M-H, Liu Y-C, Yang S-T, Lee G-H (2009) Inorg Chem 48:7604
204. Liu Z-P, Hu P (2002) J Am Chem Soc 124:5175
205. Fan H-J, Hall MB (2001) J Am Chem Soc 123:3828
206. Zampella G, Greco C, Fantucci P, De Gioia L (2006) Inorg Chem 45:4109
207. Greco C, Bruschi M, De Gioia L, Ryde U (2007) Inorg Chem 46:5911
208. Ohki Y, Yasumura K, Kuge K, Tanino S, Ando M, Li Z, Tatsumi K (2008) Proc Natl Acad Sci USA 105:7652
209. Li Z, Ohki Y, Tatsumi K (2005) J Am Chem Soc 127:8950
210. Tanino S, Li Z, Ohki Y, Tatsumi K (2009) Inorg Chem 48:2358
211. Perra A, Davies ES, Hyde JR, Wang Q, McMaster J, Schröder M (2006) Chem Commun 1103
212. Barton BE, Whaley CM, Rauchfuss TB, Gray DL (2009) J Am Chem Soc 131:6942
213. Yang X, Hall MB (2009) J Am Chem Soc 131:10901
214. Kirch M, Lehn J-M, Sauvage J-P (1979) Helv Chim Acta 62:1345
215. Moradpour A, Amouyal E, Keller P, Kagan H (1978) Nouv J Chim 2:547
216. Kalyanasundaram K, Kiwi J, Grätzel M (1978) Helv Chim Acta 61:2720

217. Krishnan CV, Sutin N (1981) J Am Chem Soc 103:2141
218. Goldsmith JI, Hudson WR, Lowry MS, Anderson TH, Bernhard S (2005) J Am Chem Soc 127:7502
219. Tinker LL, McDaniel ND, Curtin PN, Smith CK, Ireland MJ, Bernhard S (2007) Chem Eur J 13:8726
220. Cline ED, Adamson SE, Bernhard S (2008) Inorg Chem 47:10378
221. Tinker LL, Bernhard S (2009) Inorg Chem 48:10507
222. Curtin PN, Tinker LL, Burgess CM, Cline ED, Bernhard S (2009) Inorg Chem 48:10498
223. Du P, Schneider J, Luo G, Brennessel WW, Eisenberg R (2009) Inorg Chem 48:4952
224. Du P, Knowles K, Eisenberg R (2008) J Am Chem Soc 130:12576
225. Zhang J, Du P, Schneider J, Jarosz P, Eisenberg R (2007) J Am Chem Soc 129:7726
226. Lazarides T, McCormick T, Du P, Luo G, Lindley B, Eisenberg R (2009) J Am Chem Soc 131:9192
227. Rau S, Schäfer B, Gleich D, Anders E, Rudolph M, Friedrich M, Görls H, Henry W, Vos JG (2006) Angew Chem 118:6361, Angew Chem Int Ed 45:6215
228. Elvington M, Brown J, Arachchige SM, Brewer KJ (2007) J Am Chem Soc 129:10644
229. Ozawa H, Haga M, Sakai K (2006) J Am Chem Soc 128:4926
230. Palmans R, Frank AJ (1991) J Phys Chem 95:9438
231. Na Y, Wang M, Pan J, Zhang P, Åkermark B, Sun L (2008) Inorg Chem 47:2805
232. Lomoth R, Ott S (2009) Dalton Trans 9952
233. Gärtner F, Sundararaju B, Surkus A-E, Boddien A, Loges B, Junge H, Dixneuf PH, Beller M (2009) Angew Chem Int Ed 48:9962
234. Kluwer AM, Kapre R, Hartl F, Lutz M, Spek AL, Brouwer AM, van Leeuwen PWNM, Reek JNH (2009) Proc Natl Acad Sci 106:10460
235. Knölker H-J, Goesmann H, Klauss R (1999) Angew Chem Int Ed 38:702
236. Mercury 2.2 for Windows. This material is available free of charge via the Internet at http://www.ccdc.cam.ac.uk/mercury/

Top Organomet Chem (2011) 33: 83–109
DOI: 10.1007/978-3-642-14670-1_3

Fe-Catalyzed Oxidation Reactions of Olefins, Alkanes, and Alcohols: Involvement of Oxo- and Peroxo Complexes

Kristin Schröder, Kathrin Junge, Bianca Bitterlich, and Matthias Beller

Abstract In this review, recent developments of iron-catalyzed oxidations of olefins (epoxidation), alkanes, arenes, and alcohols are summarized. Special focus is given on the ligand systems and the catalytic performance of the iron complexes. In addition, the mechanistic involvement of high-valent iron–oxo species is discussed.

Keywords Homogeneous catalysis · Iron · Oxidation

Contents

1 Introduction

Iron is an essential element for the proper function of nearly all known biological systems. In living organism, iron is generally stored in the center of metalloproteins, most important is the incorporation into heme complexes. These complexes

---

K. Schröder, K. Junge, B. Bitterlich, and M. Beller (✉)
Leibniz-Institut für Katalyse an der Universität Rostock e.V., Albert-Einstein-Straße 29a, D-18059 Rostock, Germany
e-mail: Matthias.Beller@catalysis.de

are an essential part of cytochromes, e.g., P450 cytochromes, which mediate numerous redox reactions, and of oxygen carrier proteins, for example hemoglobin. Nonheme iron-based enzymes include, for example, methane monooxygenase and hemerythrins, which regulate oxygen transport and fixation in marine invertebrates. The key active species in numerous biological oxidation reactions in which activation of oxygen is involved are known to be high-valent iron–oxo intermediates of heme and nonheme complexes [1, 2]. In the past two decades, significant advancements towards the direct characterization of Fe(IV)O-, Fe(V)O-, and even Fe(VI)O-species have been made [3]. Hence, detailed knowledge on the structure of model complexes such as $[Fe^{IV}(TMCS)(=O)](PF_6)$ (TMCS = 1-mercaptoethyl-4,8,11-trimethyl-1,4,8,11-tetraza cyclotetradecane) [4] and $Fe^{V}TAML(=O)$ (TAML = tetraamido macrocyclic ligand) [5] are nowadays available. There is no doubt that such basic knowledge is important for the understanding of biologically relevant actions and the more rational design of artificial catalysts. However, so far the catalytic performance of most of these model complexes is far away from the efficiency of biologically active systems and sometimes with model complexes no catalytic behavior is observed at all.

Because there exist a number of reviews which deals with the structural and mechanistic aspects of high-valent iron–oxo and peroxo complexes [6, 7], we focus in this report on the application and catalysis of iron complexes in selected important oxidation reactions. When appropriate we will discuss the involvement and characterization of Fe–oxo intermediates in these reactions.

2 Epoxidation of Olefins

Among the different oxidation reactions, both from an academic as well as industrial point of view, the epoxidation of olefins is of considerable interest [8, 9]. Subsequent ring opening reactions make epoxides versatile building blocks for large-scale materials, bulk, and fine chemicals as well as agrochemicals and pharmaceuticals [10]. Due to their low cost, benignancy to the environment and biological relevance, there is an increasing interest to use iron complexes as catalysts for epoxidation reactions [11, 12]. Notably, in addition to the catalysts, the applied oxidant determines the value of the oxidation system to a significant extent (for a list of common oxidants, their active oxygen contents and waste products see [13]). From ecological and economical points of view, molecular oxygen [14, 15] and hydrogen peroxide are the oxidants of choice with regard to waste and by-products.

As an example heme-models have been reported to catalyze the epoxidation of olefins to the corresponding epoxides in good yield [16, 17]. In particular, $[Fe^{III}(TPP)Cl]$ (TPP = 5,10,15,20-meso-tetraphenylporphyrin) was reported to oxidize naturally occurring propenylbenzenes to the corresponding epoxides up to 98% selectivity (conversion 98%) using H_2O_2 as oxidant [16]. The major drawback for these heme-model systems is the low tunability of the catalysts for different olefins. Mechanistic insights into the iron porphyrin-catalyzed epoxidation using hydrogen peroxide were published by Bell and coworkers [18, 19].

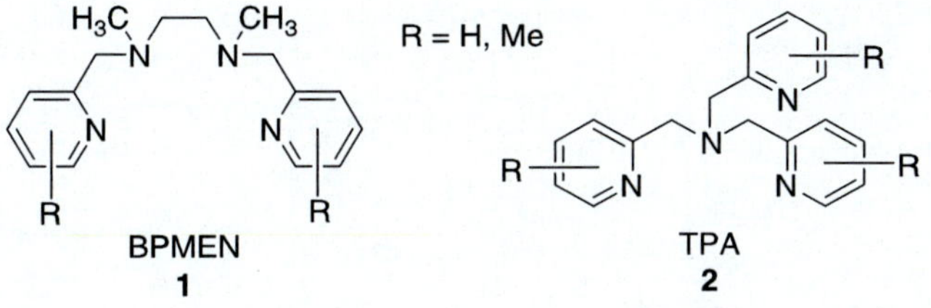

Scheme 1 BPMEN and TPA ligands for the iron-catalyzed epoxidation of olefins

In addition, also nonheme iron catalysts containing BPMEN **1** and TPA **2** as ligands are known to activate hydrogen peroxide for the epoxidation of olefins (Scheme 1) [20–26]. More recently, especially Que and coworkers were able to improve the catalyst productivity to nearly quantitative conversion of the alkene by using an acetonitrile/acetic acid solution [27–29]. The carboxylic acid is required to increase the efficiency of the reaction and the epoxide/diol product ratio. The competitive dihydroxylation reaction suggested the participation of different active species in these oxidations (Scheme 2).

Scheme 2 Epoxidation of cyclooctene catalyzed by Fe(BPMEN) and Fe(TPA) complexes

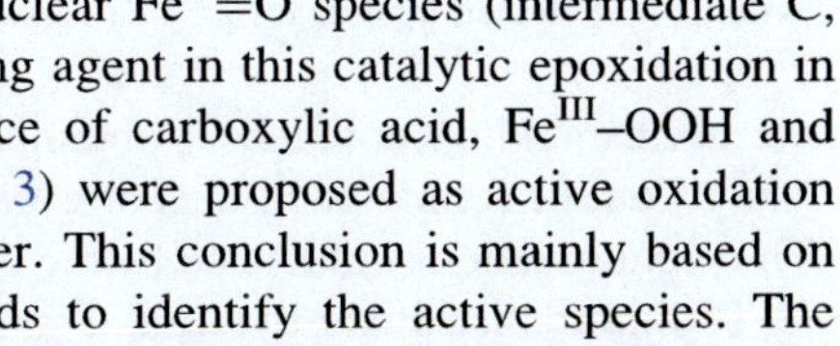

Mechanistic studies revealed a mononuclear $Fe^{V}{=}O$ species (intermediate C, Scheme 3) to be most likely the epoxidizing agent in this catalytic epoxidation in the presence of acetic acid. In the absence of carboxylic acid, Fe^{III}–OOH and $Fe^{V}{=}O$ (intermediates A and B, Scheme 3) were proposed as active oxidation species depending on the presence of water. This conclusion is mainly based on labeling studies and spectroscopic methods to identify the active species. The reported results are quite controversial to the results of the groups of Talsi and Comba. Talsi and coworkers reported the formation of an $Fe^{IV}{=}O$ intermediate supported by NMR- and EPR-spectroscopy [30], which is in agreement with the results obtained by Comba and coworkers [31]. In addition, other groups explored similar high-valent iron–oxo species, which shed light on this catalytic mechanism [32, 33].

Iron complexes with the pentadentate ligand **3** derived from pyridyl and prolinol building blocks containing a stereogenic center were reported from the group of Klein Gebbink (Scheme 4) [34]. In alkene oxidations with hydrogen peroxide,

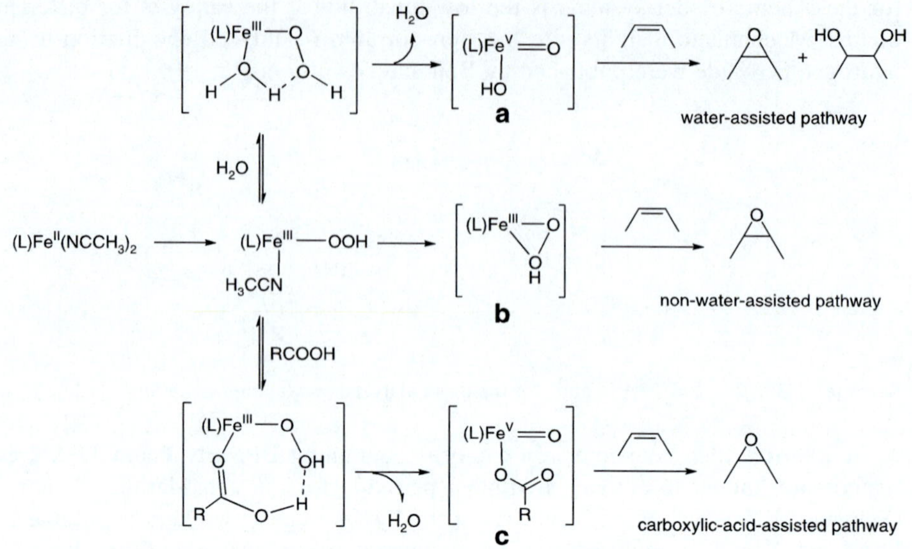

Scheme 3 Proposed mechanism of iron-catalyzed oxidation by Que and coworkers

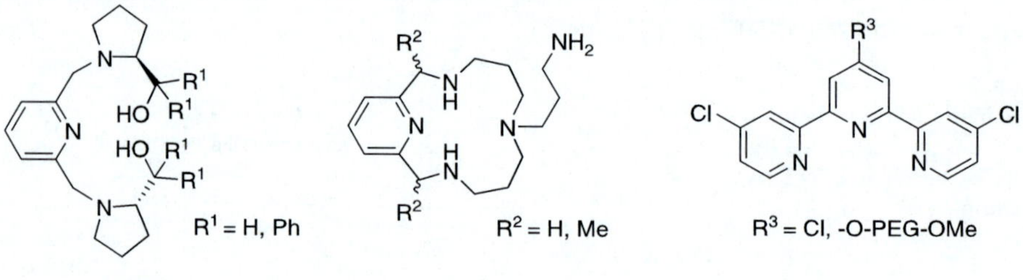

3 4 5

Scheme 4 Selected ligands for iron-catalyzed epoxidation of alkenes

the corresponding epoxide is obtained only in low yield as a racemic mixture. Moreover, the formation of the corresponding allylic alcohol and ketone is observed. The product ratio is influenced by the nature of the coordinating anions (Cl or OTf).

To mimic the square-pyramidal coordination of iron bleomycin, a series of iron (II)complexes with pyridine-containing macrocycles **4** was synthesized and used for the epoxidation of alkenes with H_2O_2 (Scheme 4) [35]. These macrocycles bear an aminopropyl pendant arm and in presence of poorly coordinating acids like triflic acid a reversible dissociation of the arm is possible and the catalytic active species is formed. These complexes perform well in alkene epoxidations (66–89% yield with 90–98% selectivity in 5 min at room temperature). Furthermore, recyclable terpyridines **5** lead to highly active Fe^{II}-complexes, which show good to excellent results (up to 96% yield) for the epoxidation with oxone at room temperature (Scheme 4) [36].

With respect to generality, it still remains challenging to discover iron-catalyzed epoxidations, which allow efficient and selective reactions for both aromatic and aliphatic olefins. A convenient and efficient method for the fast epoxidation of a variety of olefins was developed by our group. The simple and practical in situ catalyst system consists of iron trichloride hexahydrate, pyridine-2,6-dicarboxylic acid (H_2pydic), and an organic amine [37–39]. By modifying the organic amine almost all classes of olefins are accessible for epoxidation with hydrogen peroxide under mild conditions. Pyrrolidine **6** and benzylamine derivatives **7** turned out to be advantageous as coligands (Scheme 5). The development of a second generation catalyst in the absence of pyridine-2,6-dicarboxylic acid was achieved by using iron trichloride hexahydrate in combination with bio-inspired imidazole derivatives **8** [40, 41]. Selected results are shown in Table 1.

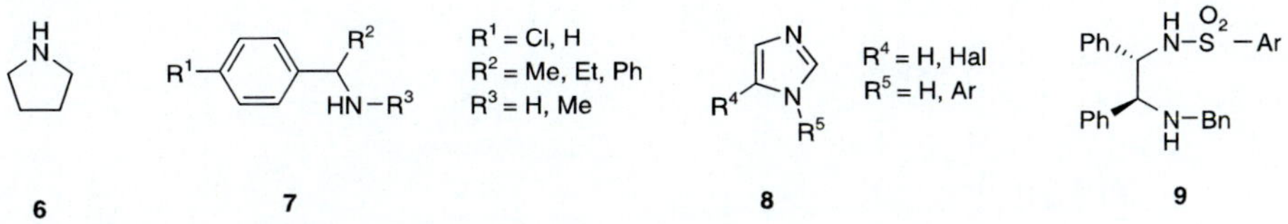

Scheme 5 Organic coligands used in the epoxidation of alkenes

Until recently only few examples on asymmetric epoxidation using iron-based catalysts were reported in the literature (Scheme 6) [42–44]. With [Fe(BPMCN)($CF_3SO_3)_2$] **10**, 58% of the epoxide with 12% *ee* was obtained in the oxidation of *trans*-2-heptene [42].

By elaborating 5,760 metal–ligand combinations, Francis and Jacobsen identified three Fe-complexes with peptide-like ligands, which gave the epoxide in 15–20% *ee* in the asymmetric epoxidation of *trans*-β-methylstyrene utilizing 30% H_2O_2. The homogeneous catalyst **11** derived from this study gave 48% *ee* with 100% conversion of *trans*-β-methylstyrene [43]. Aerobic epoxidation of styrene derivatives with an aldehyde coreductant catalyzed by tris(δ,δ-dicampholylmethanato) iron(III) complex **12** was also reported.

A breakthrough in iron-catalyzed asymmetric epoxidation of aromatic alkenes using hydrogen peroxide has been reported by our group in 2008. Good to excellent isolated yields of aromatic epoxides are obtained with *ee*-values up to 97% for stilbene derivatives using diphenylethylenediamines **9** as ligands (Scheme 5) [45, 46].

Additional recent ligand developments to be mentioned are the chiral bipyridine ligand **13** and the polypyridine ligand **14** (Scheme 7) [47, 48]. The μ-oxo-dinuclear iron complex of **13** performed the enantioselective epoxidation with peracetic acid in high conversion and yield. Unsymmetrical alkenes were oxidized with enantiomeric excess ranging from 9 to 63%, whereas symmetrical *trans*-stilbene led to racemic mixtures. The importance for the dinuclearity of the catalyst has been pointed out since the mononuclear complexes of the ligand showed lower rate, yield, and lower enantiomeric excess. Disadvantage of the reaction is the use of peracetic acid as oxidant which leads to significant amounts of acetic acid as

Table 1 Selected result of the epoxidation using iron trichloride hexahydrate, pyridine-2,6-dicarboxylic acid, and an organic amine

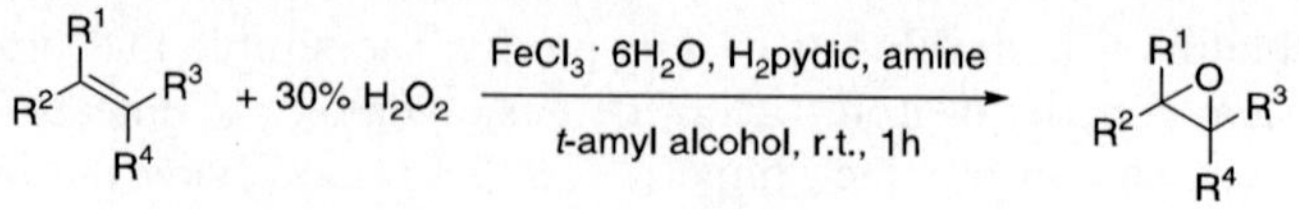

R^1, R^2, R^3, R^4 = H, aryl, alkyl

Entry	Substrate	Amine	Yield[a]
1	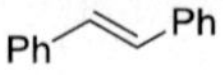	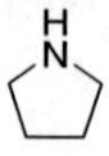	97
2	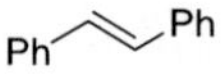		93
3	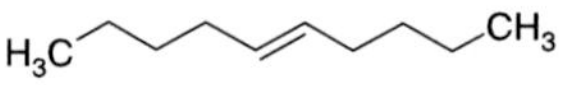	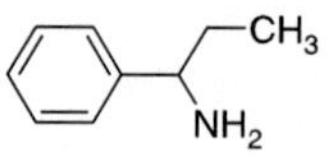	96
4	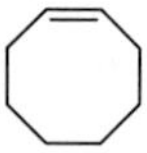	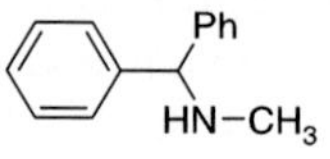	89
8[b]	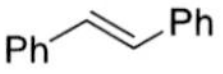	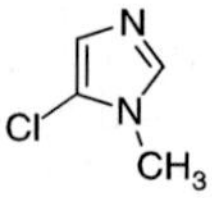	87
9[b]	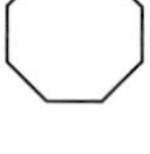	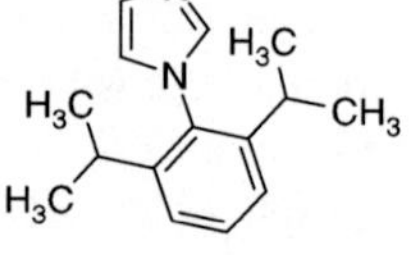	65

Reaction conditions: 5 mol% $FeCl_3 \cdot 6H_2O$, 5 mol% H_2pydic, 10–12 mol% amine, 3 equiv. H_2O_2
[a]Yield was determined by GC analysis;
[b]H_2pydic was omitted

by-product. Hence, acid labile epoxides are not accessible. The μ-oxo-dinuclear iron complex of **14** showed excellent reactivity and selectivity towards terminal and 1,2-substituted aromatic alkenes in the epoxidation with H_2O_2 in acetonitrile/acetic acid. Enantiomeric excess up to 43% was achieved.

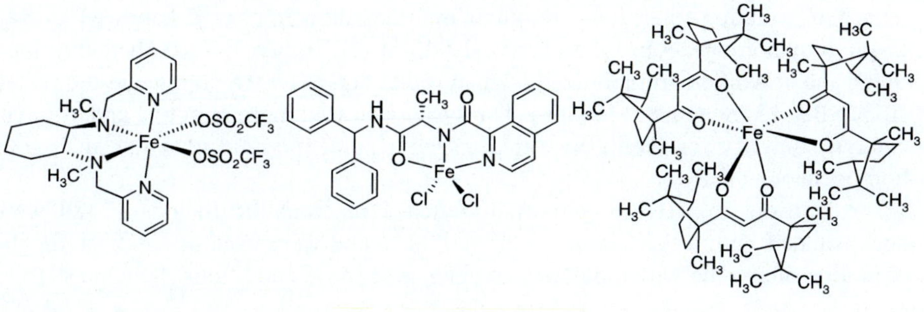

10 11 12

Scheme 6 Chiral iron complexes for the asymmetric epoxidation of olefins

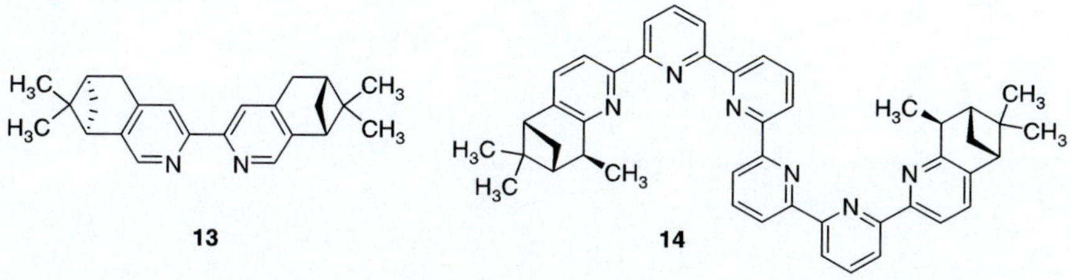

13 14

Scheme 7 Recently reported ligands for iron-catalyzed asymmetric epoxidations

Asymmetric epoxidation systems using iron porphyrin heme-mimics are also known, however the labor-intensive and expensive syntheses is limiting their applications [49].

The immobilization of metal catalysts onto solid supports has become an important research area, as catalyst recovery, recycling as well as product separation is easier under heterogeneous conditions. In this respect, the iron complex of the Schiff base HPPn **15** (HPPn = *N,N′*-bis(*o*-hydroxyacetophenone) propylene diamine) was supported onto cross-linked chloromethylated polystyrene beads. Interestingly, the supported catalyst showed higher catalytic activity than the free metal complex (Scheme 8) [50, 51]. In terms of chemical stability, particularly with

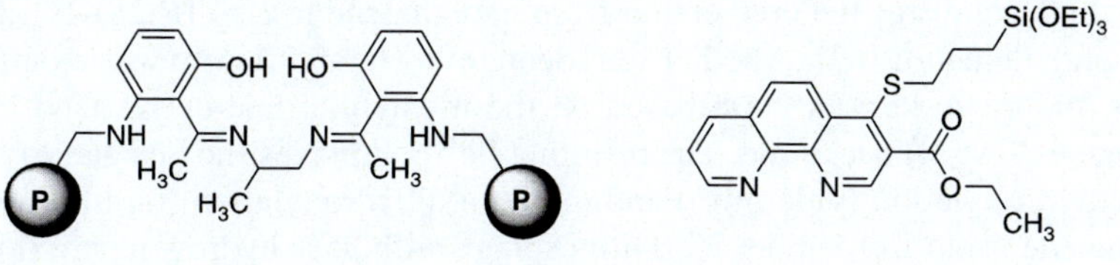

15 16

Scheme 8 Immobilized HPPn and Stacks phenantroline ligand regard to oxidizing conditions, inorganic matrices show improved stability [52–54]. Inspired by homogeneous $[\{Fe(phen)_2(H_2O)\}_2\ \mu\text{-}O]^{4+}$ (phen = 1,10-phenantroline), Stack and coworkers immobilized phenantroline derivative **16** on micelle-templated silica SBA-15 (Scheme 8) [55, 56]. The system showed more selective and efficient catalytic activity for olefin epoxidations with peracetic acid than the analogous homogeneous catalyst.

In addition, iron(II) complexes of tetraaza macrocyclic ligands **17–20** were encapsulated within the nanopores of zeolite-Y and were used as catalysts for the oxidation of styrene with molecular oxygen under mild conditions (Scheme 9) [57].

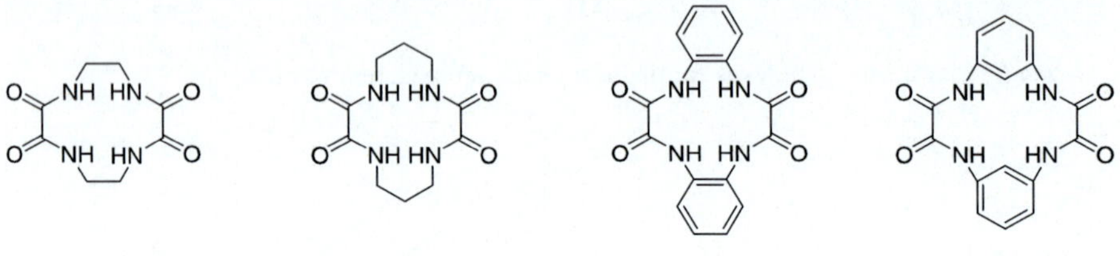

17 **18** **19** **20**

Scheme 9 Tetraaza macrocyclic ligands for supported iron catalysts

3 Dihydroxylations

1,2-Diols are applied on a multimillion ton scale as antifreezing agents and polyester monomers (ethylene and propylene glycol) [58]. In addition, they are starting materials for various fine chemicals. Intimately connected with the epoxidation-hydrolysis process, dihydroxylation of C=C double bonds constitutes a shorter and more atom-efficient route to 1,2-diols. Although considerable advancements in the field of biomimetic nonheme complexes have been achieved in recent years, still osmium complexes remain the most efficient and reliable catalysts for dihydroxylation of olefins (reviews: [59]).

So far, biomimetic and bio-inspired approaches mimicking the nonheme oxygenases have been studied with ligands such as the tetradentate N_4 ligand BPMEN **1** also named as mep or the tripodal ligand TPA **2** (reviews: [60–62], Scheme 1). Lately, TPA ligands were used also as model for NDO (naphthalene 1,2-dioxygenase), which catalyzes the conversion from naphthalene to *cis*-(1*R*,2*S*)-1,2-dihydro-1,2-naphthalenediol [63]. Another tetradentate N_4-ligand family was examined by Costas and coworkers in 2008 based on the methylpyridine-derivatized triazacyclononane (TACN) backbone. The resulting Fe^{II} complexes showed activity both in alkane hydroxylation (vide infra) and in olefin dihydroxylation [64]. For example, cyclooctene is oxidized under inert atmosphere with 35% hydrogen peroxide in the presence of catalytic amounts of different iron complexes to attain the corresponding epoxide and *cis*-diol in 19% and 62% yield, respectively, based on the oxidant (Table 2, entry 1). The unprecedented high water incorporation into the

Table 2 Epoxidation of cyclooctene with iron catalysts and H_2O_2

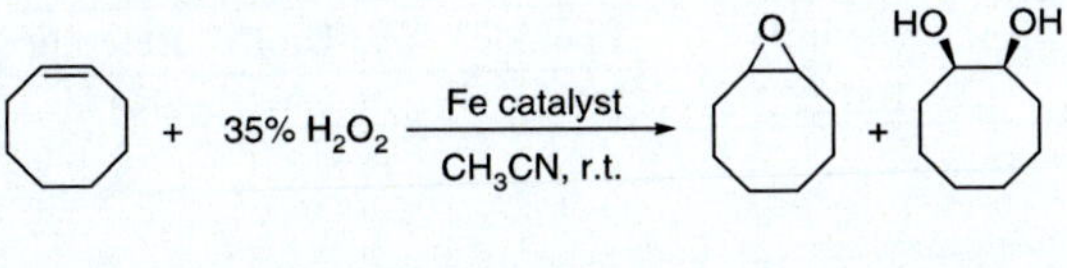

Entry	Catalyst	Olefin:H_2O_2	Epoxide[a]	*cis*-Diol[a]	References
1[b]	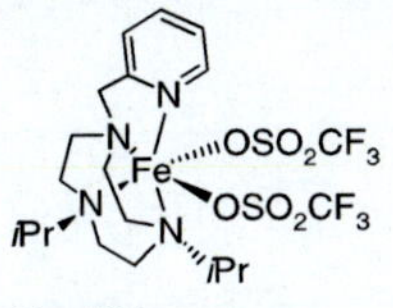 10 mol%	100:1	19	62	[64]
2[b]	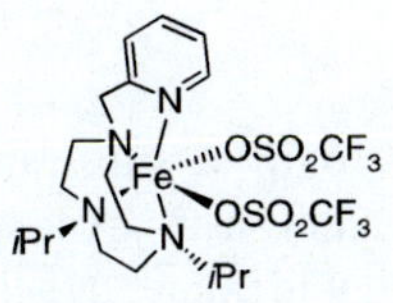 1 mol%	10:1	12	73	[64]
3[c]	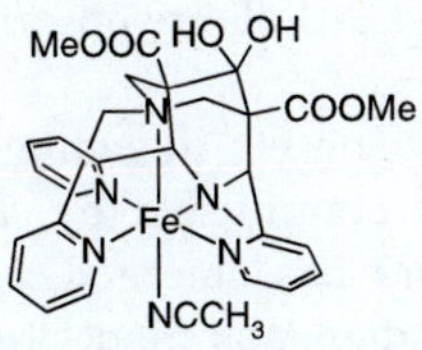 10 mol%	100:1	50	0	[29, 31]
4[b,c]	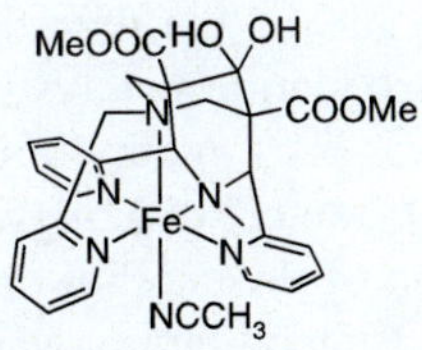 10 mol%	100:1	10	10	[29, 31]
5	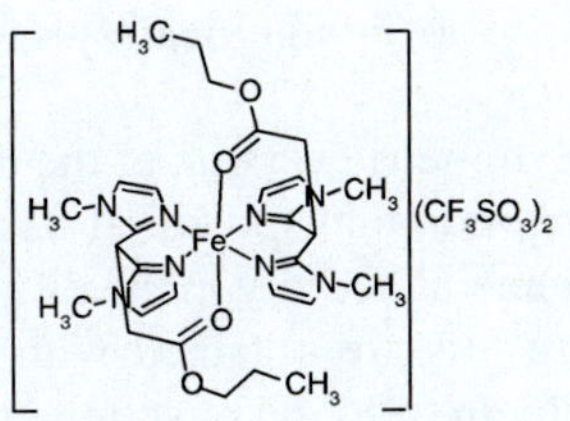 10 mol%	100:1	28	11	[69]

(*continued*)

Table 2 (continued)

Entry	Catalyst	Olefin:H_2O_2	Epoxide[a]	*cis*-Diol[a]	References
6	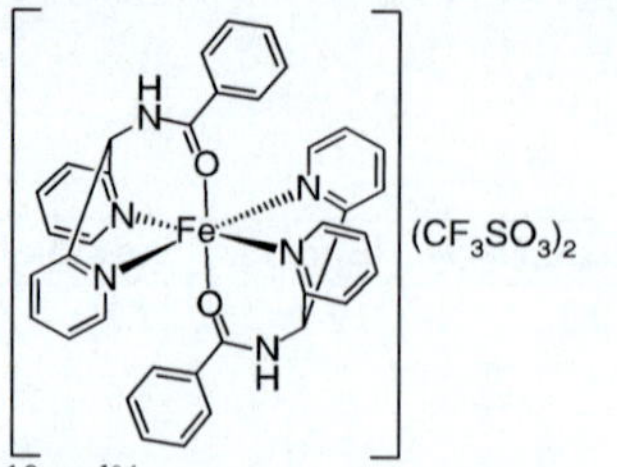 10 mol%	100:1	5%	70	[70]

[a]Yield based on the limiting reagent
[b]Reactions carried out in inert atmosphere
[c]The corresponding anions were not mentioned

corresponding epoxides and diols was later explained by a substrate-dependent interplay between two isomerically related high-valent iron oxo species [65]. Further improvements were achieved using a pentadentate N_5-ligand motif known as bispidine ligands [29, 31]. This Fe^{II} complex catalyzed the oxidation of cyclooctene with hydrogen peroxide mainly to afford the epoxide as product. However, under anaerobic conditions *cis*- and *trans*-1,2-diols are observed (Table 2, entries 3 and 4).

In a specific subset of the nonheme iron oxygenases including the Rieske dioxygenases, the mononuclear iron(II) center is coordinated by the so-called 2-his-1-carboxylate triad [66]. This ligand scaffold has emerged as a versatile platform for oxidative transformations and as a fundamental model for the design of new ligands. For example, Klein Gebbink and coworkers introduced bis(1-alkylimidazol-2-yl)propionates as a new tridentate, tripodal N, N, O ligand family [67], which were applied as model of the active site of the extradiol cleaving catechol dioxygenases [68] and as epoxide/dihydroxylation catalysts mimicking the Rieske oxygenases [69]. Whereas propyl 3,3,-bis(1-methylimidazol-2-yl)-propionate favored epoxidation (Table 2, entry 5), novel (di-(2-pyridyl)methyl) benzamide ligands showed predominantly formation of the *cis*-diol product in the oxidation of cyclooctene. This member of the N,N,O-ligand family was introduced by Que and coworkers in 2005 [70]. The iron(II) complexes of this ligand catalyzed the dihydroxylation of various aliphatic and aromatic olefines (Table 2, entry 6). It should be noted that the *cis*-diol to epoxide ratio is significantly increased and mainly the *cis*-diol is formed.

In 2008, Que and coworkers reported an asymmetric version of the dihydroxylation with a new type of ligands bearing bipyrrolidine as the chiral backbone [71]. The corresponding iron(II) complex showed general activity in the dihydroxylation of various olefins using H_2O_2. Satisfactory results are obtained with aliphatic as well as with aromatic olefins. For example, dihydroxylation of styrene gave styrene oxide and 1-phenylethane-1,2-diol in <1% and 65% yield, respectively (Scheme 10).

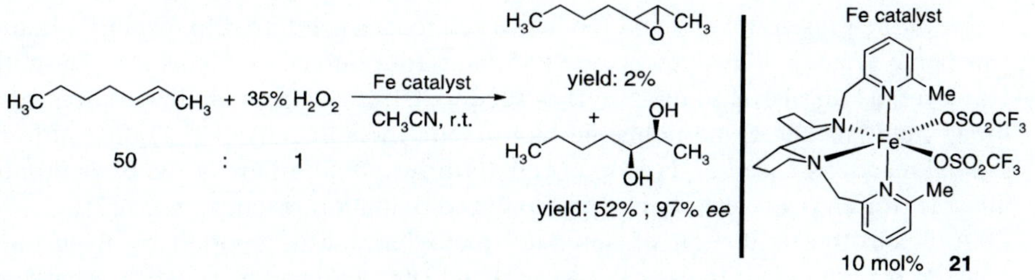

Scheme 10 Asymmetric epoxidation of *trans*-2-heptene with the chiral iron complex and H_2O_2

Most striking result goes to the oxidation of 2-heptene. In this system, the *cis*-diol is obtained in 55% yield with 97% *ee* (Scheme 10).[1] However, due to the unusual reaction conditions (large excess of olefin), this method is unlikely to be applied in organic synthesis. In-between homogeneous and heterogeneous iron catalysts, unsupported "free" nano-γ-Fe_2O_3 displays a unique opportunity to combine reusability with activity and selectivity. The nano-γ-Fe_2O_3 catalyst with a particle size of 20–50 nm showed oxidative cleavage of aromatic olefins to the corresponding aldehydes applying hydrogen peroxide as oxidant (Scheme 11) [72, 73]. Various aromatic olefins with different substituents were successfully oxidized into aldehydes with high selectivity although only low conversion was achieved. Interestingly, the catalytic activity can be tuned by changing the particle size of nano-iron oxide.

Scheme 11 Selective cleavage of olefins to aldehydes

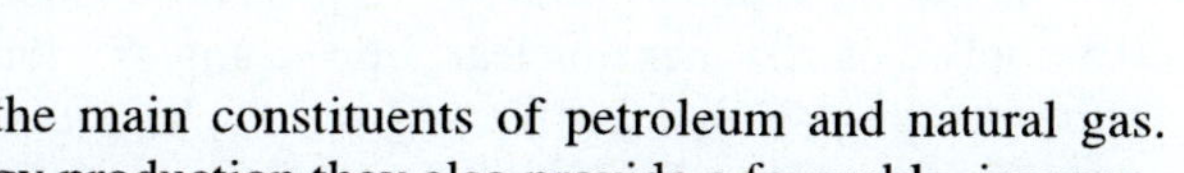

4 Alkane Oxidation

Saturated hydrocarbons are the main constituents of petroleum and natural gas. Mainly used as fuels for energy production they also provide a favorable, inexpensive feedstock for chemical industry [74]. Unfortunately, the inertness of alkanes renders their chemical conversion challenging with respect to selectivity. Clearly, the development of new and improved methods for the selective transformation of alkanes belongs to the central goals of catalysis. Iron-catalyzed processes might be a smart tool for such transformations (for reviews see [75–77]).

Remarkably, alkanes are oxidatively transformed by biological organisms at benign temperatures and pressures. Clearly, enzymatic transformations of alkanes and their well studied mechanisms (e.g., for cytochrome P450) are beyond the

---

[1]Yield and catalyst concentration based on the limiting reagent.

scope of the present review and the interested reader is referred to recent literature (for heme systems like Cytochrome P450 and for nonheme systems see [78–80]). However, as discussed in the previous section, these enzymes are inspiration for a broad range of heme and nonheme biomimetic model catalysts [81]. In addition, radical reactions based on Fenton chemistry offer possibilities for the oxidation of alkanes (for an overview about iron-catalyzed oxidation reactions see [82]).

A ligand-free oxidation of activated methylenes was reported by Bolm and coworkers with $Fe(ClO_4)_2$ as catalyst [83] and later with an $FeCl_3{\cdot}6H_2O$ as catalyst source in an improved system [84]. In the latter system, no acid was needed under $GoAgg^V$-type conditions, where pyridine was used as solvent. Benzylic oxidation was achieved to the corresponding carbonyl compounds (up to >99% yield) with aqueous TBHP (=*t*-butyl hydroperoxide) as terminal oxidant. Besides, alcohols are converted to the corresponding ketones. In 2008, we introduced an $FeCl_3{\cdot}6H_2O$ ligand-free catalyst system for the α-oxidation of β-ketoesters (Scheme 12) [85]. By using cyclic β-ketoesters as starting material 75–90% yield of the hydroxylation products are obtained with H_2O_2 as oxidant.

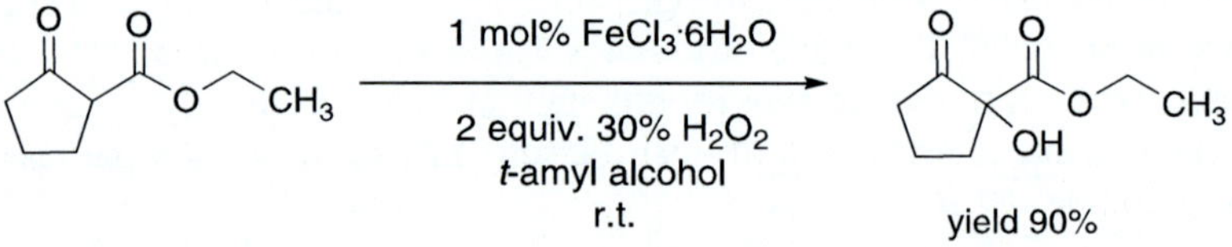

Scheme 12 Selective hydroxylation of 2-ethoxycarbonyl-1-oxo-cyclopentane

Inspired by $Gif^{III,IV}$ or $GoAgg^{III}$ type chemistry [77], iron carboxylates were investigated for the oxidation of cyclohexane, recently. For example, Schmid and coworkers showed that a hexanuclear iron *p*-nitrobenzoate $[Fe_6O_3(OH)(p\text{-}NO_2C_6H_4COO)_{11}(dmf)_4]$ with an unprecedented $[Fe_6^{III}O_3(\mu_3\text{-}O)(\mu_2\text{-}OH)]^{11+}$ core is the most active catalyst [86]. In the oxidation of cyclohexane with only 0.3 mol% of the hexanuclear iron complex, total yields up to 30% of the corresponding alcohol and ketone were achieved with 50% H_2O_2 (5.5–8 equiv.) as terminal oxidant. The ratio of the obtained products was between 1:1 and 1:1.5 and suggests a Haber–Weiss radical chain mechanism [87, 88] or a cyclohexyl hydroperoxide as primary oxidation product.

In contrary to the not easy adjustable ligand-free systems, systems with ligands provide possibilities in tuning activity and selectivity. Shul'pin et al. examined the addition of bipyridine to $FeCl_3$ and hydrogen peroxide as oxidant [89]. They monitored a 35 times enhanced reactivity (TON up to 400) for oxidation of cyclohexane. They observed the predominant formation of cyclohexyl hydroperoxide and the corresponding transformations into cyclohexanol and cyclohexanone. Another radical-based system involving cyclohexyl hydroperoxide was developed by Pombeiro and coworkers. They used an FeN_2S_2 center bearing a noninnocent ligand (Scheme 13) [90]. In the presence of pyrazinecarboxylic acid (Hpca) at room temperature within 6 h, yields up to 13% are achieved (alcohol as the major product).

Scheme 13 The FeN_2S_2 center bearing a coordinatively unsaturated iron(III) with a noninnocent ligand

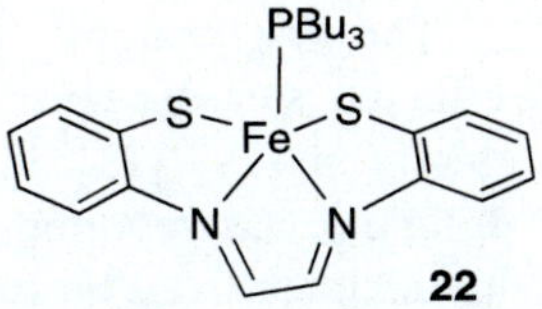

22

Based on Shul'pin's examinations on the rate enhancement by adding Hpca [91] and in contrast to the inhibition while adding Hpca [110], Reedjik and coworkers investigated the role of Hpca with defined iron complexes [92]. In their studies, [Fe $(pca)_2(py)_2]\cdot py$ showed moderate activity with a maximum yield of 31% based on hydrogen peroxide.

A major challenge in the application of "biomimetic" or "bio-inspired" ligands is to direct the unselective radical pathways of simple Fe salts and H_2O_2 into chemo-, regio-, and stereoselective transformations. Thus, homolytic cleavage of the iron peroxo bond has to be omitted. For this purpose, various N- and N,O-ligands were developed (for a review about "Biologically inspired oxidation catalysis" see [93]). Tridentate ligands like bis(imino)pyridine or bis(amino)pyridine turned out to be not appropriate. Here, two ligands can cover all coordination sites and the Fe complexes showed only moderate oxidation reactivity [94, 95]. More suitable ligands are those of tetradentate nitrogen-donor ligands. An arrangement that allows two *cis*-oriented coordination sites for peroxide binding is also necessary as stated by Que [96]. During the past decade, tetradentate N_4-ligands and recently N,O-ligands showed interesting activity not only in olefin oxidation but also in more challenging alkane oxidation. Scheme 14 shows important N-ligands, which have been used in such reactions.

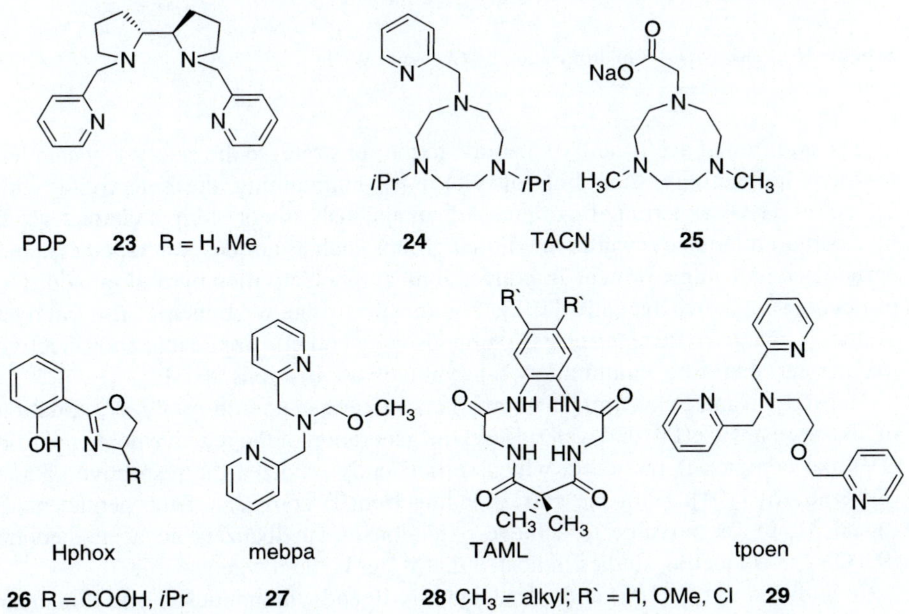

Scheme 14 Biomimetic ligands used in iron-catalyzed alkane oxidation

The first example of a nonheme iron catalyst with the TPA ligand **2**, which effected stereospecific alkane hydroxylation, was developed by Que and coworkers in 1997 [97]. In the same year, the linear N_4 tetradentate mep or BPMEN **1** ligand was reported by Nishida and coworkers [98]. [Fe(**1**)(OTf)$_2$] complexes are characterized by two essential pyridine donors connected with an ethylendiamine bridge, two labile *cis*-coordination sites at the metal center and single *cis-a* coordination geometry [99]. Therefore, iron complexes of **1** are quite stable against the addition of excessive H_2O_2 (up to 100 equiv. excess) and show a reactivity that is distinct from Fenton-type chemistry. Hence, the catalyst converts 65% of the H_2O_2 into oxygenated products using 10 equiv. of H_2O_2 [100]. A breakthrough in selective alkane oxidation was reported in 2007 by Chen and White, who developed a catalyst system based on a modified mep ligand. The resulting hydroxylation catalyst [Fe(*S*,*S*-**30**)(CH$_3$CN)$_2$](SbF$_6$)$_2$ reacts with electron-rich, tertiary C–H bonds using H_2O_2 in good yields with predictable selectivity (Scheme 15) [101].

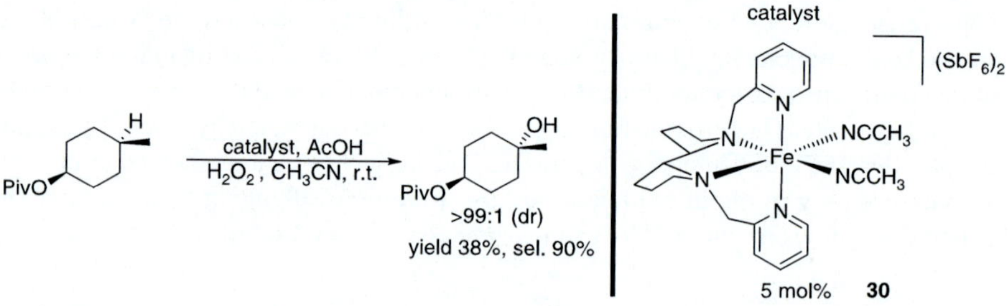

Scheme 15 Selective C–H hydroxylation by Chen and White

The addition of acetic acid (0.5 equiv. to the substrate) to the catalyst system led to increased activity (doubling of yield) by maintaining the selectivity with 1.2 equiv. H_2O_2 as terminal oxidant. Advantageously, the system is characterized by a certain tolerance towards functional groups such as amides, esters, ethers, and carbonates. An improvement in conversions and selectivities by a slow addition protocol was shown recently [102]. For the first time, a nonheme iron catalyst system is able to oxidize tertiary C–H bonds in a synthetic applicable and selective manner and therefore should allow for synthetic applications [103].

Recently, Nam, Fukuzumi, and coworkers succeed in an iron-catalyzed oxidation of alkanes using Ce(IV) and water. Here, the generation of the reactive nonheme iron (IV) oxo complex is proposed, which subsequently oxidized the respective alkane (Scheme 16) [104]. With the corresponding iron(II) complex of the pentadentate ligand **31**, it was possible to achieve oxidation of ethylbenzene to acetophenone (9 TON). ^{18}O labeling studies indicated that water is the oxygen source.

In addition to tri- [105] and tetradentate N-ligands, mononuclear and dinuclear iron complexes with pentadentate N,N,N,N,O-ligands were applied to alkane

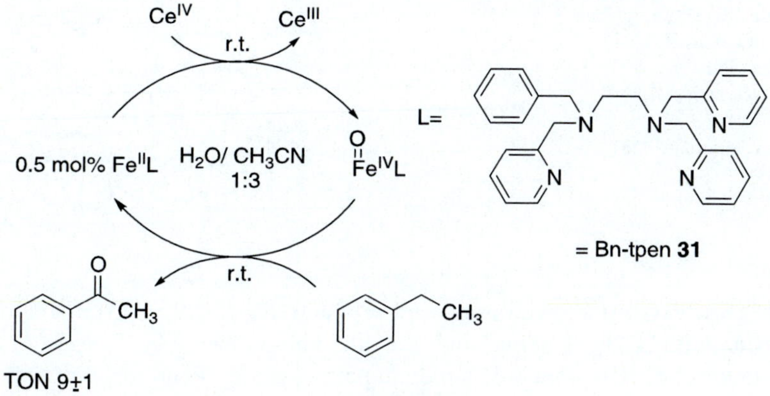

Scheme 16 Oxidation of ethylbenzene with Ce(IV)

oxidations. As an example Sun and Wang used tpoen, **29** in the oxidation of cyclohexane [106]. Mediocre conversion (18%) based on the oxidant H_2O_2 are observed. Another multidentate ligand (mebpa, **27**) was reported by Reedjik (first report on the mebpa ligand: [107, 108]). At low H_2O_2 concentration in the presence of additives, they succeeded in 54% conversion with an alcohol/ketone ratio up to 3.5.

A mononuclear diastereopure high-spin Fe^{III} alkylperoxo complex with a pentadentate N,N,N,O,O-ligand **33** (Scheme 17) was reported by Klein Gebbink and coworkers [109, 110]. The complex is characterized by unusual seven-coordinate geometry. However, in the oxidation of ethylbenzene the iron complex with **33** and TBHP yielded with large excess of substrate only low TON's (4) and low ee (6.5%) of 1-phenylethanol.

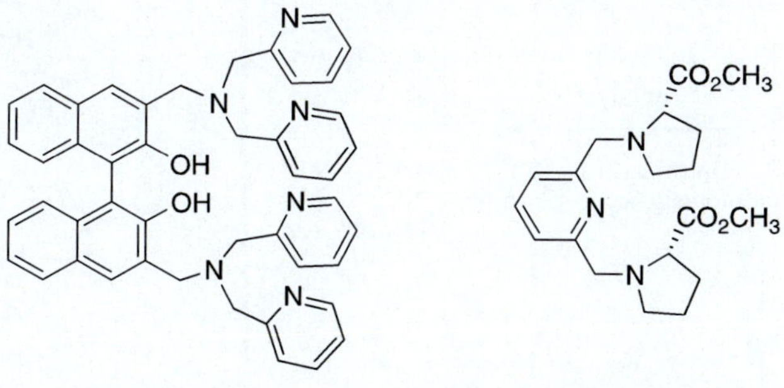

32 **33**

Scheme 17 Biomimetic ligands used in Fe-catalyzed alkane oxidations

For nearly two decades TACN-type ligands are of continuing interest for oxidation chemistry. A more recent example is described by Shul'pin and coworkers, who prepared novel di- and tetranuclear complexes with TACN ligands **25** bearing pendant acetato arms bridging the iron(III) centers (Scheme 18) [111]. The

Scheme 18 Tetranuclear $[Fe_4(N_3O_2\text{-}L_4(\mu\text{-}O)_2)]^{4+}$ complex with 1-carboxymethyl-4,7-dimethyl-1,4,7-triazacyclononane (**25**) ligands

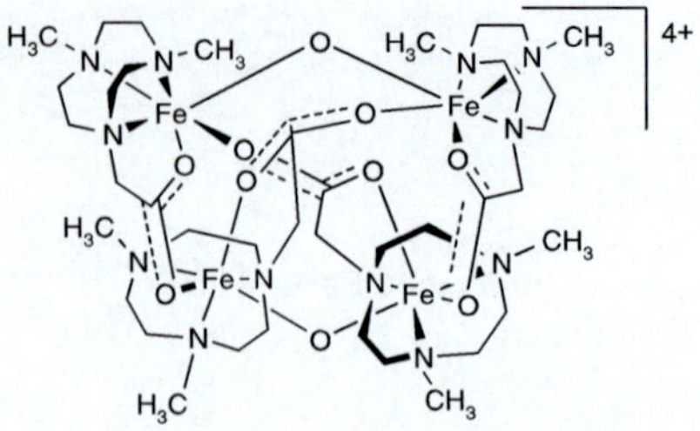

tetranuclear catalyst showed improved productivity (TON = 43) in cyclohexane oxidation with H_2O_2 as oxidant with an alcohol/ketone (A/K) ratio of 6. Similarities in the A/K ratio conclude a radical pathway, e.g., Fenton chemistry, for this reaction.

Apart from selective organic synthesis, there exists a significant interest in the nonselective iron-catalyzed oxidations of all kinds of organic compounds. In this respect, Collins and coworkers developed and examined tetraamido macrocyclic ligands (TAML, **28**) for the treatment of waste water from paper and textile mills with H_2O_2 [112]. During their investigations, Fe(V)-oxo species and Fe(IV) complexes were proposed and observed as key intermediates [113]. Notably, the isolated Fe (V)-oxo species gave substoichiometric reactions with olefins and with ethylbenzene [4]. For hydrocarbon decompositions also photocatalytic oxidations were reported in 2006 by Nocera and coworkers. More specifically, they used fluorinated Pacman bisporphyrin ligands bridged by a dibenzofuran scaffold with visible light and oxygen as terminal oxidant (Scheme 19) [114].

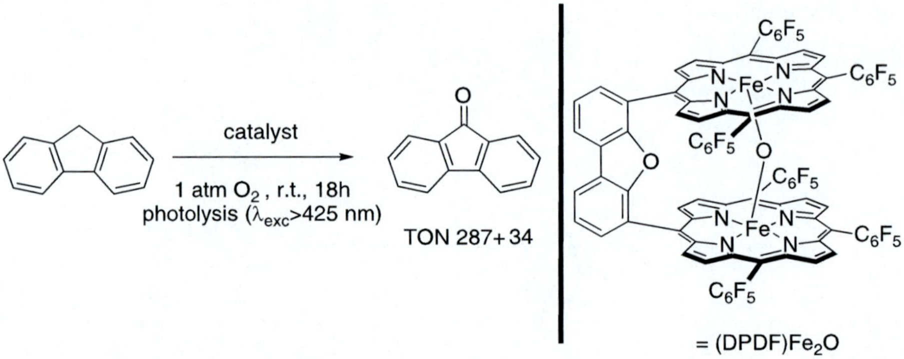

Scheme 19 Photocatalytic hydroxylation of fluorene with a Pacman system

The Pacman catalyst selectively oxidized a broad range of organic substrates including sulfides to the corresponding sulfoxides and olefins to epoxides and ketones. However, cyclohexene gave a typical autoxidation product distribution yielding the allylic oxidation products 2-cyclohexene-1-ol (12%) and 2-cyclohexene-1-one (73%) and the epoxide with 15% yield [115].

In addition to porphyrin-type ligands, also porphyrazine complexes show interesting properties such as activation of H_2O_2 and pH-dependent decomposition of dyes [116]. Recently, Sorokin and coworkers applied binuclear phthalocyanine complexes for the oxidation of methane [117]. With their μ-nitrido-bridged phthalocyanine complexes (Scheme 20), they were able to perform homogenous oxidation in CH_3CN to give formic acid with H_2O_2 as oxidant.

Scheme 20 μ-Nitrido-bridged iron phthalocyanine complexes

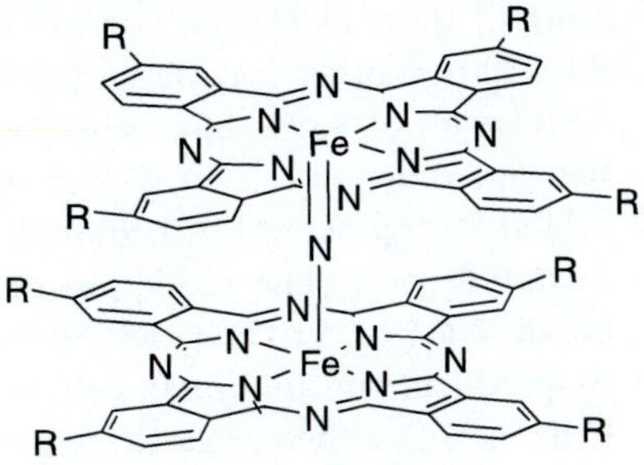

34 R = H, $C(CH_3)_3$

Labeling studies indicated that the obtained formic acid was originated from both substrate and solvent. When the catalyst was supported onto silica to provide a heterogeneous catalyst, methane is oxidized at 80 °C and 32 bars CH_4 to CH_2O (up to 1.1 TON) and HCOOH (up to 27.3 TON).

Metal-oxygen cluster species such as polyoxometalates (POM's) represent an interesting and interdisciplinary field in oxidation catalysis. Especially, high-valent iron–oxo species of POM's should be highly active catalysts [118]. Unfortunately, until today experimental investigations did not prove the existence of this type of powerful oxidants. Novel protocols for the oxidation of alkanes with Fe-containing POM's include the use of iron-supported polyoxotungstates ($FeSiW_{11}$) for the oxidation of cyclohexane in the presence of microwave-induced heating [119, 120]. More specifically, tetrairon(III)-substituted polytungstates immobilized on (3-aminopropyl)triethoxysilane (apts)-modified SBA-15 showed high catalytic activity in the oxidation of long chain alkanes [121]. For example, *n*-hexadecane gave 18% conversion (TOF = 2,043 h^{-1}) using air as oxidant at 150 °C. Selectivities of 50% C_{16} ketones and 28% C_{16} alcohols were obtained. Mechanistic investigations showed that the reaction occurred via a free-radical chain autooxidation.

5 Oxidation of Aromatic C–H Bonds

Selective C–H hydroxylation on arenes to give the corresponding phenols displays an attractive tool for the chemical industry and organic synthesis. Unfortunately, the desired phenolic product is more electron rich than the substrate and therefore tends to over-oxidation resulting in catechols, hydroquinones, benzoquinones, and finally tars.

Already in the beginning of the nineteenth century, radical reactions on arenes with iron(II)sulfate and H_2O_2 were discovered, known as Fenton reaction. Later on, numerous attempts were undertaken to direct the unselective radical reactions to more selective ones by employing different oxidants, ligands, biphasic systems, photochemistry, or electro-catalytic methods. More recent examples include the work of Bianchi et al. who reported enhanced selectivity for benzene oxidation by the addition of a pyrazine-carboxylate ligand under biphasic conditions [122]. At a conversion of 8.6%, 85% selectivity based on H_2O_2 and 97% selectivity based on benzene were observed. Pombeiro and coworkers improved the conversion and selectivity by using Fe^{III}(gma)(PBu_3) as catalyst, which was also active in alkane oxidation [90]. The oxidation of benzene proceeded under rather mild reaction conditions (r.t., 6 h) and yielded 20% (110 TON) phenol without any other observed by-products and hydrogen peroxide as terminal oxidant (4 equiv. with respect to benzene). Another biphasic approach made use of iron(II)sulfate as catalyst and H_2O_2 as oxidant in a system with a polypropylene hydrophobic porous support as separation barrier [123]. The performance in terms of conversion is low (up to 1.2%). Not surprisingly, the selectivities to corresponding phenol (99.9%) and H_2O_2 conversions to phenol (96.8%) were excellent. It is well-known from heterogeneous catalyst systems that at higher conversion the selectivities will be lower. A micro-emulsion catalytic system consisting of water, benzene, acetic acid, ferric dodecane sulfonate as catalyst and sodium dodecylbenzene sulfonate as surfactant was shown to be also applicable for hydroxylation of benzene with H_2O_2 as oxidant [124]. With low H_2O_2/benzene ratio phenol selectivities up to 92.9% could be achieved, at 21.9% benzene conversion. At higher H_2O_2 concentration, benzene conversion is enhanced but with the disadvantage of a more unselective reactions.

In the field of nonheme aromatic hydroxylations the oxidation of benzoic acids to salicylic acid was investigated. In 2005, the group of Rykbak-Akimova achieved stoichiometric hydroxylation of benzoic acid with an $[Fe^{II}(\mathbf{1})(CH_3CN)_2](ClO_4)_2$ complex and H_2O_2 as oxidant to yield exclusively *o*-hydroxylated salicylate complex $[Fe^{III}(\mathbf{1})(OOC(C_6H_4)O)](ClO_4)_3$ (up to 84% at r.t.) [125]. In the same year, Nam and coworkers reported that perbenzoic acid is converted into salicylate complexes while reacting with $[Fe^{II}(\mathbf{2})(CH_3CN)_2]^{2+}$ [126]. Several investigations were performed with nonheme iron(II)complexes of Bn-tpen **31** and N_4Py. They resulted in the conclusion that the iron(IV)-oxo group attacks the aromatic ring via an electrophilic pathway to produce either a tetrahedral radical or cationic σ-complex [127]. A catalytic transformation of arenes was achieved with nonheme iron complexes of tpen ligands **35** (Scheme 21) [128]. Addition of a reducing agent (1-naphthol) enhanced in most cases the yields of substrates; e.g., 59% yield of phenol.[2]

[2]Yields are based on the limiting reagent.

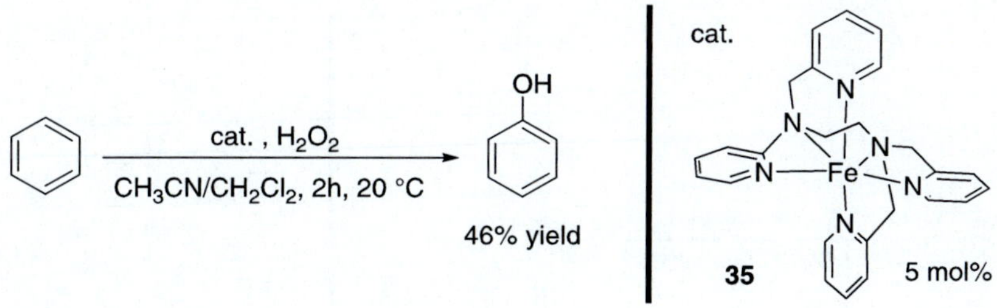

Scheme 21 Hydroxylation of benzene to phenol with nonheme iron complex **35** [142]

In addition to nonheme iron complexes also heme systems are able to catalyze the oxidation of benzene. For example, porphyrin-like phthalocyanine structures were employed to benzene oxidation (see also alkane hydroxylation) [129]. Mechanistic investigations of this type of reactions were carried out amongst others by Nam and coworkers resulting in similar conclusions like in the nonheme case [130]. More recently, Sorokin reported a remarkable "biological" aromatic oxidation, which occurred via formation of benzene oxide and involves an NIH shift. Here, phenol is obtained with a TON of 11 at r.t. with 0.24 mol% of the catalyst.

Compared with the selective hydroxylation of arenes or polyarenes to phenols, similar reactions to quinones were so far of limited interest. Again, the stability of the corresponding products under the reaction conditions is often problematic. Nevertheless, this type of reaction is of industrial interest for the preparation of vitamin intermediates. Here, menadione (vitamin K_3) and 2,3,6-trimethylquinone (key intermediate in vitamin E synthesis) represent the most important intermediates. Moreover, polynuclear partly functionalized aromatic hydrocarbons (PAHs) are a central class of environmental carcinogens and the complete oxidation and decomposition is important for environmentally benign waste disposals (mostly wet air oxidations are used for decomposition of PAH's; for iron-catalyzed examples see [131, 132]). For the oxidation of such compounds, metalloporphyrins are applied as models for cytochrome P450. Thus, 1 mol% *o*-substituted tetraarylporphyrinatoiron(III)chloride with electron withdrawing groups at the porphyrin ring as well as at the phenyl rings gave with H_2O_2 as oxidant high conversion for different PAH's. For example, the oxidation of pyrene proceeded in up to 92% conversion and resulted in two major products (pyrene-4,5-dione, pyrene-4,5,9,10-tetrone) (Scheme 22) [133].

Previous studies by Sorokin with iron phthalocyanine catalysts made use of oxone in the oxidation of 2,3,6-trimethylphenol [134]. Here, 4 equiv. $KHSO_5$ were necessary to achieve full conversion. Otherwise, a hexamethyl-biphenol is observed as minor side-product. Covalently supported iron phthalocyanine complexes also showed activity in the oxidation of phenols bearing functional groups (alcohols, double bonds, benzylic, and allylic positions) [135]. Besides, silica-supported iron phthalocyanine catalysts were reported in the synthesis of menadione [136].

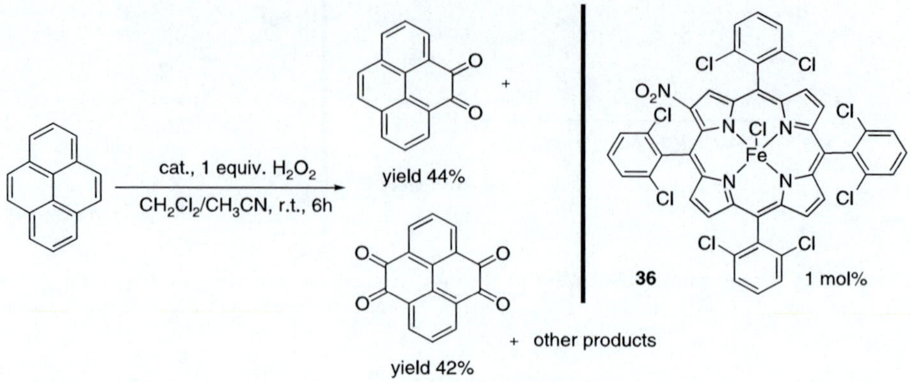

Scheme 22 Oxidation of pyrene with Fe-porphyrin catalysts

6 Alcohol Oxidations

A number of easily available iron salts and complexes can be used for the selective oxidation of alcohols to carbonyl compounds. Often iron oxo species are proposed as catalytic intermediates. Examples include the simple combination of $FeCl_3$ and hydrogen peroxide, which is known to be of moderate activity in alcohol oxidations as shown in the case of 2-cyanoethanol [137]. An $FeCl_3$–TEMPO–$NaNO_2$ catalyst system using air as oxidant was introduced in 2005 [138]. At ambient temperature various alcohols including sulfur-containing compounds were converted to the corresponding aldehydes and ketones with high conversion and excellent selectivities. Pearson and coworkers presented an iron carbonyl precatalyst (1,3-cyclohexadiene) $Fe(CO)_3$ in the presence of triethylamine-*N*-oxide as oxidant yielding the corresponding carbonyl compounds in 88–98% yield [139].

Under solvent-free conditions in the presence of stoichiometric amounts of iron nitrate good to very good yields for the oxidation of benzyl and various secondary alcohols were obtained [140]. Despite these examples, controlling iron-catalyzed oxidation reactions of alcohols with air or hydrogen peroxide remain difficult. In 2008, we demonstrated the possibility to switch between nonselective radical pathways and selective nonradical reactions by tuning the absolute pH of the reaction system [141]. As a benchmark reaction, the oxidation of benzyl alcohol was studied, which is catalyzed by various iron salts (mainly $Fe(NO_3)_3$) to give benzaldehyde. A constant pH value resulted in high conversion (pH value close to 1.00), whereas the chemoselectivity is controlled by the change of the pH.

In the field of nonheme iron complexes, Münck, Collins, and Kinoshita reported the oxidation of benzylic alcohols via stable μ-oxo-bridged diiron(IV) TAML complexes, which are formed by the reaction of iron-**28** complexes with molecular oxygen (Scheme 23) [142].

Remarkably, the shown Fe^{III} complexes reacted directly with oxygen to afford high-valent oxo–iron species. In addition, Kim, Nam, and coworkers explored

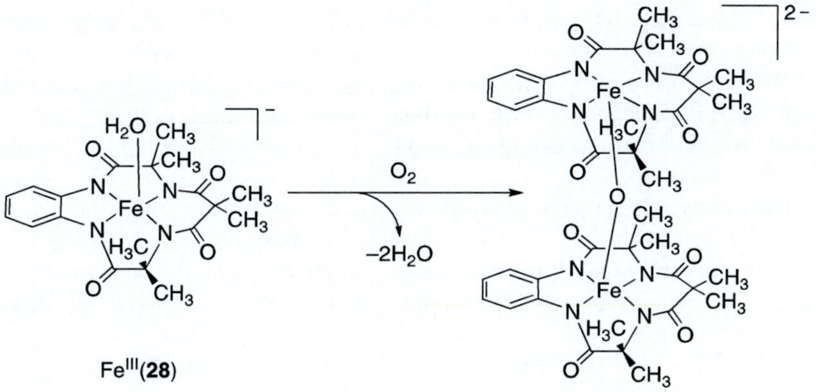

Scheme 23 Formation of the catalytic active μ-oxo-bridged diiron(IV) complex

mechanistic details of the alcohol oxidation with heme and nonheme complexes [143]. They suggested that this oxidation occurred via a α-CH hydrogen atom abstraction followed by an electron transfer. Later Kim and coworkers reported an iron(III) complex with tetradentate ligands bearing amide moieties (Scheme 24) [32]. These complexes oxidized alcohols as well as olefins to the corresponding carbonyls and epoxides on treatment with *m*CPBA as oxidant. However, low TON's between 50 and 90 were achieved using unusual substrate:oxidant:catalyst ratio of 100:10:0.5.

Scheme 24 Biomimetic ligands **37** and **38**

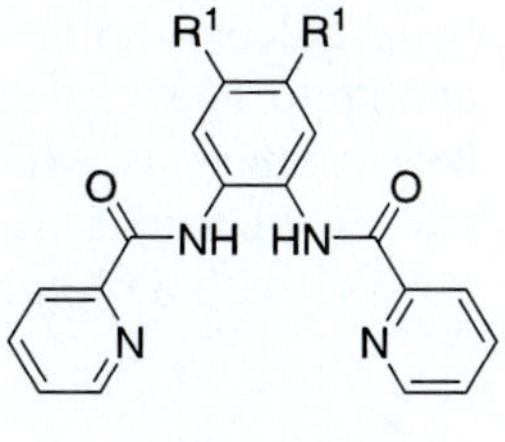

37: $R^1 = CH_3$
38: $R^1 = Cl$

A biomimetic oxidation with perfluorinated porphyrin complexes [(F_{20}TPP) FeCl] showed high catalytic activity with secondary alcohols with over 97% yield in all cases [144]. Furthermore, this catalyst is able to oxidize a broad range of alcohols under mild conditions with *m*CPBA as terminal oxidant. Here, an α-hydroxyalkyl radical species is proposed as central intermediate.

Another iron porphyrin complex with 5,10,15,20-tetrakis(2′,6′-dichloro-3′-sulfonatophenyl)porphyrin was applied in ionic liquids and oxidized veratryl alcohol (3,4-dimethoxybenzyl alcohol) with hydrogen peroxide in yields up to 83% to the aldehyde as the major product [145]. In addition, TEMPO was incorporated via a phenyl linker into the porphyrin scaffold [146]. The resulting catalyst showed a slight increase of the activity in the oxidation of benzyl alcohol using bleach as oxidant. However, similar Mn-based complexes showed distinct better activity.

Finally, phthalocyanine iron catalysts were also used for the oxidation of alcohols to yield corresponding carbonyl compounds with nonbenign hypervalent iodine oxidants [147].

Interestingly, "free" nano-iron oxide particles are active catalysts for the selective oxidation of alcohols to yield the corresponding aldehydes/ketones [72, 73]. Different aromatic alcohols and secondary aliphatic alcohols were oxidized with high selectivity, but at low conversion. Here, further improvement should be possible (Scheme 25).

Scheme 25 Selective oxidation of alcohols to aldehydes and ketones

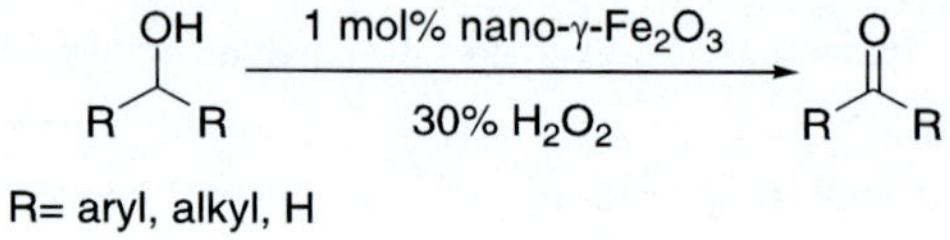

Under microwave irradiation and applying MCM-41-immobilized nano-iron oxide higher activity is observed [148]. In this case also, primary aliphatic alcohols could be oxidized. The TON for the selective oxidation of 1-octanol to 1-octanal reached to 46 with 99% selectivity. Hou and coworkers reported in 2006 an iron coordination polymer $\{[Fe(fcz)_2Cl_2]\cdot 2CH_3OH\}_n$ with fcz = 1-(2,4-difluorophenyl)-1,1-bis[(1*H*-1,2,4-triazol-l-yl)methyl]ethanol which catalyzed the oxidation of benzyl alcohol to benzaldehyde with hydrogen peroxide as oxidant in 87% yield and up to 100% selectivity [149]. An alternative approach is based on the use of heteropoly acids, whereby the incorporation of vanadium and iron into a molybdophosphoric acid catalyst led to high yields for the oxidation of various alcohols (up to 94%) with molecular oxygen [150].

7 Conclusions

Iron-based redox reactions are of key importance for the correct function of biological systems. Often in these systems, high-valent iron–oxo species are proposed as crucial intermediates. In the past decade, significant advancements have been reported with respect to the understanding and characterization of such active iron oxo species. However, despite all this work still the development of synthetically useful iron catalysts is significantly driven by the synthesis of new ligands and catalytic testing. Notably, selective catalytic oxidations with iron complexes allow to "streamline" organic synthesis and to perform the synthesis of advanced intermediates with improved efficiency.

References

1. Sheldon RA (1994) Metalloporphyrins in catalytic oxidations. Marcel Dekker, New York
2. Trautheim AX (1997) Bioinorganic chemistry: transition metals in biology and their coordination chemistry. Wiley-VCH, Weinheim
3. Berry JF, Bill E, Bothe E, George SD, Miener B, Neese F, Wieghardt K (2006) Science 312:1937–1941
4. Bukowski MR, Koehntop KD, Stubna A, Bominaar EL, Halfen JA, Munck E, Nam W, Que L (2005) Science 310:1000–1002
5. de Oliveira FT, Chanda A, Banerjee D, Shan X, Mondal S, Que L, Bominaar EL, Münck E, Collins TJ (2007) Science 315:835–838
6. Krebs C, Fujimori DG, Walsh CT, Bollinger JM (2007) Acc Chem Res 40:484–492
7. Makris TM, von Koenig K, Schlichting I, Sligar SG (2006) J Inorg Biochem 100:507–518
8. Yudin AK (2006) Aziridines and epoxides in organic synthesis. Wiley-VCH, Weinheim, pp 229–269
9. Larrow JF, Jacobsen EN (2004) Topics Organomet Chem 6:123–152
10. Jacobsen EN, Pfaltz A, Yamamoto H (1999) Comprehensive asymmetric catalysis, vol 3., pp 1309–1326
11. Plietker B (2008) Iron catalysis in organic chemistry. Wiley-VCH, Weinheim, pp 73–123
12. Enthaler S, Junge K, Beller M (2008) Angew Chem 120:3363–3367, Angew Chem Int Ed 47:3317–3321
13. Bäckvall J-E (2004) Modern oxidation methods. Wiley-VCH, Weinheim, pp 22–50
14. Punniyamurthy T, Velusamy S, Iqbal J (2005) Chem Rev 105:2329–2363
15. Simándi LI (2003) Advances in catalytic activation of dioxygen by metal complexes. Kluwer, Dordrecht
16. Souza DPB, Fricks AT, Alvarez HM, Salomão GC, Olsen MHN, Filho LC, Fernandes C, Atunes OAC (2007) Catal Commun 8:1041–1046
17. Traylor TG, Tsuchiya S, Byun YS, Kim C (1993) J Am Chem Soc 115:775–2781
18. Stephenson NA, Bell AT (2007) J Mol Catal A Chem 275:54–62
19. Stephenson NA, Bell AT (2005) J Am Chem Soc 127:8635–8643
20. Bassan A, Blomberg MRA, Siegbahn PEM, Que L (2005) Angew Chem 117:2999–3001, Angew Chem Int Ed 44:2939–2941
21. Kodera M, Itoh M, Kano K, Funabiki T, Reglier M (2005) Angew Chem 117:7266–7268, Angew Chem Int Ed 44:7104–7106
22. Fujita M, Que L (2004) Adv Synth Catal 346:190–194
23. Chen K, Costas M, Kim J, Tipton AT, Que L (2002) J Am Chem Soc 124:3026–3035
24. White MC, Doyle AG, Jacobsen EN (2001) J Am Chem Soc 123:7194–7195
25. Costas M, Tipton AK, Chen K, Jo D-H, Que L (2001) J Am Chem Soc 123:6722–6723
26. Chen K, Que L (1999) Chem Commun 1375–1376
27. Mas-Ballesté R, Que L (2007) J Am Chem Soc 129:15964–15972
28. Mas-Ballesté R, Costas M, Van den Berg T, Que L (2006) Chem Eur J 12:7489–7500
29. Bukowski MR, Comba P, Lienke A, Limberg C, De Laorden CL, Mas-Ballesté R, Merz M, Que L (2006) Angew Chem 118:3524–3528, Angew Chem Int Ed 45:3446–3449
30. Duban E A, Bryliakov K P, Talsi E P (2007) Eur J Inorg Chem 852–857
31. Bautz J, Comba P, De Laorden CL, Menzel M, Rajaraman G (2007) Angew Chem 119:8213–8216, Angew Chem Int Ed 46:8067–8070
32. Lee SH, Han JH, Kwak H, Lee SJ, Lee EY, Kim HJ, Lee JH, Bae C, Lee CN, Kim Y, Kim C (2007) Chem Eur J 13:9393–9398
33. Suh Y, Seo MS, Kim KM, Kim YS, Jang HG, Tosha T, Kitagawa T, Kim J, Nam W (2006) J Inorg Biochem 100:627–633
34. Gosiewska S, Lutz M, Spek AL, Klein Gebbink RJM (2007) Inorg Chim Acta 360:405–417
35. Taktak S, Ye W, Herrera AM, Rybak-Akimova EV (2007) Inorg Chem 46:2929–2942
36. Liu P, Wong EL-M, Yuen AW-H, Che C-M (2008) Org Lett 10:3275–3278

37. Bitterlich B, Schröder K, Tse MK, Beller M (2008) Eur J Org Chem 29:4867–4870
38. Bitterlich B, Anilkumar G, Gelalcha FG, Spilker B, Grotevendt A, Jackstell R, Tse MK, Beller M (2007) Chem Asian J 2:521–529
39. Gopinathan A, Bitterlich B, Gelalcha FG, Tse MK, Beller M (2007) Chem Commun 289–291
40. Schröder K, Enthaler S, Bitterlich B, Schulz T, Spannenberg A, Tse MK, Junge K, Beller M (2009) Chem Eur J 15:5471–5481
41. Schröder K, Tong X, Bitterlich B, Tse MK, Gelalcha FG, Brückner A, Beller M (2007) Tetrahedron Lett 48:6339–6342
42. Costas M, Tipton AK, Chen K, Jo D-H, Que L (2001) J Am Chem Soc 123:6722–6723
43. Francis MB, Jacobsen EN (1999) Angew Chem Int Ed 38:937–941
44. Cheng QF, Xu XY, Ma WX, Yang SJ, You TP (2005) Chin Chem Lett 16:1467–1470
45. Gelalcha FG, Anilkumar G, Tse MK, Brückner A, Beller M (2008) Chem Eur J 14:7687–7698
46. Gelalcha FG, Bitterlich B, Gopinathan A, Tse MK, Beller M (2007) Angew Chem 119:1–6, Angew Chem Int Ed 46:1–5
47. Marchi-Delapierre C, Jorge-Robin A, Thibon A, Ménage S (2007) Chem Commun 1166–1168
48. Yeung H-L, Sham K-C, Tsang C-S, Lau T-C, Kwong H-L (2008) Chem Commun 3801–3803
49. Rose E, Andrioletti B, Zrig S, Quelquejeu-Ethève M (2005) Chem Soc Rev 34:573–583
50. Gupta KC, Sutar AK (2008) Polym Adv Technol 19:186–200
51. Gupta KC, Sutar AK (2008) J Appl Polym Sci 108:3927–3941
52. Wang Y, Zhang Q, Shishido T, Takehira K (2002) J Catal 209:186–196
53. Nozaki C, Lugmair CG, Bell AT, Tilley TD (2002) J Am Chem Soc 124:13194–13203
54. Duma V, Hönicke D (2000) J Catal 191:93–104
55. Terry TJ, Dubois G, Murphy A, Stack TD (2007) Angew Chem 119:963–965, Angew Chem Int Ed 46:945–947
56. Dubois G, Murphy A, Stack TD (2003) Org Lett 5:2469–2472
57. Salavati-Niasari M (2007) J Mol Catal 278:22–28
58. Weissermehl K, Arpe H J (2003) Worldwide production capacities for ethylene glycol in 2000: 13.6 Mio to/a; worldwide production of 1,2-propylene glycol in 1996: 1.4 Mio to/a. In: Ind Org Chem, Vol 4. Wiley-VCH, Weinheim, p 152 and p 277
59. Zaitsev AB, Adolfsson H (2006) Synthesis 1725–1756
60. Costas M, Mehn MP, Jensen MP, Que L (2004) Chem Rev 104:939–986
61. Oldenburg PD, Que L (2006) Catal Today 117:15–21
62. Bruijnincx PCA, Van Koten G, Klein Gebbink RJM (2008) Chem Soc Rev 37:2716–2744
63. Feng Y, Ke C-Y, Xue G, Que L (2009) Chem Commun 50–52
64. Company A, Gómez L, Fontrodona X, Ribas X, Costas M (2008) Chem Eur J 14:5727–5731
65. Company A, Feng Y, Güell M, Ribas X, Luis JM, Que L, Costas M (2009) Chem Eur J 15:3359–3362
66. Koehntop KD, Emerson JP, Que L (2005) J Biol Inorg Chem 10:87–93
67. Bruijnincx PCA, Lutz M, Spek AL, Van Faasen EL, Weckhuysen BM, Van Koten G, Klein Gebbink RJM (2005) Eur J Inorg Chem 779–787
68. Bruijnincx PCA, Lutz M, Spek AL, Hagen WR, Weckhuysen BM, Van Koten G, Klein Gebbink RJM (2007) J Am Chem Soc 129:2275–2286
69. Bruijnincx PCA, Buurmans ILC, Gosiewska S, Moelands MAH, Lutz M, Spek AL, Van Koten G, Klein Gebbink RJM (2008) Chem Eur J 14:1228–1237
70. Oldenburg PD, Shteinman AA, Que L (2005) J Am Chem Soc 127:15672–15673
71. Suzuki K, Oldenburg PD, Que L (2008) Angew Chem 120:1913–1915, Angew Chem Int Ed 47:1887–1889
72. Shi F, Tse MK, Pohl M-M, Brückner A, Zhang S, Beller M (2007) Angew Chem Int Ed 46:8866–8868

73. Shi F, Tse MK, Pohl M-M, Radnik J, Brückner A, Zhang S, Beller M (2008) J Mol Catal 292:28–35
74. Brazdil JF (2006) Top Catal 38:289–294
75. Bauer EB (2008) Current Org Chem 12:1341–1369
76. Shilov AE, Shul'pin GB (1997) Chem Rev 97:2879–2932
77. Labinger JA, Bercaw JE (2002) Nature 417:507–514
78. Ortiz de Montellano PR (2005) Cytochrome P450: structure, mechanism and biochemistry, 3rd edn. Kluwer/Plenum, New York
79. Abu-Omar MM, Loaiza A, Hontzeas N (2005) Chem Rev 105:2227–2252
80. Chakrabarty S, Austin RN, Deng D, Groves JT, Lipscomb JD (2007) J Am Chem Soc 129:3514–3515
81. Meunier B (2000) Biomimetic oxidation catalyzed by transition metal complexes. College Press, London
82. Plietker B (2008) Iron catalysis in organic chemistry. Wiley-VCH, Weinheim, pp 73–91
83. Pavan C, Legros J, Bolm C (2005) Adv Synth Catal 347:703–705
84. Nakanishi M, Bolm C (2007) Adv Synth Catal 349:861–864
85. Li D, Schröder K, Bitterlich B, Tse MK, Beller M (2008) Tetrahedron Lett 49:5976–5979
86. Trettenhahn G, Nagl M, Neuwirth N, Arion VB, Jary W, Pöchlauer P, Schmid W (2006) Angew Chem 118:2860–2865, Angew Chem Int Ed 45:2794–2798
87. Haber F, Willstätter R (1931) Ber Dtsch Chem Ges 64:2844–2856
88. Haber F, Weiss J (1932) Naturwissenschaften 20:948–950
89. Shul'pin GP, Golfeto CC, Süss-Fink G, Shul'pina LS, Mandelli D (2005) Tetrahedron Lett 46:4563–4567
90. Fernandes RR, Kirillova MV, da Silva JAL, Fraústo da Silva JJR, Pombeiro AJL (2009) Appl Catal A Gen 353:107–112
91. Shul'pin GP, Nizova GV, Kozlov YN, Cuervo LG, Süss-Fink G (2004) Adv Synth Catal 346:317–332
92. Tanase S, Marques-Gallego P, Browne WR, Hage R, Bouwman E, Feringa BL, Reedijk J (2008) Dalton Trans 2026–2033
93. Que L, Tolman WB (2008) Nature 455:333–340
94. Britovsek GJP, England J, Spitzmesser SK, White AJP, Williams DJ (2005) Dalton Trans 945–955
95. Carvalho NMF, Horn A, Antunes OAC (2006) Appl Catal A Gen 305:140–145
96. Oldenburg PD, Ke C-Y, Tipton AA, Shteinman AA, Que L (2006) Angew Chem 118:8143–8146, Angew Chem Int Ed 45:7975–7978
97. Kim C, Chen K, Kim J, Que L (1997) J Am Chem Soc 119:5964–5965
98. Okuno T, Ito S, Ohba S, Nishida Y (1997) J Chem Soc Dalton Trans 3547–3551
99. England J, Davies CR, Banaru M, White AJP, Britovsek GJP (2008) Adv Synth Catal 350:883–897
100. Britovsek GJP, England J, White AJP (2005) Inorg Chem 44:125–8134
101. Chen MS, White C (2007) Science 318:783–787
102. Vermeulen NA, Chen MS, White MC (2009) Tetrahedron 65:3078–3081
103. Christmann M (2008) Angew Chem 120:2780–2783, Angew Chem Int Ed 47:2740–2742
104. Lee Y-M, Dhuri SN, Sawant SC, Cho J, Kubo M, Ogura T, Fukuzumi S, Nam W (2009) Angew Chem 121:1835–1838, Angew Chem Int Ed 48:1803–1806
105. Godbole MD, Puig MP, Tanase S, Kooijman H, Spek AL, Bouwman E (2007) Inorg Chim Acta 360:1954–1960
106. Li F, Wang M, Ma C, Gao A, Chen H, Sun L (2006) Dalton Trans 2427–2434
107. Ito S, Okuno T, Itoh H, Ohba S, Matsushima H, Tokii T, Nishida Y (1997) Z Naturforsch B 52:719–727
108. Tanase S, Foltz C, De Gelder R, Hage R, Bouwman E, Reedijk J (2005) J Mol Catal A Chem 225:161–167

109. Gosiewska S, Cornlissen JLM, Lutz M, Spek AL, Van Knoten G, Klein Gebbink RJM (2006) Inorg Chem 45:4214–4227
110. Gosiewska S, Permentier HP, Bruins AP, Van Knoten G, Klein Gebbink RJM (2007) Dalton Trans 3365–3368
111. Romakh VB, Therrien B, Süss-Fink G, Shul'pin GB (2007) Inorg Chem 46:3166–3175
112. Collins TJ (2002) Acc Chem Res 35:782–790
113. Chanda A, Shan X, Chakrabarti M, Ellis WC, Popescu DL, Tiago de Oliveira F, Wang D, Que L, Collins TJ, Münck E, Bominaar EL (2008) Inorg Chem 47:3669–3678
114. Rosenthal J, Luckett TD, Hodgkiss JM, Nocera DG (2006) J Am Chem Soc 128:6546–6547
115. Rosenthal J, Pistorio BJ, Cheng LJ, Nocera DG (2005) J Org Chem 70:1885–1888
116. Theodoridis A, Maigut J, Puchta R, Kudrik EV, Van Eldik R (2008) Inorg Chem 47: 2994–3013
117. Sorokin AB, Kudrik EV, Bouchu D (2008) Chem Commun 2562–2564
118. Kumar D, Derat E, Khenkin AM, Neumann R, Shaik S (2005) J Am Chem Soc 127:17712–17718
119. Bonchio M, Carraro M, Scorrano G, Kortz U (2005) Adv Synth Catal 347:1909–1912
120. Bonchio M, Carraro M, Sartorel A, Scorrano G, Kortz U (2006) J Mol Catal A Chem 251:93–99
121. Chen L, Zhu K, Bi L-H, Suchopar A, Reicke M, Mathys G, Jaensch H, Kortz U, Richards RM (2007) Inorg Chem 46:8457–8459
122. Bianchi D, Bortolo R, Tassinari R, Ricci M, Vignolo R (2000) Angew Chem 112:4491–4493, Angew Chem Int Ed 39:4321–4323
123. Molinari R, Poerio T, Argurio P (2006) Catal Today 118:52–56
124. Liu H, Fu Z, Yin D, Yin D, Liao H (2005) Catal Commun 6:638–643
125. Taktak S, Flook M, Foxman BM, Que L, Rybak-Akimova EV (2005) Chem Comm 5301–5303
126. Oh NY, Seo MS, Lim MH, Consugar MB, Park MJ, Rohde J-U, Han J, Kim KM, Kim J, Que L, Nam W (2005) Chem Comm 5644–5646
127. de Visser SP, Oh K, Han A-R, Nam W (2007) Inorg Chem 46:4632–4641
128. Thibon A, Bartoli J-F, Guillot R, Sainton J, Martinho M, Mansuy D, Banse F (2008) J Mol Catal A Chem 287:115–120
129. Kudrik EV, Sorokin AB (2008) Chem Eur J 14:7123–7126
130. Kang M-J, Song WJ, Han A-R, Choi YS, Jang HG, Nam W (2007) J Org Chem 72: 6301–6304
131. Quintanilla A, Casas JA, Zazo JA, Mohedano AF, Rodríguez JJ (2006) Appl Catal B Environ 62:115–120
132. Abussaud BA, Ulkem N, Berk D, Kubes GJ (2008) Ind Eng Chem Res 47:4325–4331
133. Giri NG, Chauhan SMS (2009) Catal Commun 10:383–387
134. Çimen Y, Türk H (2008) Appl Catal A Gen 340:52–58
135. Zalomaeva OV, Sorokin AB (2006) New J Chem 30:1768–1773
136. Kholdeeva OA, Zalomaeva OV, Sorokin AB, Ivanchikova ID, Della Pina C, Rossi M (2007) Catal Today 121:58–64
137. Shul'pina L, Veghini D, Kudinov AR, Shul'pina GB (2006) React Kinet Catal Lett 88: 157–163
138. Wang N, Liu R, Chen J, Liang X (2005) Chem Commun 5322–5324
139. Pearson AJ, Kwak Y (2005) Tetrahedron Lett 46:5417–5419
140. Namboodiri VV, Polshettiwar V, Varma RS (2007) Tetrahedron Lett 48:8839–8842
141. Shi F, Tse MK, Li Z, Beller M (2008) Chem Eur J 14:8793–8797
142. Ghosh A, Tiago de Oliveira F, Yano T, Nishioka T, Beach ES, Kinoshita I, Münck E, Ryabov AD, Horwitz CP, Collins TJ (2005) J Am Chem Soc 127:505–2513
143. Oh NY, Suh Y, Park MJ, Seo MS, Kim J, Nam W (2005) Angew Chem 117:4307–4311, Angew Chem Int Ed 44:4235–4239
144. Han JH, Yoo S-K, Seo JS, Hong SJ, Kim SK, Kim C (2005) Dalton Trans 402–406

145. Kumar A, Jain N, Chauhan SMS (2007) Synlett 411–414
146. Huang J-Y, Li S-J, Wang Y-G (2006) Tetrahedron Lett 47:5637–5640
147. Geraskin IM, Luedtke MW, Neu HM, Nemykin VN, Zhdankin VV (2008) Tetrahedron Lett 49:7410–7412
148. González-Arellano C, Campelo JM, Macquarrie DJ, Marinas JM, Romero AA, Luque R (2008) ChemSusChem 1:746–750
149. Han H, Zhang S, Hou H, Fan Y, Zhu Y (2006) Eur J Inorg Chem 1594–1600
150. Nagaraju P, Pasha N, Prasad PSS, Lingaiah N (2007) Green Chem 9:1126–1129

Top Organomet Chem (2011) 33: 111–138
DOI: 10.1007/978-3-642-14670-1_4

Catalysis by Fe=X Complexes (X = NR, CR_2)

Chi-Ming Che, Cong-Ying Zhou*, and Ella Lai-Ming Wong

Abstract Iron–nitrene/imido and carbene complexes are reactive chemical species capable of performing C–N and C–C bond formation, respectively. They have been widely utilized in the synthesis of natural and unnatural bioactive organic compounds and there is a growing interest in developing iron-catalyzed organic transformation reactions involving these chemical species as reaction intermediates. In the first part of this chapter, the organometallic chemistry of iron–nitrene/imido and carbene complexes including their synthesis, structures, spectroscopic properties and reactivity are presented. In the second part, a variety of iron-catalyzed nitrene and carbene transfer reactions are discussed.

Keywords Carbene · Catalysis · Complex · Iron · Nitrene

Contents

C.-M. Che, C.-Y. Zhou (✉), and E.L.M. Wong
Department of Chemistry and Open Laboratory of Chemical Biology of the Institute of Molecular Technology for Drug Discovery and Synthesis, The University of Hong Kong, Pokfulam Road, Hong Kong, China
e-mail: cmche@hku.hk

Abbreviations

Ac	Acetyl
acac	Acetylacetonate
Ar	Aryl
Cp	Cyclopentadienyl
DCM	Dichloromethane
DMF	*N,N*-Dimethylformamide
dppe	Bis(diphenylphosphino)ethane
dr	Diastereomeric ratio
ee	Enantiomeric excess
equiv	Equivalent(s)
Et	Ethyl
h	Hour(s)
i-Pr	Isopropyl
Me	Methyl
Mes	Mesityl, 2,4,6-trimethylphenyl
min	Minute(s)
mol	Mole(s)
NHC	*N*-Heterocyclic carbene
Nu	Nucleophile
Ph	Phenyl
rt	Room temperature
t-Bu	*tert*-Butyl
THF	Tetrahydrofuran
Ts	Tosyl, *p*-toluenesulfonyl

1 Introduction

Transition metal-catalyzed carbenoid transfer reactions, such as alkene cyclopropanation, C–H insertion, X–H insertion (X = heteroatom), ylide formation, and cycloaddition, are powerful methods for the construction of C–C and C–heteroatom bonds [1–6]. In contrast to a free carbene, metallocarbene-mediated reactions often proceed stereo- and regioselectively under mild conditions with tolerance to a wide range of functionalities. The reactivity and selectivity of metallocarbenes can be tuned by changing metal ion or auxiliary ligands. The decomposition of diazo compounds by transition metal complexes is a convenient way to generate metal–carbene intermediates.

Nitrene, a nitrogen analogue of carbene, can undergo alkene aziridination, C–H insertion, and imination of sulfur compounds, straightforwardly affording amino compounds, which are ubiquitous in natural products and biologically active molecules. Transition metal complexes can efficiently and stereoselectively effect these nitrene transfer reactions via intermediacy of metallonitrenes under mild conditions [7–16]. PhI=NR is the most commonly used nitrene precursor, which can be formed in situ by reaction of RNH_2 with $PhI(OAc)_2$ or PhIO. Other nitrene sources such as chloramine-T, bromamine-T, and organic azides can also be used for metal-catalyzed nitrene transfer reactions.

Although complexes of platinum group metals such as that of rhodium are well documented to efficiently catalyze carbene and nitrene transfer reactions [1–16], the development of new catalysts based on inexpensive and biocompatible metals is an appealing subject in this area of research. Iron is inexpensive due to its natural abundance and easy accessibility. Iron complexes are biocompatible in comparison to that of other transition metals as they are involved in a wide range of biological processes in the human body. These features make iron an ideal choice of element for the development of transition metal-catalyzed reactions. This chapter summarizes the catalysis through the intermediacy of iron–nitrene and carbene complexes. The first part presents organometallic aspects of these complexes, including their synthesis, structures, spectroscopic properties, and reactivity. The organometallic chemistry of iron porphyrin carbene complexes is discussed, since iron porphyrins have been successfully applied for catalytic carbenoid transfer reactions. The Fisher-type iron carbene complexes such as half-sandwich iron carbene complexes, which are commonly employed in stoichiometric reactions, are not covered in this chapter. The second part summarizes the relevant catalysis.

2 Iron Carbene Complexes

2.1 Synthesis

In 1964, Fischer and coworkers reported the first stable transition metal carbene complex [17]. Thirteen years later, Mansuy reported the first iron porphyrin carbene complex, which is also the first iron carbene complex isolated [18]. Treatment of [Fe(TPP)] with CCl_4 in the presence of an excess of iron powder under argon gave [Fe(TPP)CCl_2] in 90% yield (Scheme 1). With this method, various iron porphyrin carbene complexes [Fe(Por)C(X)R] have been synthesized by the reaction of iron porphyrins with polyhalogenated compounds (RCX_3) in the presence of excess reducing agents such as iron powder or sodium dithionite ($Na_2S_2O_4$) [19, 20]. These halogenated carbenes include CCl_2, CBr_2, CF_2, C(Cl)F, C(Br)F, C(Cl)CO_2Et, C(Cl)CN, and C(Cl)CF_3. Five-coordinate halocarbene complexes [Fe(Por)C(X)R] can bind donor solvent molecule or organic base (such as pyridine) to form six-coordinate species [(L)Fe(Por)C(X)R]. [Fe(TPP)(CCl_2)(H_2O)] is the first iron porphyrin carbene complex characterized by single-crystal X-ray diffraction analysis [21].

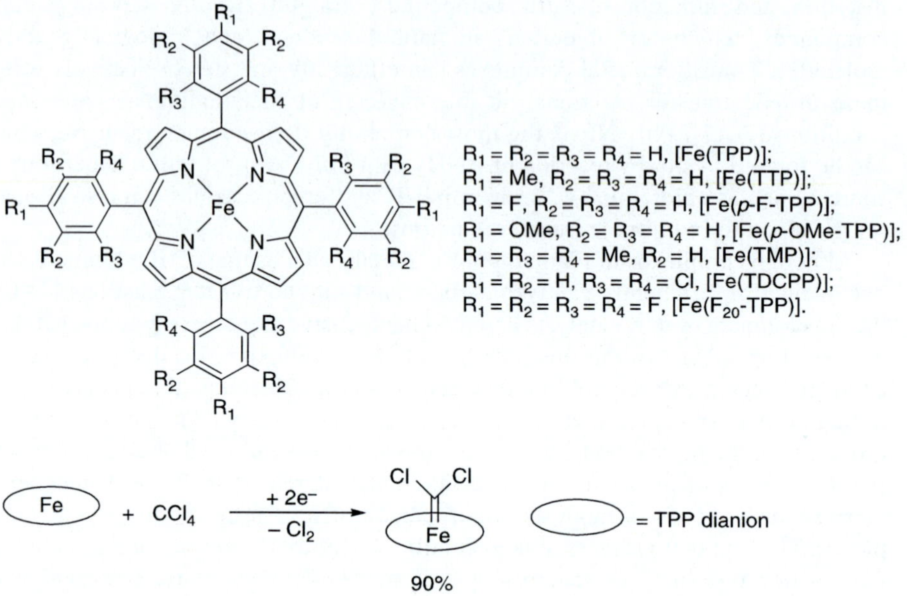

Scheme 1 Schematic structures of iron *meso*-tetraarylporphyrins and preparation of Fe-porphyrin carbene complexes

Decomposition of diazo compounds by iron porphyrins is a convenient method for the synthesis of non-heteroatom carbene–iron porphyrins [22]. Reaction of [Fe(F_{20}-TPP)] [F_{20}-TPP = *meso*-tetrakis(pentafluorophenyl)porphyrinato dianion] with diazo compounds $N_2C(Ph)R$ (R = Ph, CO_2Et, $CO_2CH_2CH{=}CH_2$) under an inert atmosphere afforded complexes [Fe(F_{20}-TPP)C(Ph)R] in 65–70% yields (Scheme 2). Like the halocarbene complex [Fe(TPP)(CCl_2)], [Fe(F_{20}-TPP)CPh_2] reacted with MeIm to afford six-coordinate species [(MeIm)Fe(F_{20}-TPP)CPh_2] in 65% isolated yield.

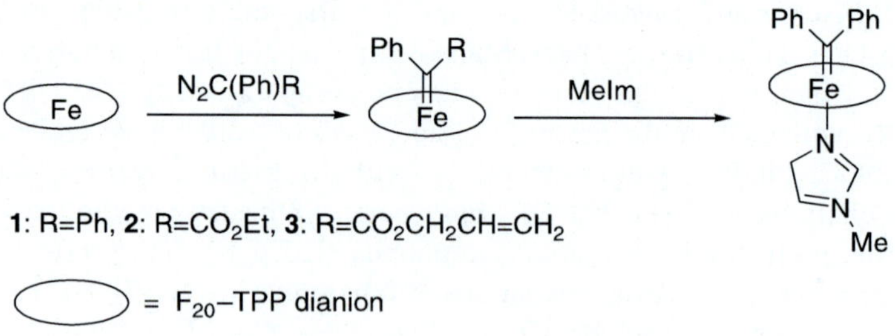

Scheme 2 Synthesis of non-heteroatom carbene-iron porphyrins

2.2 *Spectroscopic Feature and X-Ray Structure*

All iron porphyrin carbene complexes are diamagnetic and can be characterized by NMR spectroscopy. These complexes are low-spin iron(II) species. The ^{13}C NMR data for iron porphyrin carbene complexes are summarized in Table 1. As shown in this table, the resonances of Fe=*C*(carbene) are deshielded with chemical shifts at 220–390 ppm range. For halocarbene complexes [Fe(TPP)C(X)R], electron-rich substituent (R = alkyl) leads to Fe=*C*(carbene) signal at 310 ppm, whereas electron-withdrawing substituent (R=CN, CO_2Et) upfields the shift of Fe=*C*(carbene) signal to 220 ppm. For non-heteroatom carbene–iron porphyrins [Fe(F_{20}-TPP)C(Ph)R], the arylcarbene complexes [Fe(F_{20}-TPP)CPh_2] and [(MeIm)Fe(F_{20}-TPP)CPh_2] display downfield Fe=*C*(carbene) signals at 358.98 ppm and 385.44 ppm, respectively, compared to the (alkoxycarbonyl) carbene complexes [Fe(F_{20}-TPP)C(Ph)CO_2Et] (δ = 327.47 ppm) and [Fe(F_{20}-TPP)C(Ph)$CO_2CH_2CH{=}CH_2$] (δ = 325.67 ppm). Notably, coordination of MeIm to [Fe(F_{20}-TPP)CPh_2] downfield shifts the Fe=*C*(carbene) signal by 26 ppm.

Table 1 ^{13}C NMR data for iron porphyrin carbene complexes

Complex	^{13}C NMR (ppm) Fe=C	References
[Fe(TPP)C(Cl)$CHMe_2$]	324.0	[23, 24]
[Fe(TPP)C(Cl)CH(OH)Me]	312.0	[23, 24]
[Fe(TPP)C(Cl)CH(OH)Ph]	303.0	[23, 24]
[Fe(TPP)C(Cl)CO_2Et]	234.0	[19]
[Fe(TPP)CCl_2]	224.7	[18]
[Fe(TPP)C(Cl)CN]	210.0	[19]
[Fe(F_{20}-TPP)CPh_2]	358.98	[22]
[Fe(F_{20}-TPP)C(Ph)CO_2Et]	327.47	[22]
[Fe(F_{20}-TPP)C(Ph)$CO_2CH_2CH{=}CH_2$]	325.67	[22]
[(MeIm)Fe(F_{20}-TPP)CPh_2]	385.44	[22]

Although a large number of iron porphyrin carbene complexes have been reported in the literature, only a few of these complexes have been characterized by single-crystal X-ray diffraction analysis [21, 22]. Table 2 shows the structural data of the iron porphyrin carbene complexes reported in the literature. The Fe=C (carbene) distance of [Fe(F_{20}-TPP)CPh_2] is 1.767(3) Å. Coordination of MeIm to [Fe(F_{20}-TPP)CPh_2] to form [(MeIm)Fe(F_{20}-TPP)CPh_2] results in lengthening of the Fe=CPh_2 bond to 1.827(5) Å, which is similar to the Fe=CCl_2 distance in [Fe (TPP)(CCl_2)(H_2O)] [1.83(3) Å].

Table 2 X-ray structural data for iron porphyrin carbene complexes

Complex	Fe=C(Å)	Fe–N(Å)	References
[Fe(TPP)(CCl_2)(H_2O)]	1.83(3)	1.984(4)	[21]
[Fe(F_{20}-TPP)CPh_2]	1.767(3)	1.966(3)	[22]
[(MeIm)Fe(F_{20}-TPP)CPh_2]	1.827(5)	1.973(4)	[22]

The porphyrin ring of [Fe(F_{20}-TPP)CPh_2] exhibits a marked ruffling distortion with a mean deviation of 0.138 Å from the mean plane of the 24 component atoms. The porphyrin ring in [(MeIm)Fe(F_{20}-TPP)CPh_2] shows a even more severe ruffling distortion with a mean deviation of 0.198 Å from the mean plane of the 24 component atoms (Fig. 1). In contrast, the porphyrin ring of [Fe(TPP)(CCl_2)(H_2O)] is essentially planar (mean deviation: 0.03 Å).

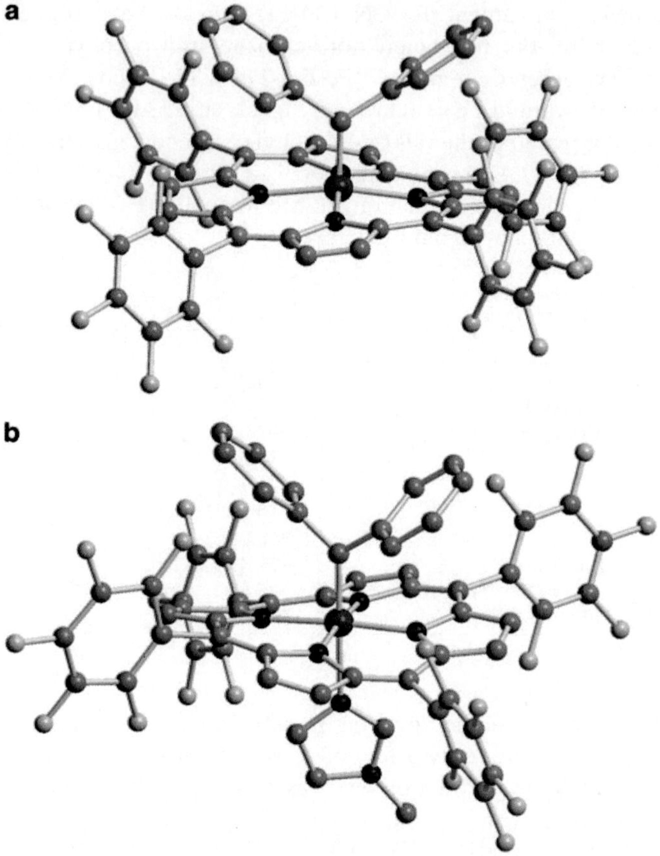

Fig. 1 Ball and stick drawings of the structures of (**a**) [Fe(F_{20}-TPP)(CPh_2)]·0.5C_6H_6·0.5CH_2Cl_2 and (**b**) [Fe(F_{20}-TPP)(CPh_2)(MeIm)]. The solvent molecules and/or hydrogen atoms are not shown [22]

The iron atom in [Fe(F_{20}-TPP)CPh_2] is 0.294 Å out of the mean porphyrin plane toward the carbene group, whereas [(MeIm)Fe(F_{20}-TPP)CPh_2] exhibits a considerably smaller displacement of the iron atom from the mean plane of the porphyrin ring (0.122 Å toward the CPh_2 group). In contrast, the iron atom of

$[Fe(TPP)(CCl_2)(H_2O)]$ is not significantly displaced from the porphyrin plane. Despite the different structures of $[Fe(F_{20}\text{-}TPP)CPh_2]$, $[(MeIm)Fe(F_{20}\text{-}TPP)CPh_2]$, and $[Fe(TPP)(CCl_2)(H_2O)]$, the average Fe–N(pyrrole) distances in these three complexes are similar.

2.3 *Reactivity*

$[Fe(TPP)CCl_2]$ reacts with primary amine RNH_2 to form isonitrile complex (eq. 1 in Scheme 3) [25, 26]. Visible light can activate iron porphyrin dihalocarbene complex to generate free carbene which reacts with alkenes to produce cyclopropanes in good yields (eq. 2 in Scheme 3) [27, 28]. In contrast, $[Fe(F_{20}\text{-}TPP)C(Ph)CO_2Et]$ can undergo stoichiometric cyclopropanation of alkenes at 60°C without the need of light irradiation (eq. 3 in Scheme 3) [22]. However, $[Fe(F_{20}\text{-}TPP)(CPh_2)]$ is unreactive toward stoichiometric cyclopropanation of styrene, even though it can catalyze intermolecular cyclopropanation of styrene with EDA and intramolecular cyclopropanation of allylic diazoacetates. Interestingly, the six-coordinate carbene complex $[(MeIm)Fe(F_{20}\text{-}TPP)CPh_2]$ can react with styrene at 80°C to give 1,1,2-triphenyl cyclopropane in 53% yield. This result can be rationalized by the *trans* effect of MeIm, which weakens the Fe=C bond as evident from the longer Fe=C distance in $[(MeIm)Fe(F_{20}\text{-}TPP)CPh_2]$ than in $[Fe(F_{20}\text{-}TPP)CPh_2]$ [22].

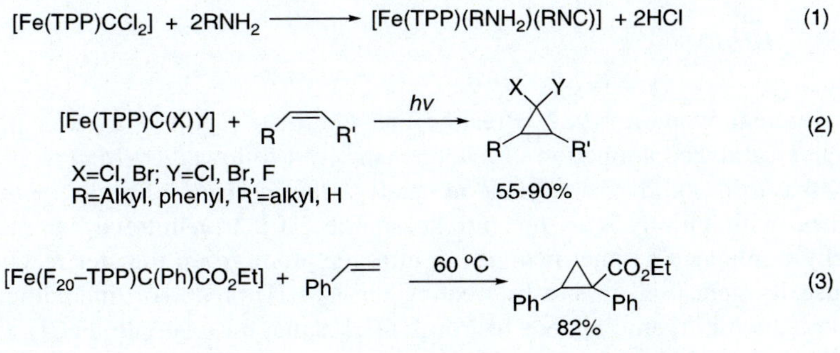

Scheme 3

Interestingly, $[Fe(F_{20}\text{-}TPP)C(Ph)CO_2Et]$ and $[Fe(F_{20}\text{-}TPP)CPh_2]$ can react with cyclohexene, THF, and cumene, leading to C–H insertion products (Table 3) [22]. The carbenoid insertion reactions were found to occur at allylic C–H bond of cyclohexene, benzylic C–H bond of cumene, and α C–H bond of THF. This is the first example of isolated iron carbene complex to undergo intermolecular carbenoid insertion to saturated C–H bonds.

Table 3 Stoichiometric C–H Insertion Reactions of [Fe(F_{20}-TPP)C(Ph)CO_2Et] **2** and [(MeIm)Fe(F_{20}-TPP)CPh_2] **4** with Alkenes or Tetrahydrofuran

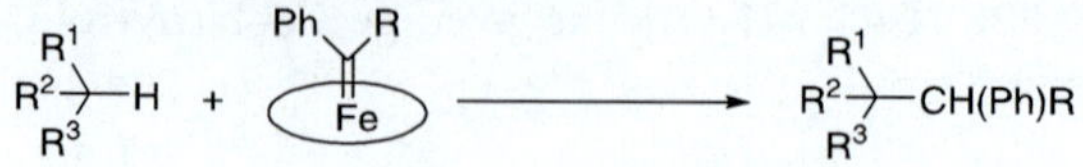

Entry	Complex	Substrate	Temperature (°C)	Time (h)	Product	Yield (%)
1	**2**	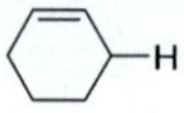	80	24	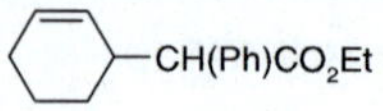	64
2	**4**	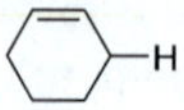	80	36	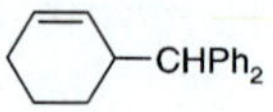	24
3	**2**	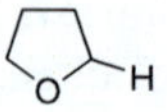	65	24	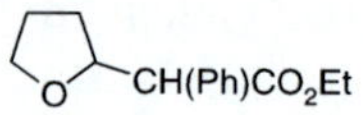	88
4	**2**	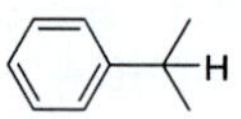	80	15	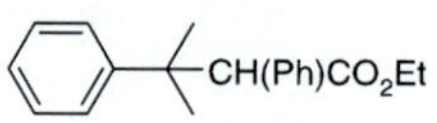	83
5	**4**	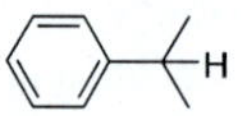	80	20	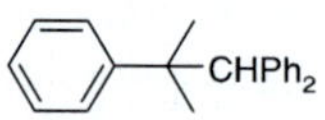	59

3 Iron Nitrene/Imido Complexes

3.1 Synthesis

In a seminal work in 1982, Breslow and Gellman reported an iron porphyrin complex catalyzed amination of cyclohexane [29] followed by Mansuy's work in 1984 that iron porphyrins are able to catalyze the insertion of a nitrene moiety to alkanes, with PhI=NTs as the nitrene source [30]. Iron nitrene complexes are usually implicated as intermediates in nitrogen atom/group transfer reactions and are usually generated in situ by treating the iron(II) precursor, including iron(II) supported with organic ligands or iron(II)/(III) salts, for example, $FeCl_2$, $FeCl_3$ Fe$(OTf)_2$ [31, 32], in appropriate solvent (e.g., CH_3CN) with solid sample of PhI=NTs. Within minutes of the addition, the color change of the solution can be observed showing the formation of nitrene species [33]. Normally, the iron nitrene complexes generated in situ are used directly for nitrene transfer reactions, without further isolation. Although implicated as reaction intermediates in group transfer reactions, iron nitrene complexes are rare. Only a few examples have been reported in the literature and iron nitrene complexes are usually characterized by spectroscopic methods. The complex [Fe(TPP)(NTs)(Cl)] reported by Mansuy and coworkers [34] was characterized by X-ray crystallography to contain a bridging NTs ligand, whereas the structure of $[Fe(N4Py)(NTs)]^{2+}$ reported by Que and coworkers [35] was characterized by DFT geometry optimization.

Besides, iron(IV) imides have also been implicated as reaction intermediates in nitrogen atom/group transfer reactions [36]. The first structurally characterized iron–imide complex was reported by Lee and coworkers [37] through the reaction of $FeCl_3$ with 2 equiv. of $LiNH^tBu$ in THF, giving one of the products as a stable site-differentiated cubane having three Fe(III) and one Fe(IV) centers and the Fe(IV) center has a terminally bonded imido ligand (Scheme 4).

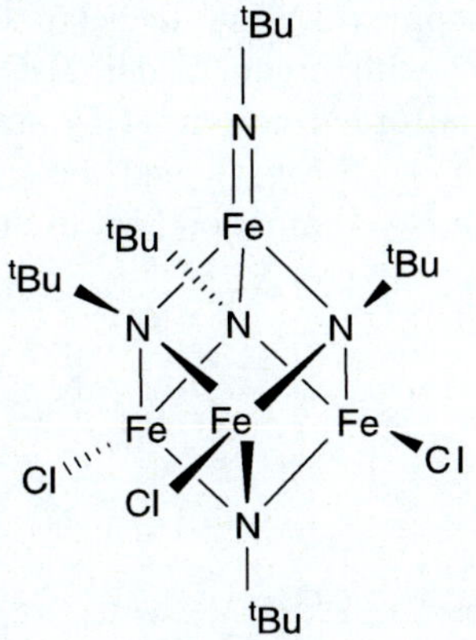

Scheme 4

Subsequently, Peters and coworkers reported the structurally characterized mononuclear iron(III) and iron(IV) imides, $[PhB(CH_2PPh_2)_3]Fe^{III}{\equiv}N$-p-tolyl and $\{[PhBP_2{}^{tBu}(pz')]Fe^{IV}{\equiv}NAd\}^+$, respectively, supported by tridentate pyrazolyl/bis (phosphino)borate ligand [38, 39] (Scheme 5). These iron–imides are prepared via oxidative nitrene transfer from organic azides, using low-valent iron(I) precursors containing phosphine or dinitrogen ligands.

$\{[PhBP^{tBu}{}_2(pz')]Fe^{III}{\equiv}NAd\}$ $L^{tBu}FeNAd$

Scheme 5

With the same tactic, a meta-stable 3-coordinate iron(III) imido complex was reported by Holland and coworkers [40]. Using the diketiminate-supported dinitrogen iron(I) complex $[L^{Me}FeNNFeL^{Me}]$ as a source of iron(I) precursor [L^{Me} = 2,4-bis (2,6-diisopropylphenylimino)pent-3-yl], the addition of adamantyl azide (AdN_3) in the presence of 4-tert-butylpyridine (^{t}Bupy, 2 equiv.) led to the formation of [L^{Me}FeNAd] as the major product in 70% yield. In the absence of pyridine additive, the AdN_3 units will be coupled to give a diiron(II) hexazene complex $\{L^{Me}Fe\}_2$(μ-AdN_6Ad). The more bulky derivative was also prepared by similar method using sterically demanding β-diketiminate ligand L^{tBu} (L^{tBu} = 2,2,6,6-tetramethyl-3,5-bis (diisopropylphenylimido)-hept-4-yl) giving [L^{tBu}FeNAd] as product [41] (Scheme 5).

In 2008, Power and coworkers reported that the reaction of a sterically encumbered iron(I) aryl/arene complex 3,5-Pr_2^{i}Ar*Fe(η^6-C_6H_6) [Ar*-3,5-Pr_2^{i}=C_6H-2,6 (C_6H_2-2,4,6-Pr_3^{i})$_2$-3,5-Pr_2^{i}] with organoazides $N_3C_6H_3$-2,6-Mes_2 (Mes = C_6H_2-2,4,6-Me_3) or N_3(1-Ad) afforded the iron(II) amido derivative {$CH_2C_6H_2$-2 (C_6H_3-2-N(H)FeAr*-3,5-Pr_2^{i})-3,5-Me_2}$_2$ and Fe(V) bis(imido) complex {3,5-Pr_2^{i}Ar*Fe[N(1-Ad)]$_2$} with N_2 elimination [42] (Scheme 6).

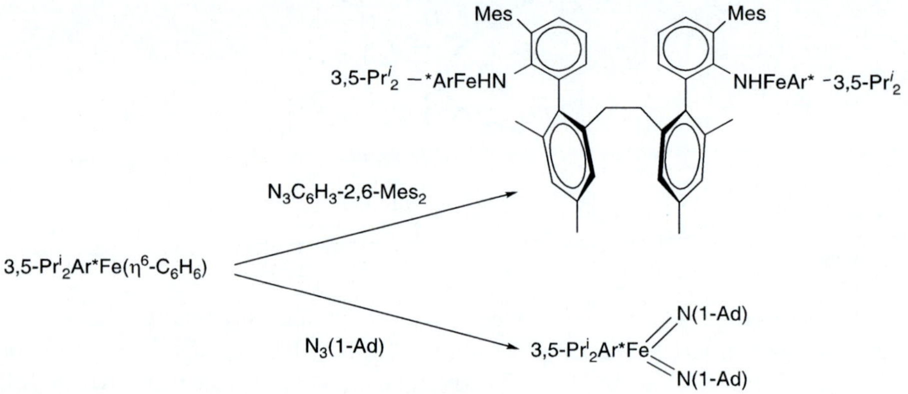

Scheme 6

High-valent iron–imido complexes have also been proposed as reaction intermediates in several reactions of the iron catalysis. Que and coworkers have provided evidence for Fe(IV) imide as a reaction intermediate in the reaction of [(6-PhTPA)-$Fe^{II}(CH_3CN)_2]^{2+}$ with PhI=NTs. Borovik and coworkers have also reported the formation of an amide product involving the generation of a putative iron(IV) imide [36] (Scheme 7).

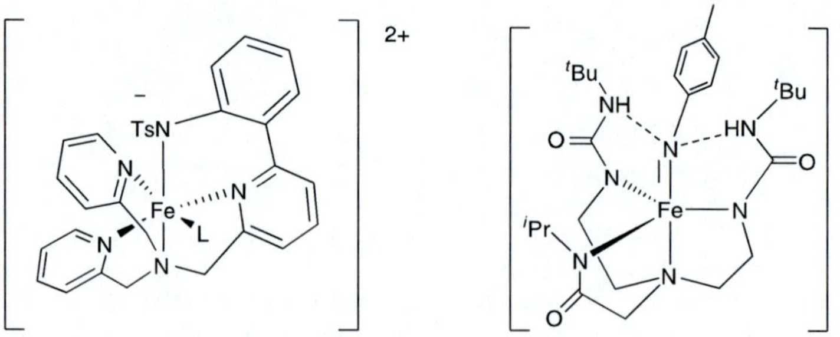

Scheme 7 Proposed iron (IV) imido reaction intermediates

3.2 Spectroscopic Feature and X-Ray Structure

Despite the analogy between iron–imido and iron–oxo complexes and extensive studies on iron(IV)–oxo complexes following the works by Que and coworkers [43, 44], well-characterized iron nitrene species are sparse in the literature. Most of the iron–imido/nitrene complexes are unstable with short lifetimes and are usually characterized by spectroscopic methods, such as UV-vis, NMR spectroscopy, and mass spectrometry at low temperature. In some instances, they have also been characterized by magnetic measurements using EPR or Mössbauer spectroscopy performed at low temperature for probing the oxidation state of the iron center. Structural information has been obtained using Fe K-edge X-ray absorption spectroscopy (XAS) and EDAX (for iron–imido species) and single X-ray crystallography (for iron–imides). Density functional theory (DFT) calculations were performed to provide insight into the electronic structures.

Lee and coworkers reported the first stable terminal iron–imide complex $[Fe_4(\mu_3\text{-}N^tBu)_4(N^tBu)Cl_3]$ in 2000 [37]. This complex has been characterized by X-ray crystallography as a cubane cluster possessing a nearly linear terminal iron–imide moiety with a short Fe–N bond distance (1.64 Å). The short Fe–N distance indicates the existence of iron–nitrogen triple bond. The IR spectrum of this complex is dominated with prominent peaks corresponding to the stretching of the bridging tert-butylimide ligands. Mössbauer measurement showed two quadrupole doublets with $\delta = 0.36$ mm/s and -0.17 mm/s, indicating the presence of high-spin iron(III) and iron(IV), respectively.

Peters and coworkers have reported a series of mononuclear iron imides stabilized by tris(phosphino)borate ligands [38, 45, 46]. These mononuclear iron imides display short Fe–N bonds (~1.65 Å) with relatively linear Fe–N–C angles (>170°) [38, 47]. The structural data are consistent with the strong dπ(Fe)-pπ(N) bonding in Fe(IV)–imides. The complexes are low-spin species with $S = 1/2$ configuration. For the iron(III) imide complex $[PhB(CH_2PPh_2)_3]Fe^{III}{\equiv}N$-*p*-tolyl, its cyclic voltammetry shows a fully reversible $Fe^{III/II}$ couple at -1.35 V, whereas an irreversible oxidation is observed at approximately -300 mV in 0.3 M [TBA][PF_6] at 30 mV/s. The other complex $\{[PhBP_2^{tBu}(pz')]Fe^{III}{\equiv}NAd\}$ exhibits a completely irreversible reduction wave at -2.20 V, showing that the iron(II) imide anion is unstable. A quasi-reversible feature at -0.72 V (at 100 mV/s; 22°C) is observed for the same complex corresponding to a $Fe^{IV/III}$ couple, and this couple becomes fully reversible when the scan rate is increased to 500 mV/s. This suggests that $\{[PhBP_2^{tBu}(pz')]Fe^{IV}{\equiv}NAd\}^+$ might be modestly stable.

Power and coworkers prepared the iron(V) bis-imide complex $\{3,5\text{-}Pr_2^iAr^*Fe[N(1\text{-}Ad)]_2\}$ [42]. This complex has been characterized by X-ray crystallography with the iron in planar three-coordinate geometry. The Fe–N bond distances are 1.642(2) and 1.619(2)Å. Magnetic studies of $\{3,5\text{-}Pr_2^iAr^*Fe[N(1\text{-}Ad)]_2\}$ reveals that this complex has a low-spin d^3 configuration with $S = 1/2$ ground state. This compound is notable as it is a stable Fe(V) imide being well characterized.

An iron(IV)–imido complex was reported by Que and coworkers in 2006 [35]. The starting iron(II) precursor is $[Fe(N_4Py)]^{2+}$ (N_4Py = *N,N*-bis(2-pyridylmethyl)bis

(2-pyridyl)methylamine) which reacts with PhINTs to give $[Fe^{IV}(NTs)(N_4Py)]^{2+}$. $[Fe^{IV}(NTs)(N_4Py)]^{2+}$ in CH_3CN exhibits an intense absorption band at $\lambda = 445$ nm ($\varepsilon = 2700\ M^{-1}\ cm^{-1}$) and a weak broad band at $\lambda = 660$ nm ($\varepsilon = 250\ M^{-1}\ cm^{-1}$), tentatively assigned as the iron(IV)-oxo contaminant. It has a half-life of 3 h at room temperature. The Mössbauer spectrum of $[Fe^{IV}(NTs)(N_4Py)]^{2+}$ obtained in CH_3CN at 4.2 K showed a doublet with a quadrupole splitting of 0.93 mms^{-1} and an isomer shift of 0.02 mms^{-1}. All these data are consistent with the formulation of nonheme iron(IV) complexes. High-resolution ESI-MS showed a predominant ion species at m/z 741.0860, corresponding to the $\{[Fe^{IV}(NTs)(N_4Py)](CF_3SO_3)\}^+$ formulation. An analysis of the EXAFS data obtained for $[Fe^{IV}(NTs)(N_4Py)]^{2+}$ shows a single scatter at 1.73 Å, attributed to the Fe–N bond distance with double bond character. This result matches with the DFT calculation in which a $(3d_{xy})^2(3d_{xz})^1(3d_{yz})^1$ ground state electronic configuration has been suggested. The Fe–NTs bond (1.73 Å) in $[Fe^{IV}(NTs)(N_4Py)]^{2+}$ is significantly longer than the Fe–N bonds observed in the X-ray crystal structures of two other iron(IV) imido complexes reported by Lee and Peters (ca. 1.65 Å). The Mössbauer spectrum of $[Fe^{IV}(NTs)(N_4Py)]^{2+}$ is consistent with an $S = 1$ ground state.

Other iron–imido complexes have also been reported. Holland and coworkers reported the synthesis of the imidoiron(III) complex $[L^{Me}FeNAd]$ [40, 41]. This imidoiron(III) complex has not been isolated and was found to convert to a purple high-spin iron(III) complex. It has an $S = 3/2$ ground state from EPR measurement. Based on the results of QM/MM computations, $[L^{Me}FeNAd]$ is a three-coordinated complex with an Fe–N distance of 1.68 Å and has a nearly linear Fe=N–C unit with Fe–N–C angle of 174.1°. Chirik and coworkers made use of liable ligands to prepare iron–imido complexes by treatment of $(^{iPr}PDI)\text{-}Fe(N_2)_2$ ($^{iPr}PDI = (2,6\text{-}{}^iPr_2C_6H_3N = CMe)_2C_5H_3N$) with a series of aryl azides [47].

Mass spectrometry is a useful tool to detect the existence of reactive iron–imido intermediates. In intramolecular aromatic aminations, Que and coworkers used electrospray ionization mass spectrometry to show the presence of a molecular ion at m/z 590.3 and 621.2, which could be attributed to the formation of $[(6\text{-}(o\text{-}TsN\text{-}C_6H_4)\text{-}TPA)Fe^{II}]^+$ and $[(6\text{-}(o\text{-}TsN\text{-}C_6H_4)\text{-}TPA)Fe^{III}(OMe)]^+$. With the isotopically enriched $PhI^{15}NTs$ and deuterated phenyl ring in iron(II) starting materials $[(d_5\text{-}PhTPA)Fe^{II}(NCCH_3)_2]^{2+}$, the mass spectrometry results are consistent with the substitution of a tosyl nitrene group for a hydrogen on the 6-phenyl ring of the 6-PhTPA ligand. Che and coworkers also made use of high-resolution ESI-MS to detect the existence of iron–imido species in the nitrene transfer reactions using $[Fe(Cl_3\text{-}terpy)_2]^{2+}$ as catalyst and PhINTs as nitrogen source [48].

3.3 *Reactivity*

Iron–nitrene/imido complexes are proposed to be the reaction intermediates in nitrogen group transfer reactions. The nitrene group can be transferred to organic substrates. Aziridination and amination are the well-known nitrogen atom/group transfer reactions catalyzed by iron complexes. This will be discussed in Sect. 5 of this chapter.

The iron(III) imides are reactive. $[PhBP_3]Fe{\equiv}N$-*p*-tolyl can readily react with carbon monoxide at room temperature to give isocyanate (O=C=N-*p*-tolyl) and the iron(I) by-product $[PhBP_3]Fe(CO)_2$ [38]. Isocyanate (AdN=C=O, 81% yield) has also been observed upon the reaction of $[L^{tBu}FeNAd]$ with CO [41]. With the same catalyst, addition of 10 equiv. of tert-butylisocyanide (^tBuNC) or cyclohexylisocyanide (CyNC) gave the carbodiimides $AdN{=}C{=}N^tBu$ and AdN=C=NCy in nearly quantitative yields. The three-coordinate iron(III) complex $[L^{tBu}FeNAd]$ can transfer the nitrene to PMe_3 (excess, 10 equiv.) at room temperature giving $AdN{=}PMe_3$ (90% yield). No transferred product $AdN{=}PPh_3$ was observed when PPh_3 was used, attributed to steric reason.

Iron–imide complexes can also undergo hydrogenation reactions [49]. $[PhBP_3]Fe{\equiv}N$-*p*-tolyl was found to give an intense red/purple color change of the reaction mixture, with anilido complex $[PhBP_3]Fe(N(H)$-*p*-tolyl) as the major product, upon exposure to 1 atm of H_2 for 3 h at room temperature. This anilido complex has a weak υ (N–H) stretch at 3,326 cm^{-1} (Nujol) with a magnetic moment of 4.92 μ_B (S = 2). Prolonged hydrogenation of $[PhBP_3]Fe{\equiv}N$-*p*-tolyl in benzene for 3 days revealed the formation of a new diamagnetic compound $[PhBP_3]Fe(\eta^5$-cyclohexadieyl) with the release of H_2N-*p*-tolyl (eq. 1 in Scheme 8). The iron–imide complex (^{iPr}PDI)-Fe=NAr also reacts with hydrogen to give the iron–dihydrogen complex (^{iPr}PDI)-Fe(H_2) with the release of free aniline (eq. 2 in Scheme 8).

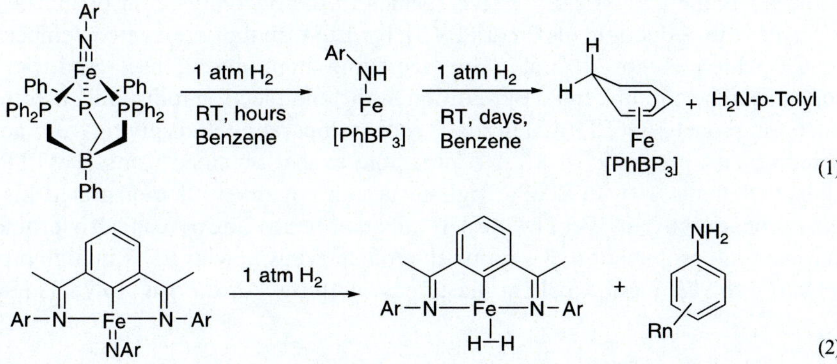

(1)

(2)

Scheme 8

4 Catalysis via Intermediacy of Iron Carbene Complexes

4.1 Cyclopropanation

Cyclopropane rings are commonly found in biologically active natural products and therapeutic drug molecules [50–52]. Iron porphyrins are active catalysts for the cyclopropanation of terminal alkenes with diazoesters with high *trans*-selectivity. In 1995, Kodadek and Woo reported an efficient [Fe(TTP)]-catalyzed cyclopropanation of styrene with EDA [53]. The reaction occurs at room temperature with high catalyst turnover (1,300), affording the corresponding cyclopropyl ester in high diastereoselectivity (Scheme 9). Decreasing the reaction temperature improves the diastereoselectivity dramatically (Scheme 9). The solvent has modest effect on the stereoselective course. Donor solvents such as THF result in higher *trans*/*cis* ratio (13:1) than that obtained in dichloromethane. The increase in *trans*/*cis* ratio could be rationalized by a *trans*-effect of the axially ligated solvent molecule on the iron porphyrin carbene intermediate, rendering the *trans*-carbene less electrophilic and hence a late transition state for the carbenoid transfer reactions.

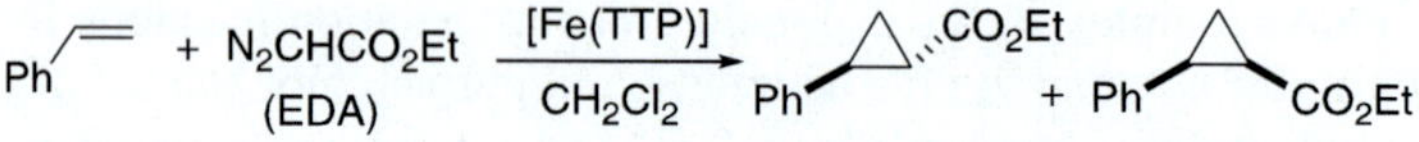

room temperature: trans:cis=8.8:1
–78 °C: trans:cis=29:1

Scheme 9

Compared with [Fe(TTP)], [Fe(TTP)Cl] is a more practical catalyst in alkene cyclopropanation due to its good air stability [53]. For the [Fe(TTP)Cl]-catalyzed alkene cyclopropanation with diazoester, it has been proposed that air-sensitive Fe (II) porphyrin generated in situ through the reduction of Fe(III) porphyrin by diazoester is the catalytically active species in the decomposition of diazoesters. However, the reduction of [Fe(TTP)Cl] by EDA requires elevated temperature (40°C), which is not favorable for achieving high diastereoselectivities. This problem can be circumvented by addition of the one-electron reductant cobaltocene which can reduce [Fe(TTP)Cl] in situ at room temperature, thus giving *trans* and *cis* cyclopropanes in a ratio of 8.7:1 comparable to that obtained using [Fe(TTP)] as catalyst (Scheme 10). In 2003, Tagliatesta and coworkers demonstrated that the bulky iron porphyrin [Fe(TDCPP)Cl] together with $CoCp_2$ can also efficiently catalyze cyclopropanation of various styrene derivatives with EDA in high product yields (up to 97%) and excellent diastereoselectivity (*trans*/*cis* up to 78:1) [54].

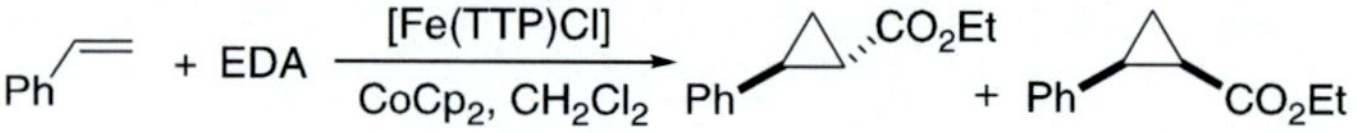

trans:cis=8.7:1

Scheme 10

To simplify the catalytic system further, Kodadek and Woo investigated the activity of [Fe(F_{20}-TPP)Cl] for alkene cyclopropanation with EDA in the absence of cobaltocene. These workers proposed that electron-deficient porphyrin would render the Fe(III) porphyrin more easily reduced by EDA. Indeed, [Fe(F_{20}-TPP)Cl] efficiently catalyzes alkene cyclopropanation with EDA with high catalyst turnover

(4,000) and good *trans*-selectivity (*trans*/*cis* = 3.3–6.0/1) without cobaltocene at room temperature [53].

Iron porphyrins display pronounced substrate preferences for alkene cyclopropanation with EDA. In general, electron-rich terminal alkenes in conjunction with aromatic moiety or heteroatoms can efficiently undergo cyclopropanation with high catalyst turnover and selectivity. In contrast, 1,2-disubstituted alkenes cannot undergo cyclopropanation with diazoesters. Alkyl alkenes are poor substrates, giving cyclopropanated products in low yields. In both cases, the dimerization product diethyl maleate was obtained in high yield [53].

Aryldiazomethane can also be used for iron porphyrin-catalyzed alkene cyclopropanation [55]. For example, the treatment of *p*-tolyldiazomethane with styrene in the presence of [Fe(TTP)] afforded the corresponding arylcyclopropapane in 79% yield with a high *trans*/*cis* ratio of 14:1 (eq. 1 in Scheme 11). Interestingly, when bulkier mesityldiazomethane was used as carbene source, *cis*-selectivity was observed (*cis*/*trans* = 2.0:1). Additionally, mesityldiazomethane was found to react with *trans*-β-styrene, the latter was found not to react with EDA or trimethylsilyldiazomethane under the similar reaction conditions, to give 1-mesityl-2-methyl-3-phenylcyclopropane in 35% yield. Trimethylsilyldiazomethane is also an active carbene source for [Fe(TTP)]-catalyzed cyclopropanation of styrene, affording 1-phenyl-2-trimethylsilylcyclopropane in 89% yield with *trans*/*cis* ratio of 10:1 (eq. 2 in Scheme 11).

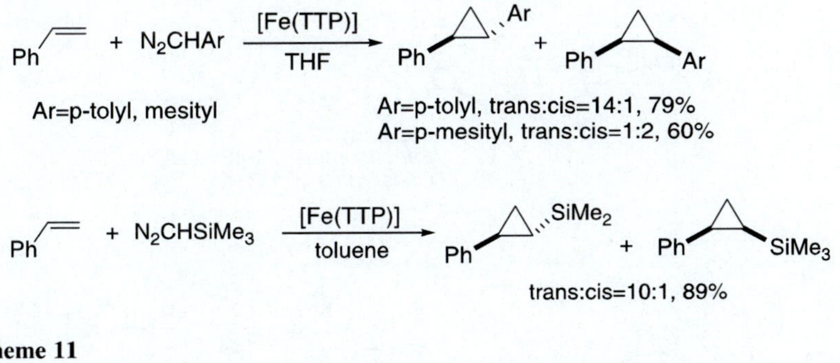

(1)

(2)

Scheme 11

More recently, Carreira reported a [Fe(TPP)Cl]-catalyzed diastereoselective synthesis of trifluoromethyl-substituted cyclopropane in aqueous media [56]. The carbene precursor trifluoromethyl diazomethane is difficult to be handled, generated in situ from trifluoroethyl amine hydrochloride, and reacts with styrene in the presence of [Fe(TPP)Cl] to give the corresponding cyclopropanes in high yields and with excellent diastereoselectivities (Scheme 12).

The complex [Fe(D_4-TmAP)Cl] with Halterman's porphyrin ligand can effect asymmetric alkene cyclopropanation with diazoacetate in high product yield and high stereoselectivity [57]. The reaction occurs smoothly at room temperature without the need for addition of $CoCp_2$, affording the cyclopropyl esters

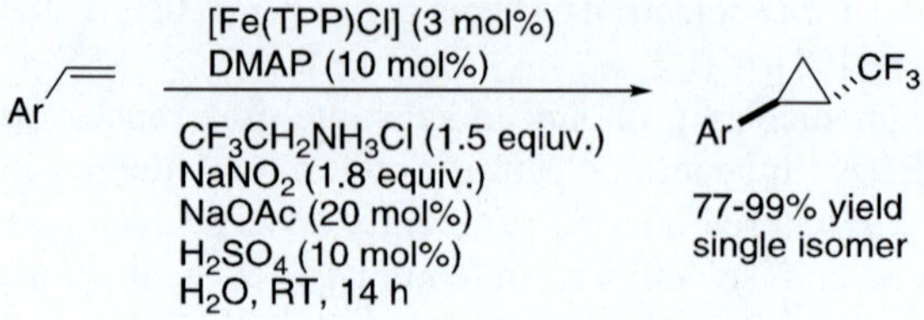

Scheme 12

with high *trans*/*cis* ratios (up to 23:1) and high enantioselectivity for the *trans*-cyclopropanes (up to 86% ee) (Scheme 13). This complex is a robust catalyst which can efficiently catalyze alkene cyclopropanation in 0.002% catalyst loading without detrimental effect on the selectivity. Like [Fe(TTP)], an axial ligation effect was found to affect the diastereoselectivity of [Fe(D_4-TmAP)Cl]-catalyzed cyclopropanations. Addition of organic base such as pyridine increases the *trans*/*cis* ratio from 12:1 to 33:1. Subsequent detection of [(py)Fe(D_4-TmAP)($CHCO_2Et$)] by ESMS provided experimental evidence for six-coordinate iron monocarbene as a catalytically active species. [Fe(D_4-TpAP)Cl] that has a D_4-symmtric porphyrin ligand similar to Halterman´s porphyrin gave high diastereoselectivity (*trans*/*cis* = 21/1) and modest enantiomeric excesses for both diastereomers 45% ee for *trans-isomer*, 21% ee for *cis-isomer*. Fe[HPhH$(dach)_2$] with tetraaza macrocycle derived from enantiomerically pure *trans*-1,2-diaminocyclohexane catalyzed the styrene cyclopropanation in modest enantioselectivity (42% ee for both isomers) [58].

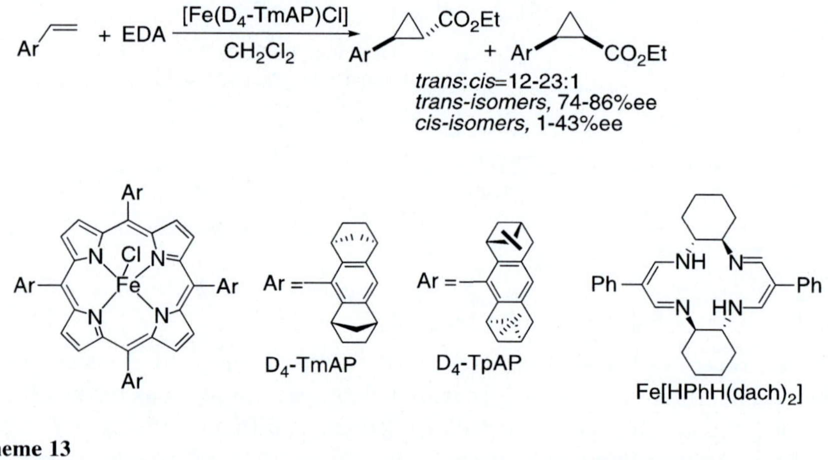

Scheme 13

μ-Oxo-bis[(salen)iron(III)] complexes (Scheme 14) reported by Nguyen are moderate active catalysts for alkene cyclopropanation, affording the corresponding cyclopropanes in moderate to high yields (28–97%) with modest stereoselectivity (*trans*/*cis* = 2–4:1) in refluxing benzene [59, 60]. Fe(III/IV) corroles (Scheme 14) reported by Gross can catalyze the cyclopropanation of styrene with EDA to give cyclopropyl esters in good yields (up to 89%) with moderate *trans/cis* ratio (up to 2.3) [61].

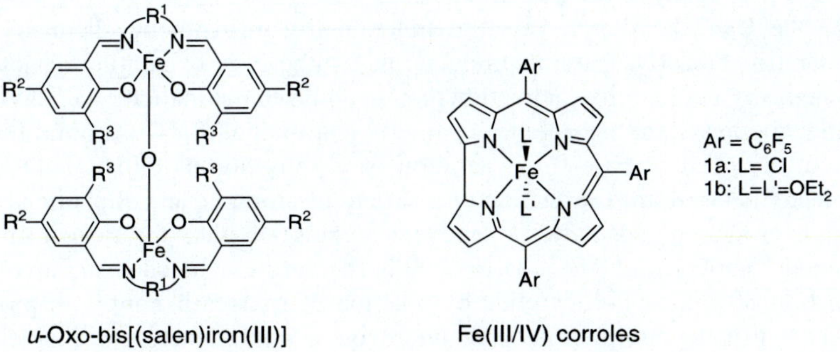

Scheme 14

Interestingly, the cyclopropanation of styrenes with EDA catalyzed by the half sandwich iron complex $[CpFe(CO)_2(THF)]^+BF_4^-$ afforded cyclopropanes in good yields and with *cis*-selectivity (*cis/trans* = 80:20) [62]. With phenyldiazomethane as a carbene source, excellent *cis*-selectivity (92–100%) was achieved (Scheme 15) [63].

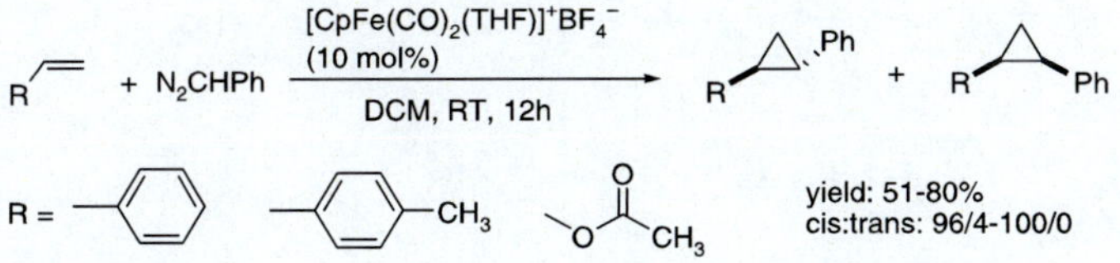

Scheme 15

4.2 *N–H Insertion and Olefination*

Carbenoid N–H insertion of amines with diazoacetates provides a useful means for the synthesis of α-amino esters. Fe(III) porphyrins [64] and Fe(III/IV) corroles [65] are efficient catalysts for N–H carbenoid insertion of various aromatic and aliphatic amines using EDA as a carbene source (Scheme 16). The insertion reactions occur at room temperature and can be completed in short reaction times and with high product yields. It is performed in a one-pot fashion without the need for slow

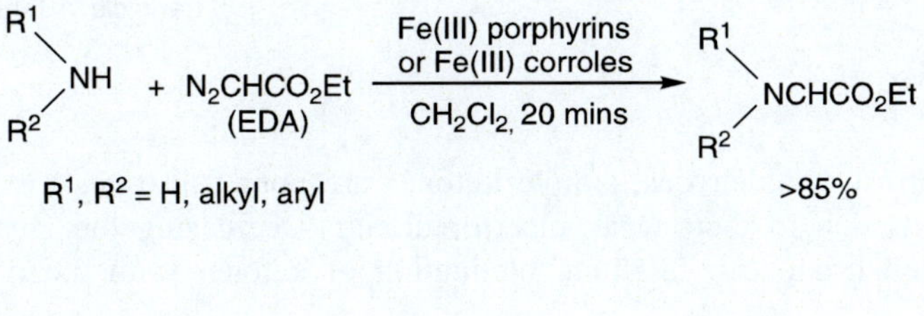

Scheme 16

addition of EDA, revealing that Fe(III) porphyrins and Fe(III/IV) corroles are not poisoned by amines. In contrast, poisoning effects by amines on ruthenium [66] and rhodium [67] catalysts have been reported.

As the C=C bond is a versatile functionality in synthetic chemistry and is prevalent in bioactive natural products, its synthesis is of continuing interest to the chemistry community. Transition metal-catalyzed olefination of aldehydes with diazo compounds and phosphines is an efficient method for C=C bond formation without the need of base for generation of a phosphorous ylide. [Fe(TTP)] can efficiently catalyze the olefination of a variety of aromatic and aliphatic aldehydes in high product yields (>85%) and excellent selectivity for *trans*-isomer was obtained (>90%) (Scheme 17) [68, 69]. The reaction mechanism involves the formation of a phosphorous ylide by reaction of iron–carbenoid with phosphine. Subsequently, the phosphorous ylide undergoes a Wittig reaction with aldehydes to generate alkenes. Besides phosphine, phosphite can also mediate Wittig olefination with diazo compounds affording alkenes in high *trans*-selectivity [70].

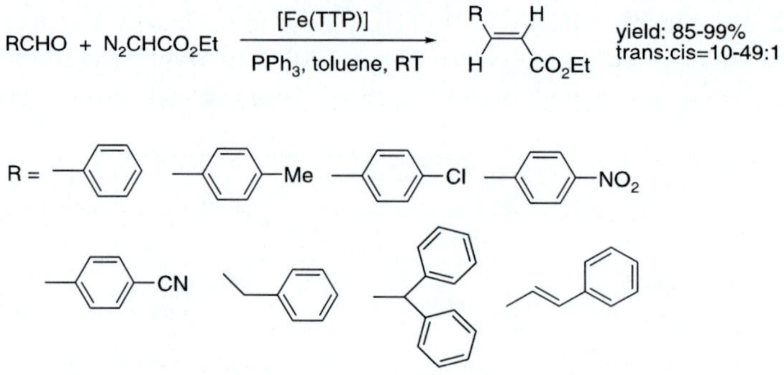

Scheme 17

A more practical catalytic system is based on the use of air-stable [Fe(TTP)Cl] as catalyst, which can efficiently catalyze the olefination reactions at 80°C in excellent yields and high *trans*-selectivity [71] (Scheme 18).

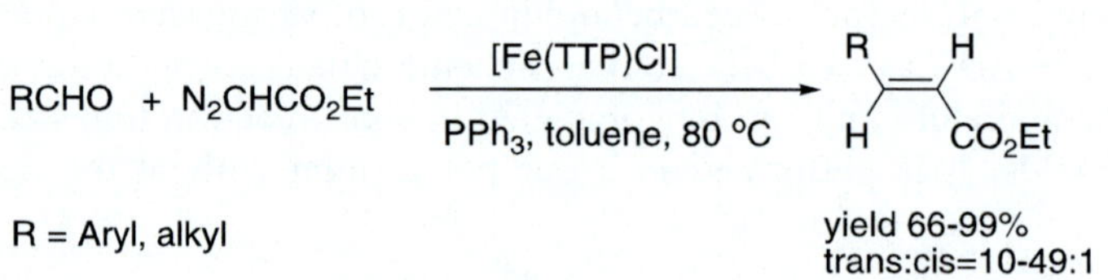

Scheme 18

In contrast to aldehydes, simple ketones are poor substrates for Fe-catalyzed olefinations due to their weak electrophilicity. Decreasing the electron density of carbonyl group can facilitate olefination of ketones with diazo compounds.

Zhang demonstrated that in the presence of [Fe(TTP)Cl], trifluoromethyl ketones can react with EDA and PPh_3 to give β-trifluoromethyl α,β-unsaturated esters in high yields (up to 95%) and high *trans* selectivity (up to 99%) [72] (eq. 1 in Scheme 19). Addition of benzoic acid was found to increase the electrophilicity of ketones through protonation of the carbonyl oxygen atom, leading to alkenes in good yields and modest selectivity [73] (eq. 2 in Scheme 19).

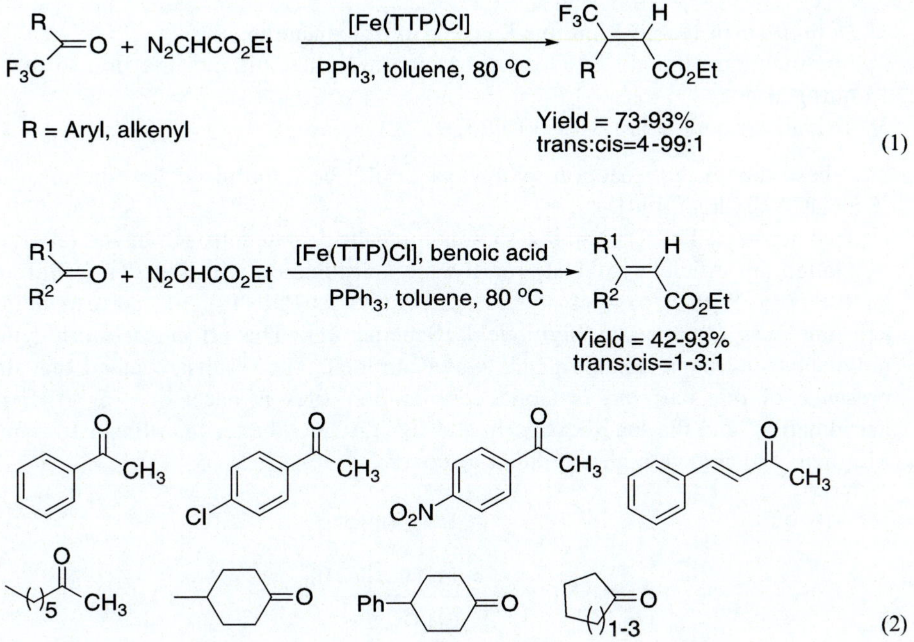

Scheme 19

5 Catalysis via Intermediacy of Iron Nitrene Complexes

5.1 Aziridination

Aziridines are versatile intermediates in organic synthesis and commonly found in bioactive molecules. The transition metal-catalyzed nitrene transfer to alkenes is an attractive method for the synthesis of aziridines [7]. In 1984, Mansuy and coworkers reported the first example of an iron-catalyzed alkene aziridination in which iron porphyrin [Fe(TTP)Cl] was used as catalyst and PhINTs was used as nitrene source [30]. Subsequently, the same authors demonstrated that [Fe(TDCPP)(ClO_4)] is a more efficient and selective catalyst than [Fe(TTP)Cl] (Scheme 20).

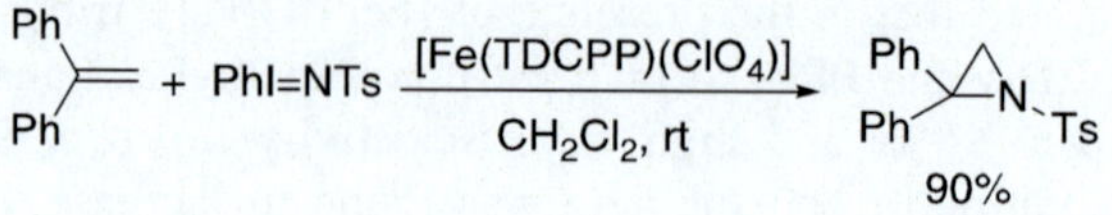

Scheme 20

Three different side reactions are found in [Fe(TTP)Cl]-catalyzed nitrene transfer reaction including

1. Hydrolysis of reactive intermediates to form tosylamide
2. Modification of [Fe(TTP)Cl] due to intramolecular nitrene insertion to iron–nitrogen bond
3. Oxidative degradation of [Fe(TTP)Cl]

but these undesired reaction pathways could be minimized by the use of [Fe(TDCPP)(ClO_4)] [74].

Two nonheme Fe (II) complexes containing a tridentate nitrogen ligand reported by Halfen are efficient catalysts for styrene aziridination with PhINTs as nitrene source [75]. With excess of styrene (25 equiv. vs PhINTs), the corresponding aziridine was obtained in high yield (Scheme 21). The effect of varying the polydentate nitrogen donor ligands was examined. The results revealed that the presence of one pair of *cis* labile coordination sites is necessary for efficient aziridination, and the macrocyclic ligand iPr_3TACN is better than linear triamine Me_5dien, which gave higher product turnovers.

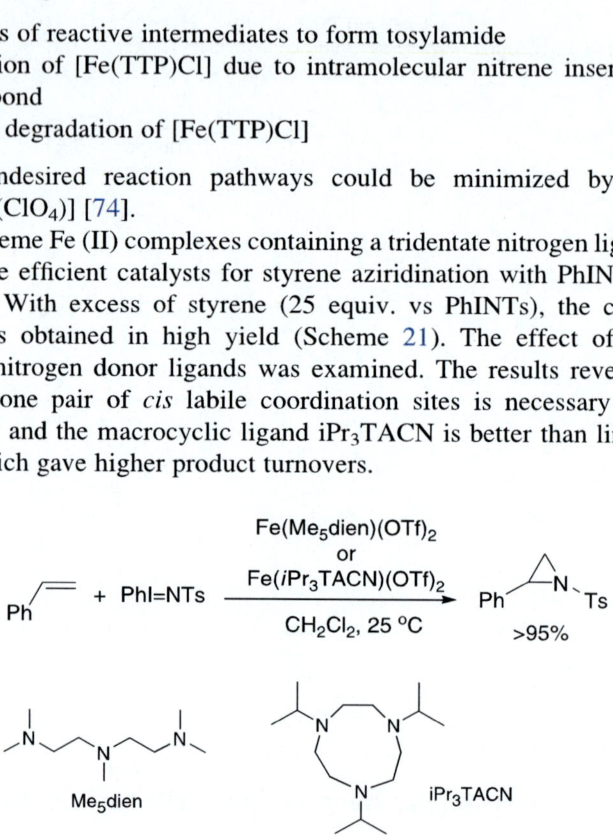

Scheme 21

A dinuclear iron(II/III) complex bearing a hexadentate phenol ligand displayed moderate activity toward aziridination of alkenes with PhINTs; a large excess of alkene (2,000 equiv. vs PhINTs) was required for good product yields (Scheme 22) [76]. It is noteworthy that complex **4** is active in the aziridination of aliphatic alkenes, affording higher product yields than copper (II) catalysts with tetradentate macrocyclic ligands [77].

The mild iron-based Lewis acid, [(η^5-C_5H_5)Fe(CO)$_2$(THF)]BF_4 reported by Hossain, catalyzed the aziridination of styrene derivatives with PhINTs with product

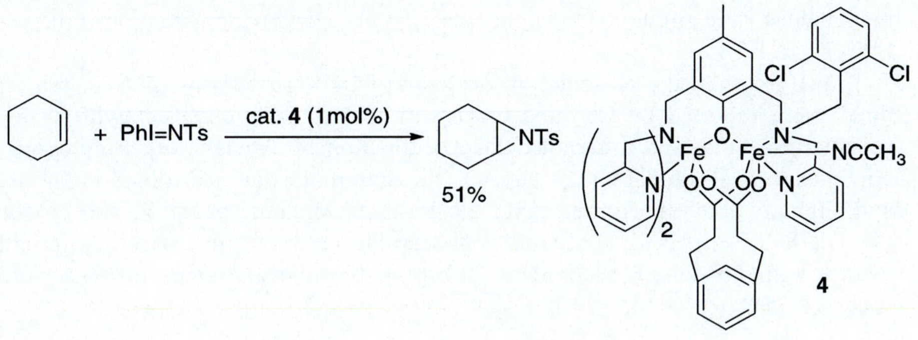

Scheme 22

yields up to 85% [78]. Importantly, the aziridination was stereospecific, affording *cis*-aziridines from *cis*-alkenes and *trans*-aziridines from *trans*-alkenes, whereas *trans*-alkenes are poor substrates, resulting in low product yields (Scheme 23).

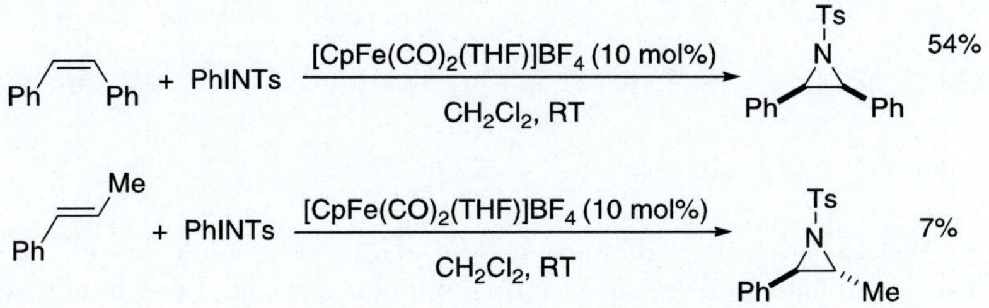

Scheme 23

A variety of alkenes can be converted to aziridines with PhINTs in good yields (48–90%) using iron phthalocyanine as catalyst (Scheme 24) [79]. It is noteworthy

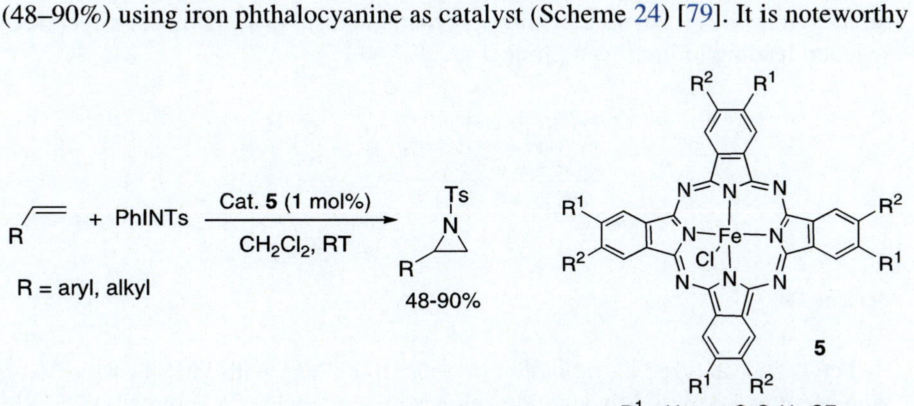

Scheme 24

that alkenes were employed in slight excess in the aziridination reactions (alkenes: PhINTs = 1.5:1).

Iron(II)-salts in the presence of the ligand "4,4′,4″-trichloro-2,2′:6′,2″-terpyridine" were reported by Che and coworkers to be an effective catalytic system for intermolecular and intramolecular aziridination of alkenes [48]. Employing a slight excess of PhINTs (1.5 equiv.), the intermolecular aziridination of aryl and aliphatic alkenes afforded good substrate conversions (63–86%) and product yields (78–97%). Alkenyl sulfamides underwent intramolecular aziridination efficiently with $PhI(OAc)_2$ as oxidant, affording bicyclic aziridines in high yields (Scheme 25).

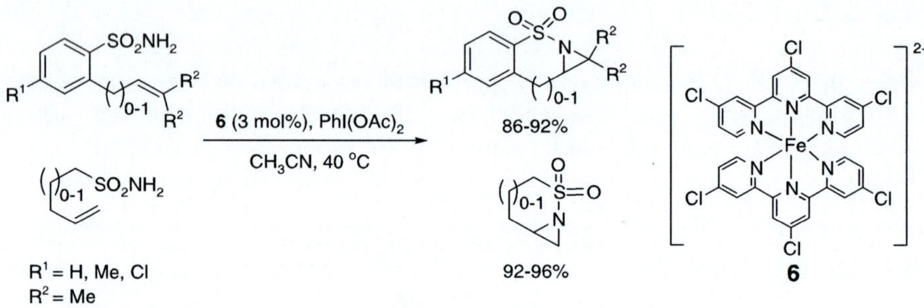

Scheme 25

The system "iron (II) triflate, quinaldic acid and an ionic liquid" efficiently catalyze aziridinations of alkenes with equimolar amounts of iminoidinane [80]. 5-Methyl-2-pyridinesulfonyliminophenyliodinane is a better nitrene source than arylsulfonyliminophenyliodinane, affording products in higher yields (Scheme 26). This result reveals that the chelating effect of pyridyl group can facilitate alkene aziridination. The additive ethylmethylimidazolium bis[(trifluoromethyl)sulfonyl] amide (emim BTA), an ionic liquid, was found to display a significant effect on the reaction leading to improved product yields.

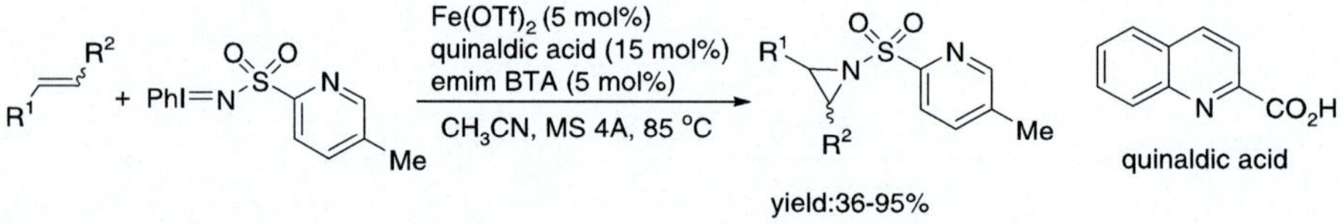

Scheme 26

$Fe(OTf)_2$-catalyzed aziridination of enol silyl ethers with PhINTs followed by ring opening led to α-N-tosylamido ketones in good yields (Scheme 27) [81]. With silyl ketene ketal (R^1 = OMe) as substrate, the N-tosyl-protected amino acid ester was obtained in 50% yield. In contrast, the copper (I) salt $CuClO_4$ was found not effective for this substrate [82].

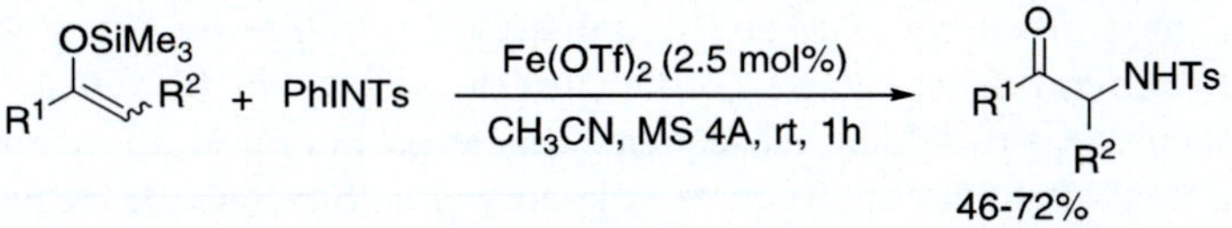

Scheme 27

A more practical, atom-economic and environmentally benign aziridination protocol is the use of chloramine-T or bromamine-T as nitrene source, which leads to NaCl or NaBr as the sole reaction by-product. In 2001, Gross reported an iron corrole catalyzed aziridination of styrenes with chloramine-T [83]. With iron corrole as catalyst, the aziridination can be performed under air atmosphere conditions, affording aziridines in moderate product yields (48–60%). In 2004, Zhang described an aziridination with bromamine-T as nitrene source and [Fe(TTP)Cl] as catalyst [84]. This catalytic system is effective for a variety of alkenes, including aromatic, aliphatic, cyclic, and acyclic alkenes, as well as α,β-unsaturated esters (Scheme 28). Moderate to low stereoselectivities for 1,2-disubstituted alkenes were observed indicating the involvement of radical intermediate.

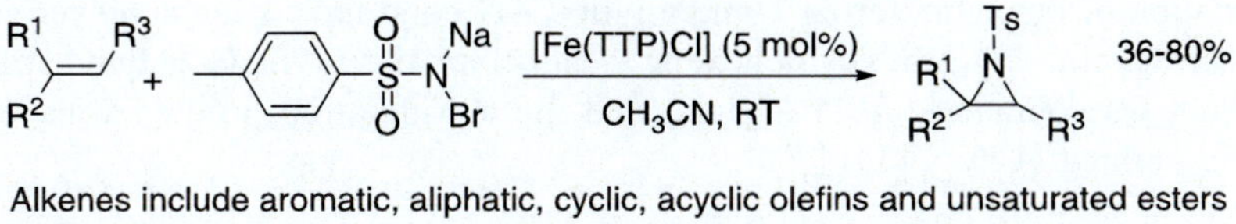

Scheme 28

5.2 C–H Insertion

In 1982, Breslow and coworkers reported the first example of iron-catalyzed nitrene C–H bond insertion [29]. They used [Fe(TTP)] as catalyst and PhINTs as nitrene precursor to achieve C–H bond amination of cyclohexane. However, the product yield was low (around 10%). Subsequently, the same authors found that iminoiodane **7** derived from 2,5-diisopropylbenzenesulfonamide underwent intramolecular C–H amination efficiently with [Fe(TPP)Cl] as catalyst at room temperature, giving the insertion product in 77% yield (Scheme 29) [85].

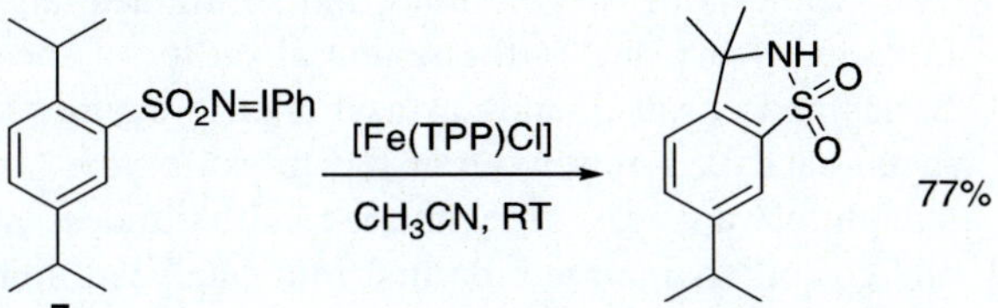

Scheme 29 **7**

With the iron complex $[Fe(Cl_3terpy)_2](ClO_4)_2$ (Cl_3terpy = 4,4′,4″-trichloro-2,2′:6′,2″-terpyridine) as catalyst, sulfamate esters react with $PhI(OAc)_2$ to generate iminoiodanes in situ which subsequently undergo intramolecular nitrenoid C–H insertion to give amidation products in good yields (Scheme 30) [48].

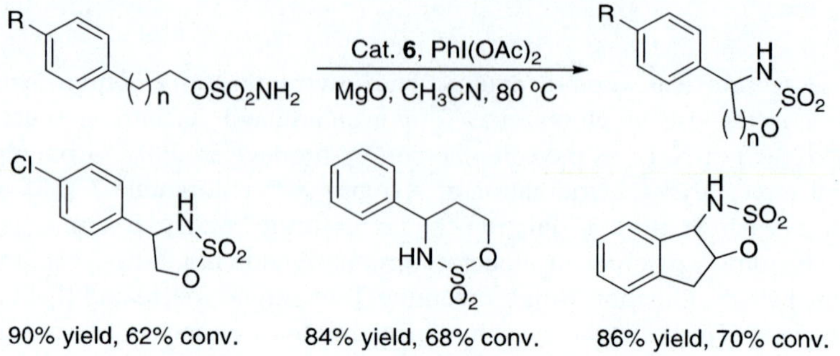

Scheme 30

Iron phthalocyanine is an efficient catalyst for intermolecular amination of saturated C–H bonds. With 1 mol% iron phthalocyanine and 1.5 equiv. PhINTs, amination of benzylic, tertiary, and allylic C–H bond have been achieved in good yields (Scheme 31). With cyclohexene as substrate, the allylic C–H bond amination product was obtained in 75% yield, and the aziridination product was found in minor amount (17% yield) [79].

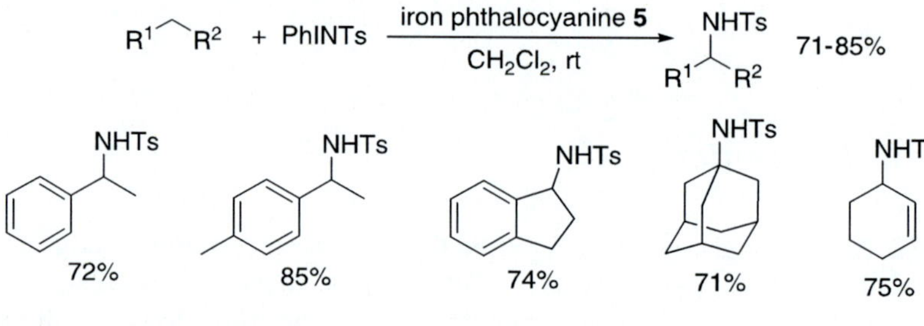

Scheme 31

5.3 *Imination of Sulfur Compounds*

The synthesis of sulfoximides and sulfimides has attracted considerable attention in recent years due to the potential utility of these compounds as efficient auxiliaries and chiral ligands in asymmetric synthesis (reviews: [86–88]). Transition metal-catalyzed nitrene transfer to sulfoxides and sulfides is an efficient and straightforward way to synthesize sulfoximides and sulfimides, respectively. Bach and coworkers reported the first iron-catalyzed imination of sulfur compounds with $FeCl_2$ as catalyst and $BocN_3$ as nitrene source. Various sulfoxides and sulfides were converted to *N*-Boc protected sulfoximides and sulfimides in good to high yields, while high catalyst loadings (25–50%) were required to guarantee the product yields (Scheme 32) [89, 90].

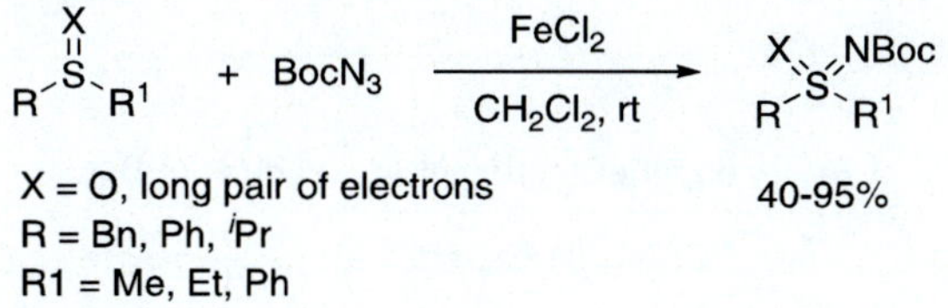

Scheme 32

To improve catalyst turnover and to avoid the use of potentially explosive $BocN_3$, Bolm and coworkers employed $Fe(acac)_2$ as catalyst and sulfonylamide+ PhI=O as nitrene source to iminate sulfur compounds [91, 92]. Using 5 mol% Fe $(acac)_2$ would allow for efficient imination of sulfur compounds to give the corresponding sulfoximides and sulfimides in high yields. The best results were obtained with *p*-nitrobenzene sulfonylamide ($NsNH_2$) as nitrene source and PhIO as terminal oxidant (eq. 1 in Scheme 33). Subsequently, Bolm demonstrated that Fe $(OTf)_2$ is a more efficient catalyst for sulfur imination [93]. Only 2.5 mol% $Fe(OTf)_2$ was needed for imination of sulfur compounds with *p*-nosylamide($NsNH_2$). Importantly, some sluggish substrates bearing benzyl, bulky alkyl, and heteroaryl substituents, which are ineffective for $Fe(acac)_2$ catalysis, could undergo sulfur imination efficiently with $Fe(OTf)_2$ as catalyst, affording corresponding products in good yields (eq. 2 in Scheme 33). Both Bach and Bolm found that iron-catalyzed sulfur imination proceeded stereospecifically with retention of configuration at sulfur providing an efficient access to enantiopure sulfoximides from corresponding chiral sulfoxides.

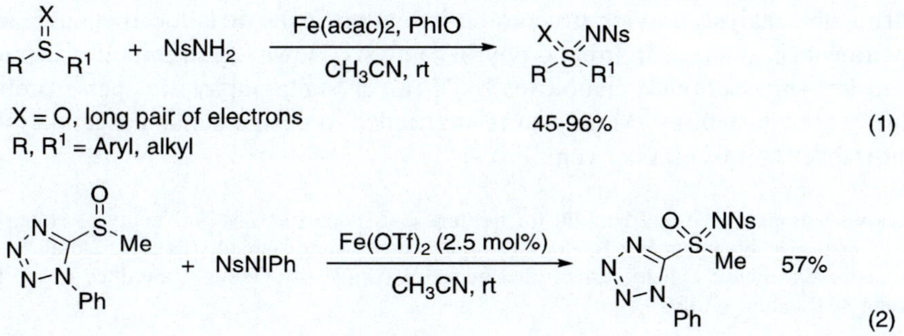

Scheme 33

When allyl sulfides are employed as substrates, they undergo imidation/[2, 3]-sigmatropic rearrangement reaction upon treatment with $BocN_3$ and a catalytic amount of $FeCl_2$, affording *N*-Boc-protected *N*-allyl sulfenamides, presumably via intermediacy of **I** (eq. 1 in Scheme 34) [94]. Propargyl sulfides could also undergo

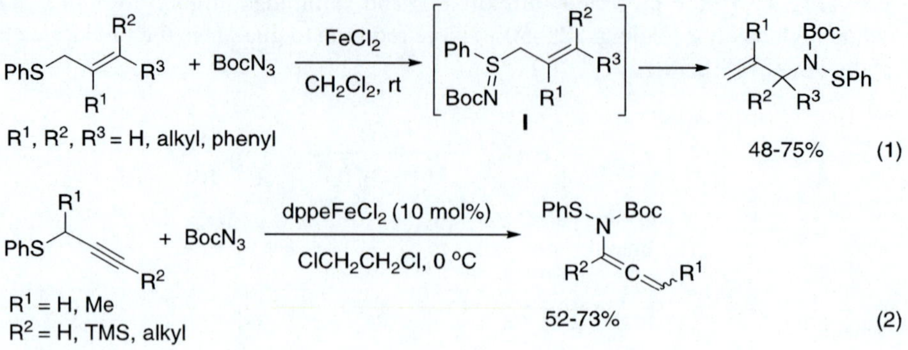

Scheme 34

this transformation to give *N*-allenyl sulfonamides when [dppe$FeCl_2$] was used as catalyst (eq. 2 in Scheme 34) [95].

6 Conclusion

There is a growing interest in iron-catalyzed carbenoid and nitrenoid transfer reactions. Iron complexes can efficiently catalyze various carbenoid transfer reactions, including alkene cyclopropanation, C–H insertion, N–H insertion, and olefination, and nitrenoid transfer reactions, including alkene aziridination, C–H insertion, and imination of sulfur compounds. More importantly, iron catalysis meets the increasing demand for environmentally benign and sustainable chemical processes in view of the low cost and biocompatibility of iron complexes. The structurally characterized iron porphyrin carbene complexes by single crystal X-ray diffraction analysis provide structural information on the metallocarbenoid reaction intermediate involved in iron porphyrin catalysis. However, structurally characterized iron–nitrene/imido complexes by X-ray crystallography are sparse probably due to their instability. More efforts are needed to design suitable ligand systems for stabilizing this class of complexes.

Acknowledgment We are thankful for the financial support of The University of Hong Kong (University Development Fund), Hong Kong Research Grant Council (HKU 7052/07P), and the Areas of Excellence Scheme established under the University Grants Committee of the Hong Kong SAR, China (AoE/P- 10/01).

References

1. Ye T, McKervey MA (1994) Chem Rev 94:1091
2. Doyle MP, Forbes DC (1998) Chem Rev 98:911
3. Doyle MP, McKervey MA, Ye T (1998) Modern catalytic methods for organic synthesis with diazo compounds. Wiley, New York

4. Davies HML, Beckwith REJ (2003) Chem Rev 103:2861
5. Godula K, Sames D (2006) Science 312:67
6. Davies HML, Manning JR (2008) Nature 451:417
7. Collet F, Dodd RH, Dauban P (2009) Chem Commun 5061
8. Fantauzzi S, Caselli A, Gallo E (2009) Dalton Trans 5434
9. Correa A, Mancheño OG, Bolm C (2008) Chem Soc Rev 37:1108
10. Halfen JA (2005) Curr Org Chem 9:657
11. Dauban P, Dodd RH (2003) Synlett 11:1571
12. Muller P, Fruit C (2003) Chem Rev 103:2905
13. Sweeney JB (2002) Chem Soc Rev 31:247
14. Jacobsen EN (1999) In: Jacobsen EN, Pfaltz A, Yamamoto H (eds) Comprehensive asymmetric catalysis, vol 2. Springer, Berlin, p 607
15. Osborn HM, Sweeney J (1997) Tetrahedron Asymmetry 8:1693
16. Tanner D (1994) Angew Chem Int Ed Engl 33:599
17. Fischer EO, Maasböl A (1964) Angew Chem 76:645
18. Mansuy D, Lange M, Chottard JC, Guerin P, Morliere P, Brault D, Rougee M (1977) J Chem Soc Chem Commun 648
19. Mansuy D, Guerin P, Chottard JC (1979) J Organomet Chem 171:195
20. Mansuy D (1980) Pure Appl Chem 52:681
21. Mansuy D, Lange M, Chottard JC, Bartoli JF, Chevrier B, Weiss R (1978) Angew Chem Int Ed Engl 17:781
22. Li Y, Huang JS, Zhou ZY, Che CM, You XZ (2002) J Am Chem Soc 124:13185
23. Guerin P, Battioni JP, Chottard JC, Mansuy D (1981) J Organomet Chem 218:201
24. Battioni JP, Dupre D, Guerin P, Mansuy D (1984) J Organomet Chem 265:53
25. Mansuy D, Lange M, Chottard JC, Bartoli JF (1978) Tetrahedron Lett 3027
26. Bruice TC, Furter PG, Ball SS (1981) J Am Chem Soc 103:4578
27. Ziegler CJ, Suslick KS (1996) J Am Chem Soc 118:5306
28. Ziegler CJ, Suslick KS (1997) J Organometal Chem 528:83
29. Breslow R, Gellman SH (1982) J Chem Soc, Chem Commun 1400
30. Mansuy D, Mahy JP, Dureault A, Bedi G, Battioni P (1984) J Chem Soc Chem Commun 1161
31. Wang Z, Zhang Y, Fu H, Jiang Y, Zhao Y (2008) Org Lett 10:1863
32. Breslow R, Gellman SH (1983) J Am Chem Soc 105:6728
33. Jensen MP, Mehn MP, Que L Jr (2003) Angew Chem Int Ed 42:4357
34. Mahy JP, Battioni P, Bedi G, Mansuy D, Fischer J, Weiss R, Morgenstern-Badarau I (1988) Inorg Chem 27:353
35. Klinker EJ, Jackson TA, Jensen MP, Stubna A, Juhász G, Bominaar EL, Münck E, Que L Jr (2006) Angew Chem Int Ed 45:7394
36. Lucas RL, Powell DR, Borovik AS (2005) J Am Chem Soc 127:11596
37. Verma AK, Nazif TN, Achim C, Lee SC (2000) J Am Chem Soc 122:11013
38. Brown SD, Betley TA, Peters JC (2003) J Am Chem Soc 125:322
39. Thomas CM, Mankad NP, Peters JC (2006) J Am Chem Soc 128:4956
40. Eckert NA, Vaddadi S, Stoian S, Lachicotte RJ, Cundari TR, Holland PL (2006) Angew Chem Int Ed 45:6868
41. Cowley RE, Eckert NA, Elhaïk J, Holland PL (2009) Chem Commun 1760
42. Ni C, Fettinger JC, Long GJ, Brynda M, Power PP (2008) Chem Commun 6045
43. Klinker EJ, Kaizer J, Brennessel WW, Woodrum NL, Cramer CJ, Que L Jr (2005) Angew Chem Int Ed 44:3690
44. Rohde JU, In JH, Lim MH, Brennessel WW, Bukowski MR, Stubna A, Munch E, Nam W, Que L Jr (2003) Science 299:1037
45. Betley TA, Peters JC (2003) J Am Chem Soc 125:10782
46. Brown SD, Peters JC (2005) J Am Chem Soc 127:1913
47. Bart SC, Lobkovsky E, Bill E, Chirik PJ (2006) J Am Chem Soc 128:5302
48. Liu P, Wong ELM, Yuen AWH, Che CM (2008) Org Lett 10:3275

49. Brown SD, Peters JC (2004) J Am Chem Soc 126:4538
50. Wong HNC, Hon MY, Tse CW, Yip YC, Tanko J, Hudlicky T (1989) Chem Rev 89:165
51. Rappoport Z (ed) (1995) The chemistry of the cyclopropyl group. Wiley, Chichester
52. Meijere A (2003) Chem Rev 103:1
53. Wolf JR, Hamaker CG, Djukic JP, Kodadek T, Woo LK (1995) J Am Chem Soc 117:9194
54. Tagliatesta P, Pastorini A (2003) J Mol Catal A Chem 198:57
55. Hamaker CG, Mirafzal GA, Woo LK (2001) Organometallics 20:5171
56. Morandi B, Carreira EM (2010) Angew Chem Int Ed 49:938
57. Lai TS, Chan FY, So PK, Ma DL, Wong KY, Che CM (2006) Dalton Trans 4845
58. Du G, Andrioletti B, Rose E, Woo LK (2002) Organometallics 21:4490
59. Edulji SK, Nguyen ST (2003) Organometallics 22:3374
60. Edulji SK, Nguyen ST (2004) Pure Appl Chem 76:645
61. Simkhovich L, Mahammed A, Goldberg I, Gross Z (2001) Chem Eur J 7:1041
62. Seitz WJ, Saha AK, Casper D, Hossain MM (1992) Tetrahedron Lett 33:7755
63. Seitz WJ, Hossain MM (1994) Tetrahedron Lett 35:7561
64. Baumann LK, Mbuvi HM, Du G, Woo LK (2007) Organometallics 26:3995
65. Aviv I, Gross Z (2006) Synlett 951
66. Lee SH, Clapham B, Koch G, Zimmerman J, Janda KD (2003) J Comb Chem 5:188
67. Aller E, Buck RT, Drysdale MJ, Ferris L, Haigh D, Moody CJ, Pearson ND, Sanghera JB (1996) J Chem Soc Perkin Trans 1:2879
68. Mirafzal GA, Cheng G, Woo LK (2002) J Am Chem Soc 124:176
69. Cheng G, Mirafzal GA, Woo LK (2003) Organometallics 22:1468
70. Aggarwal VK, Fulton JR, Sheldon CG, Vicente J (2003) J Am Chem Soc 125:6034
71. Chen Y, Huang L, Ranade MA, Zhang XP (2003) J Org Chem 68:3714
72. Chen Y, Huang L, Zhang XP (2003) J Org Chem 68:5925
73. Chen Y, Huang L, Zhang XP (2003) Org Lett 5:2493
74. Mahy JP, Bedi G, Battioni P, Mansuy D (1988) J Chem Soc Perkin Trans 2:1517
75. Klotz KL, Slominski LM, Hull AV, Gottsacker VM, Mas-Ballesté R, Que Jr. L, Halfen JA (2007) Chem Commun 2063
76. Avenier F, Latour JM (2004) Chem Commun 1544
77. Halfen JA, Uhan JM, Fox DC, Mehn MP, Que L Jr (2000) Inorg Chem 39:4913
78. Heuss BD, Mayer MF, Dennis S, Hossain MM (2003) Inorg Chim Acta 342:301
79. Yan SY, Wang Y, Shu YJ, Liu HH, Zhou XG (2006) J Mol Catal A: Chem 248:148
80. Mayer AC, Salit AF, Bolm C (2008) Chem Commun 5975
81. Nakanishi M, Salit AF, Bolm C (2008) Adv Synth Catal 350:1835
82. Evans DA, Faul MM, Bilodeau MT (1994) J Am Chem Soc 116:2742
83. Simkhovich L, Gross Z (2001) Tetrahedron Lett 42:8089
84. Vyas R, Gao GY, Harden JD, Zhang XP (2004) Org Lett 6:1907
85. Breslow R, Gellman SH (1983) J Am Chem Soc 105:6728
86. Bentley R (2005) Chem Soc Rev 34:609
87. Okamura H, Bolm C (2004) Chem Lett 33:482
88. Reggelin M, Zur C (2000) Synthesis 1
89. Bach T, Körber C (1998) Tetrahedron Lett 39:5015
90. Bach T, Körber C (1999) Eur J Org Chem 64:1033
91. Mancheño OG, Bolm C (2006) Org Lett 8:2349
92. Mancheño OG, Bolm C (2007) Chem Eur J 13:6674
93. Mancheño OG, Dallimore J, Plant A, Bolm C (2009) Org Lett 11:2429
94. Bach T, Körber C (2000) J Org Chem 65:2358
95. Bacci JP, Greenman KL, Van Vranken DL (2003) J Org Chem 68:4955

Top Organomet Chem (2011) 33: 139–175
DOI: 10.1007/978-3-642-14670-1_5

Ferrocene and Half Sandwich Complexes as Catalysts with Iron Participation

René Peters, Daniel F. Fischer, and Sascha Jautze

Abstract The unique and readily tunable electronic and spatial characteristics of ferrocenes have been widely exploited in the field of asymmetric catalysis. The ferrocene moiety is not just an innocent steric element to create a three-dimensional chiral catalyst environment. Instead, the Fe center can influence the catalytic process by electronic interaction with the catalytic site, if the latter is directly connected to the sandwich core. Of increasing importance are also half sandwich complexes in which Fe is acting as a mild Lewis acid. Like ferrocene, half sandwich complexes are often relatively robust and readily accessible. This chapter highlights recent applications of ferrocene and half sandwich complexes in which the Fe center is essential for catalytic applications.

Keywords Catalysis · Ferrocene · Half sandwich complexes · Metallacycles · Planar chirality

R. Peters (✉)
Institut für Organische Chemie, Universität Stuttgart, Pfaffenwaldring 55, D-70569 Stuttgart, Germany
e-mail: rene.peters@oc.uni-stuttgart.de

D.F. Fischer,
Laboratory of Organic Chemistry, ETH Zürich, Wolfgang-Pauli-Str. 10, CH-8093 Zürich, Switzerland

S. Jautze
Institut für Organische Chemie, Universität Stuttgart, Pfaffenwaldring 55, D-70569 Stuttgart, Germany
Laboratory of Organic Chemistry, ETH Zürich, Wolfgang-Pauli-Str. 10, CH-8093 Zürich, Switzerland

Contents

Abbreviations

Å	Ångström
Ac	Acetyl
acac	Acetylacetonate
Ar	Aryl
Bn	Benzyl
Boc	*tert*-Butyloxycarbonyl
c-	Cyclo-
cat.	Catalytic
CIP	Cahn-Ingold-Prelog
COD	Cyclooctadiene
COP	Cobalt sandwich based oxazoline palladacycle
Cp	Cyclopentadienyl
D	Deuterium
DCE	1,2-Dichloroethane
DCM	Dichloromethane
DME	1,2-Dimethoxyethane
dr	Diastereomeric ratio
ee	Enantiomeric excess
equiv.	Equivalent(s)
Et	Ethyl
h	Hour(s)
Hex	Hexyl
hν	Light irradiation
i-Bu	*iso*-Butyl
i-Pr	Isopropyl
IUPAC	International Union of Pure and Applied Chemistry
LG	Leaving group
Me	Methyl
Mes	Mesityl, 2,4,6-trimethylphenyl
min	Minute(s)

mol	Mole(s)
mp	Melting point
MTBE	*tert*-Butylmethylether
Nu	Nucleophile
Pent	Pentyl
Ph	Phenyl
PMP	*para*-Methoxyphenyl
PS	Proton sponge, 1,8-bis(dimethylamino)naphthalene
R	Residue
rds	Rate determining step
RT	Room temperature
s	Selectivity
SET	Single electron transfer
t-Bu	*tert*-Butyl
Tf	Trifluoromethylsulfonyl
THF	Tetrahydrofuran
THP	Tetrahydropyranyl
TIPS	Triisopropylsilyl
TMEDA	*N,N,N′,N′*-Tetramethyl-1,2-ethylenediamine
TMS	Trimethylsilyl
Tol	Toluyl
Ts	Toluenesulfonyl
y	Yield
Δ	Reflux conditions

1 Introduction

Ferrocene (**1**) was the first sandwich complex to be discovered, thereby opening a wide and competitive field of organometallic chemistry. The formation of ferrocene was found at almost the same time in two independent studies: on July 11, 1951, Miller, Tebboth, and Tremaine reported that on the passage of N_2 and cyclopentadiene over a freshly prepared mixture of "reduced" Fe (90%), alumina (8%), potassium oxide (1%), and molybdenum oxide (1%) at 300°C, yellow crystals identified as Cp_2Fe (Fig. 1) were obtained [1]. Due to the low yields obtained (3 g starting from 650 g ferric nitrate), doubts remain as to whether Fe(0) was the

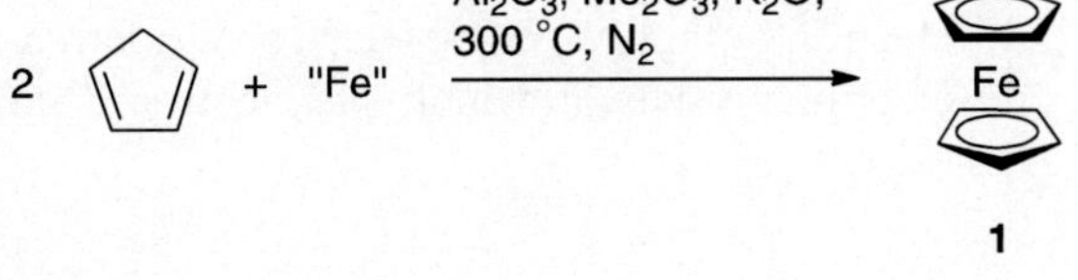

Fig. 1 Formation of Cp_2Fe by Miller, Tebboth, and Tremaine

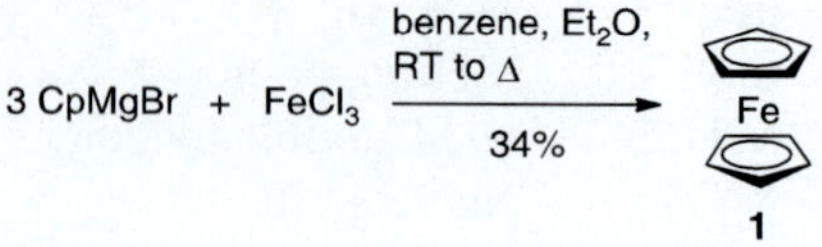

Fig. 2 Formation of Cp_2Fe by Kealy and Pauson

reactive species or if residual Fe(II) reacted with Cp^- formed in situ in the presence of basic oxides.

On August 7, 1951, Kealy and Pauson reported the formation of Cp_2Fe starting from CpMgBr and $FeCl_3$ (3.5 g from 9.05 g $FeCl_3$, Fig. 2) [2]. Initially, it was planned to form fulvalene via generation of a cyclopentadienyl radical by SET from Cp^- to Fe(III), radical combination and subsequent oxidation of Cp_2 by Fe(III). Instead, Fe(II) formed by reduction with the Grignard reagent reacted further by transmetallation.

In both original reports, the authors did not assign a sandwich structure to Cp_2Fe. However, their work immediately triggered intensive research activity by other groups to explain the unexpected properties of this new material, culminating in the structure determination as early as 1952 by E.O. Fischer (magnetic studies, X-ray crystal structure analysis) [3, 4], G. Wilkinson, and R.B. Woodward (magnetic studies, IR-spectroscopy) [5].

Due to the aromatic character of Cp_2Fe predicted by Woodward and confirmed by the reactivity toward electrophilic substitutions, which proceed with rates comparable to anisole, the name ferrocene was coined in analogy to simple aromatic systems [6].

A more efficient way to synthesize ferrocene involves a transmetallation of CpNa with $FeCl_2$ [7]. Ferrocene forms an orange crystalline solid (mp 173°C), which can be purified, for example, by sublimation, and which is relatively inert toward air and hydrolysis [1, 2].

The covalent part of the bonding character of both Cp^- ligands with Fe(II) can be summarized by σ-bond interactions ($Cp^- \rightarrow s/p_z/d_{z^2}$), strong π-interactions ($Cp^- \rightarrow d_{xz}/d_{yz}/p_x/p_y$), and weak retrodative δ-interactions ($d_{xy}/d_{x2-y2} \rightarrow Cp^-$) [8]. The molecular orbital interactions are almost independent of the conformation of the sandwich complex with respect to rotation of the Cp rings. Both the staggered (D_{5d}) as well as the eclipsed conformation (D_{5h}) possess similar binding energies [9]. The same holds true for all conformations between these two extremes meaning that the activation energy for ligand rotation is very low (ca. 0.9 ± 0.3 kcal/mol) [10, 11]. In the gas phase the eclipsed conformation is preferred, while for solid state structures of substituted derivatives, preference for one conformation is often due to packing forces or interactions of the various substituents.

All bonding or nonbonding orbitals are filled resulting in a stable diamagnetic 18-electron complex. Single-electron oxidation to a ferrocenium cation provides a 17-electron species, in which one electron is unpaired.

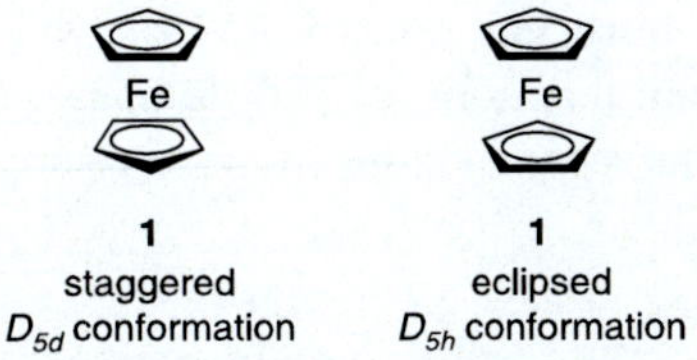

As a consequence of the molecular orbital interactions, ferrocene adopts an axially symmetrical sandwich structure with two parallel Cp ligands with a distance of 3.32 Å (eclipsed conformation) and ten identical Fe–C distances of 2.06 Å as well as ten identical C–C distances of 1.43 Å [12]. Deviation of the parallel Cp arrangement results in a loss of binding energy owing to a less efficient orbital overlap [8]. All ten C–H bonds are slightly tilted toward the Fe center, as judged from neutron-diffraction studies [13].

As mentioned above, ferrocene is amenable to electrophilic substitution reactions and acts like a typical activated electron-rich aromatic system such as anisole, with the limitation that the electrophile must not be a strong oxidizing agent, which would lead to the formation of ferrocenium cations instead. Formation of the σ-complex intermediate **2** usually occurs by *exo*-attack of the electrophile (from the direction remote to the Fe center, Fig. 3) [14], but in certain cases can also proceed by precoordination of the electrophile to the Fe center (*endo* attack) [15].

Due to the pronounced electron donating character of ferrocene, α-ferrocenyl carbocations **3** possess a remarkable stability and can therefore be isolated as salts [16]. They can also be described by a fulvene-type resonance structure **3'** (Fig. 4) in which the Fe center and the α-center are significantly shifted toward each other as revealed by crystal structure analysis, indicating a bonding interaction [17].

Fig. 3 Intermolecular electrophilic substitution reactions of ferrocene via an *exo*-attack

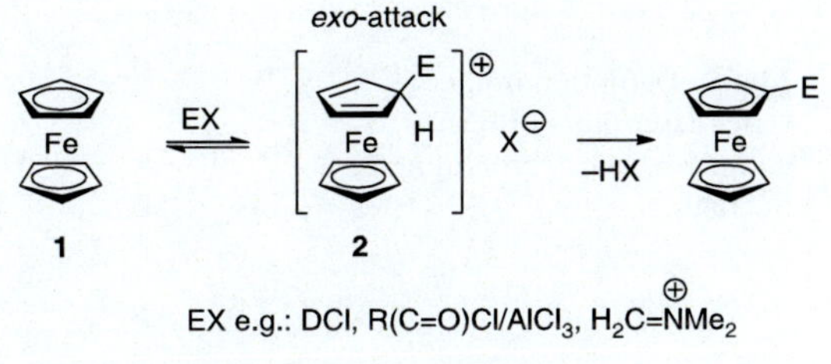

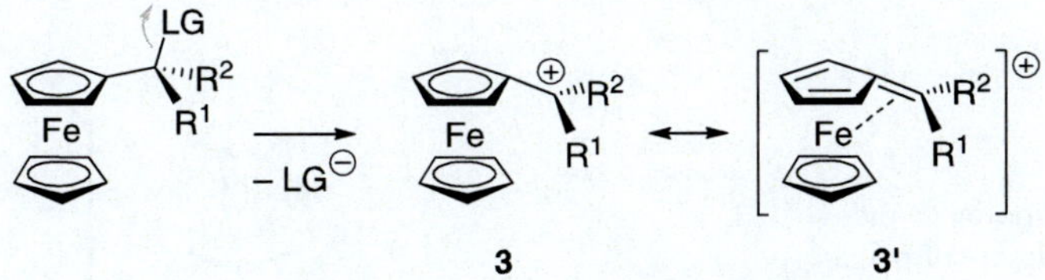

Fig. 4 Formation and stabilization of α-ferrocenyl carbocations **3**

Recently, this concept could be transferred to the homologous silylium ions **3-Si** (Si in α-position to the ferrocene core) [18, 19], which were found to be potent catalysts for Diels–Alder reactions at low temperature [19]. The electron-rich ferrocene core buffers the Lewis acidity, thus avoiding the irreversible formation of Lewis pairs.

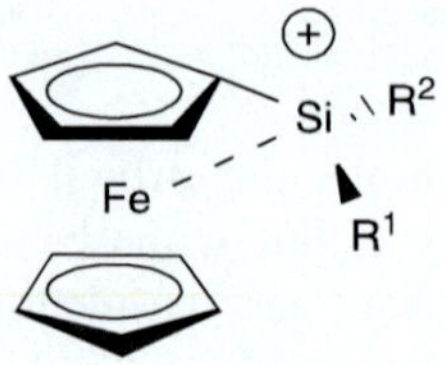

3-Si

Ferrocene derivatives **4** bearing two different substituents X,Y $\neq$ H at the same Cp ring can be planar chiral. Schlögl defined a simple rule to determine the stereodescriptors [20]: looking at the double substituted Cp ring from above, the configuration is defined as R_p if the shortest way through the space starting from the substituent of the highest priority (according to the CIP rules) to the second substituent is clockwise (Fig. 5).

Unfortunately, the IUPAC recommended rule by Cahn, Ingold, and Prelog usually leads to the opposite stereodescriptors [21]. Assuming imaginative Fe–C single bonds between the Fe center and all C atoms, the latter can be treated as distorted tetrahedrons allowing to apply the CIP rules for stereocenters (Fig. 6). The stereodescriptor of the C-atom carrying the substituent of the highest priority is used as stereodescriptor of the whole Cp ring or molecule. As the Schlögl system is more commonly used in literature, it will also be employed in this chapter.

Planar chirality has proven to be a very potent means in asymmetric catalysis to achieve high levels of stereocontrol (see Sect. 3) because planar chiral systems offer

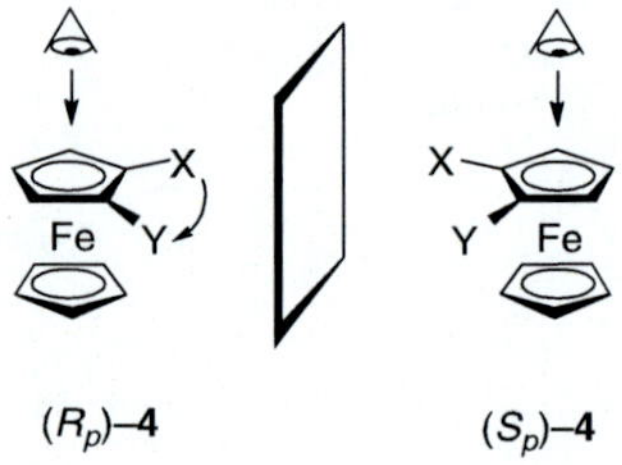

Fig. 5 Definition of the stereodescriptors of planar chiral ferrocenes according to Schlögl

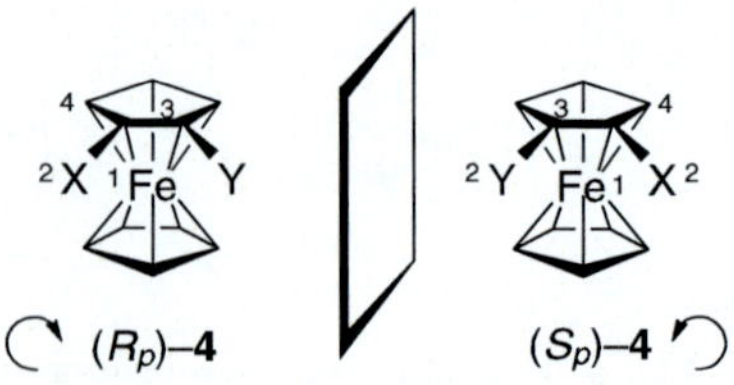

Fig. 6 Definition of the stereodescriptors of planar chiral ferrocenes according to Cahn, Ingold, and Prelog

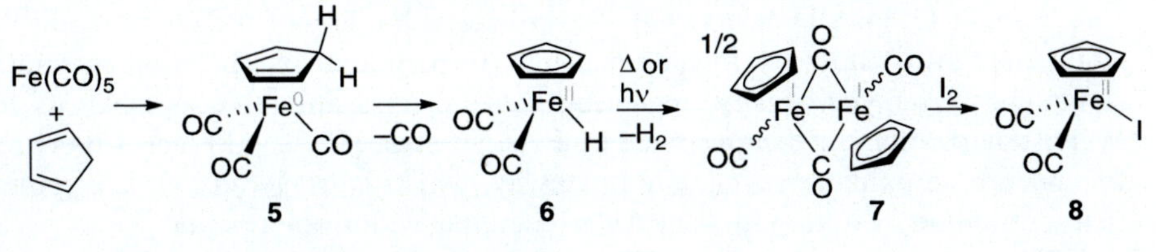

Fig. 7 Formation of the half sandwich complex $CpFe(CO)_2I$ (**8**)

a high degree of spatial control around the active site. The unique and readily tunable electronic and spatial characteristics of ferrocene and ferrocenium systems have been widely exploited in the field of asymmetric catalysis. This is showcased by selected applications in Sect. 3, in which the Fe center communicates with the catalytic site.

Of increasing importance are also Fe half sandwich complexes in which Fe is acting as a mild Lewis acid, presented in Sect. 2. Like ferrocene, half sandwich complexes such as **8** are often relatively robust and readily accessible starting from cyclopentadiene (Fig. 7). Reaction of cyclopentadiene with $Fe(CO)_5$ initially generates a diene complex **5** which further reacts to give a metal hydride species **6** [22, 23]. The latter can be transformed into dimer **7**, which reacts with oxidizing agents such as iodine to give the synthetically versatile iron prototype half sandwich complex **8** [24]. Section 2 is subdivided into systems with Fe in the oxidation states 0, +I, and +II.

2 Catalysis by Iron in Half Sandwich Complexes

2.1 *Fe(0)- and Fe(I)-Catalysts*

Jonas et al. have shown that individual Cp rings of ferrocene can be successively removed under reducing conditions [25–29]. Working under an ethylene atmosphere, the Fe(0) half sandwich complex **9** has been formed even on multigram scale as an air-sensitive crystalline solid upon co-complexation of the Li counterion with TMEDA. The labile ethylene ligands can be readily exchanged by the chelating 1,5-cyclooctadiene (COD) forming **10** or the closely related DME adduct **11**.

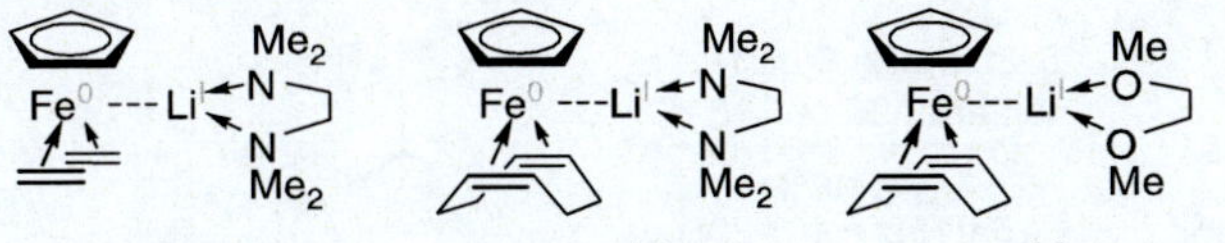

9 10 11

X-ray crystal structure analysis of **10** and **11** has revealed a close contact between Fe and the Li ion of ca. 2.5 Å. Moreover, the Li ion resides close to the olefin units. Stabilization of the electron-rich Fe center by the π-accepting character of the olefins is evidenced by the significant elongation of the C=C bonds to ca.

1.44 Å in **11** (1.34 Å in free COD). Nevertheless, binding of the olefin ligands is labile enough to allow for ligand exchange, in particular by chelating substrates containing an alkyne moiety, rendering these species attractive as catalysts for cycloisomerizations as demonstrated by Fürstner et al. [30, 31]. In fact, 1,6-enynes **12** undergo an oxidative cyclization pathway, which is triggered by the electron-rich ferrate unit, resulting in a net Alder-ene type cycloisomerization.

While the ethylene ligands in **9** are more readily displaced, COD offers the advantage that it might stabilize the resting state of the catalyst as it stays in solution serving as an ancillary protecting ligand and allowing the cycloisomerization of demanding substrates. It was found that catalyst **9** is particularly suited for the formation of bicyclic products **13** by annulation reactions, in which the size of the preexisting ring of the starting material determines the configuration at the ring junction and displays a distinct influence on the reaction rate (Fig. 8). As a general trend, substrates with larger cycloalkenes usually facilitate the reaction. In the majority of cases, the products were formed as single *trans*-diastereomers with regard to the ring-junction with the exception of [3.3.0]- and [4.3.0]-bicycles **13a** exclusively resulting in a *cis*-arrangement, while decaline products provide *cis*/*trans*-mixtures. The catalyst has a remarkable compatibility with functional groups and tolerates even terminal alkynes, aryl halides, and tertiary amino groups.

For acyclic enynes, there is a pronounced effect of the substitution pattern on the reactivity (Fig. 9). Substrates **14** with $R^2 = H$ could only be rearranged by catalyst **11**, while **9** failed. For substrates with $R^2 \neq H$, both catalysts were found to be effective providing preferentially the *trans*-configured products **15a**. Enynes carrying trisubstituted olefin moieties failed to react.

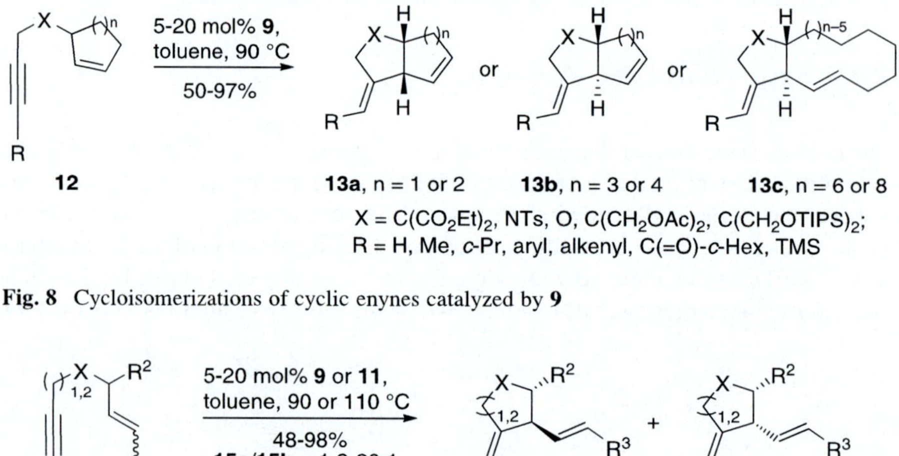

Fig. 8 Cycloisomerizations of cyclic enynes catalyzed by **9**

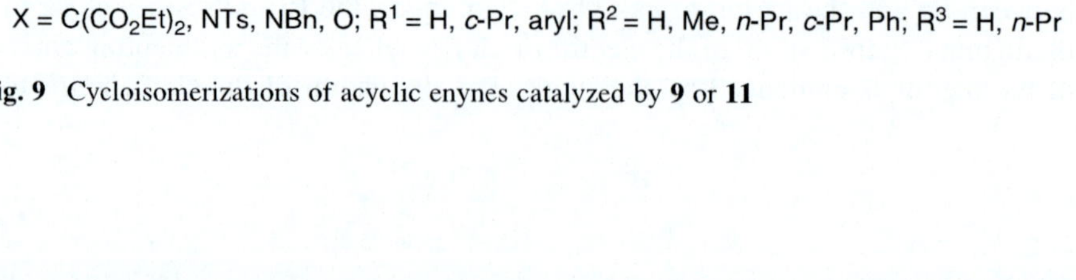

Fig. 9 Cycloisomerizations of acyclic enynes catalyzed by **9** or **11**

Various mechanistic pathways have been considered (Fig. 10), such as addition of a metal hydride species to the alkyne thus forming a vinyl metal complex **16** which subsequently reacts with the olefin part according to a Heck reaction. Alternative scenarios are either an activation of the allylic CH-bond, resulting in a π-allyl species **17**, which subsequently undergoes a metallo-ene like reaction with the alkyne, or an oxidative cyclization of the enyne to generate a metalla-cyclopentene **18**. The product **19** is then released by β-hydride elimination and reductive elimination. To get a more detailed mechanistic insight, D-labeling experiments were conducted, which provide strong evidence for the above-mentioned metallacyclic pathway.

Next to cycloisomerizations, catalysts like **11** are also useful for [4 + 2] and even more interesting for [5 + 2] cycloaddition reactions (Fig. 11), which are very

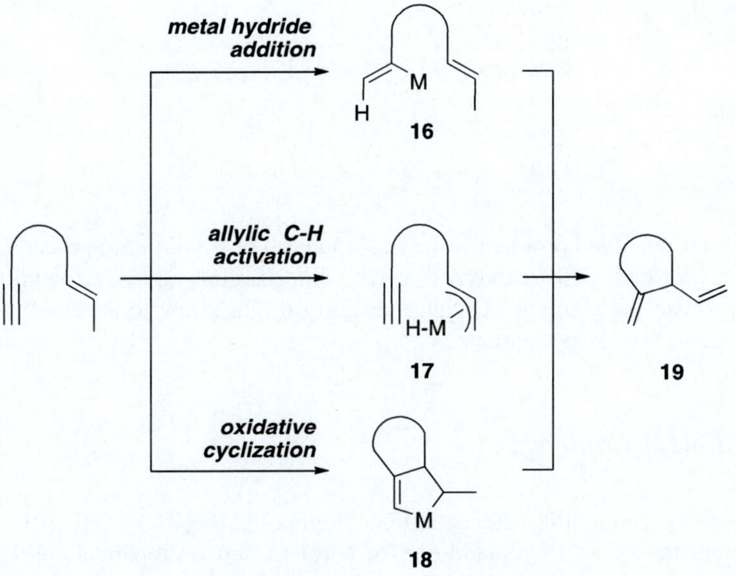

Fig. 10 Mechanistic alternatives for the cycloisomerizations of enynes

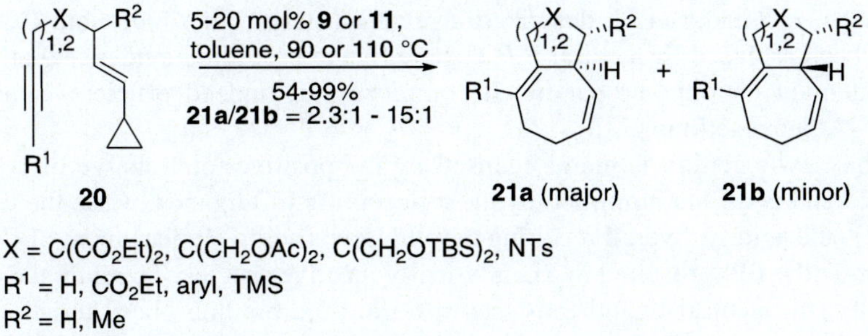

Fig. 11 [5 + 2] cycloaddition reactions catalyzed by **9** or **11**

useful for the formation of seven-membered rings **21** starting from vinyl cyclopropane **20** [31]. Again, **11** was found to be of wider applicability. The cycloheptadiene derivatives were formed with good to excellent diastereoselectivity favoring the 1,2-*trans*-disubstituted diastereomer and tolerating terminal and differently end-capped alkynes. Trisubstituted olefins failed in this approach.

In addition, complexes like **11** are also capable of catalyzing [2 + 2 + 2] cycloadditions of alkyne moieties resulting in the formation of substituted benzenes. Furthermore, Fe(I) catalysts like **22** with an odd electron count (17-electron species) have been studied in this context (Fig. 12) and the initial results demonstrate that they are catalytically relevant, uncovering a previously largely unrecognized aspect.

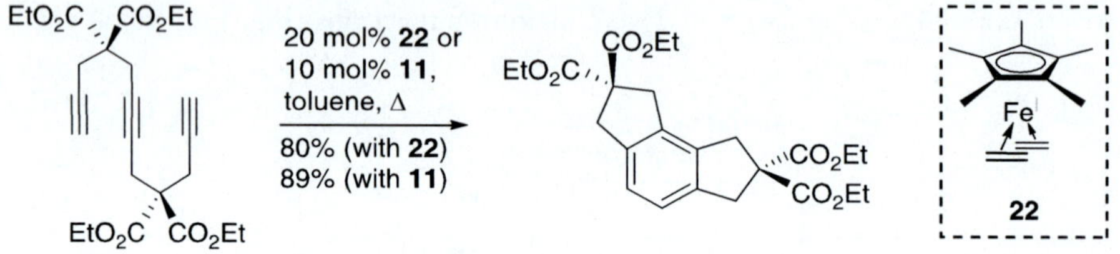

Fig. 12 [2 + 2 + 2] cycloaddition reactions catalyzed by **11** or **22**

A so far unsolved problem is the development of asymmetric procedures for the above described Fe(0)-catalyzed cycloisomerizations and cycloadditions. The option to use the element of planar chirality might allow to successfully address this issue in future applications.

2.2 Fe(II)-Catalysts

Hersh et al. found that the cationic complex [CpFe(CO)$_2$(THF)]BF$_4$ (**23**) can accelerate the [4 + 2] cycloaddition of acrolein and cyclopentadiene [32]. However, the catalytic activity was higher than expected from rate constants determined in stoichiometric experiments, indicating that a Brønsted or Lewis acid impurity might accelerate this process and generating doubts about the role of **23**.

In 1994, Kündig et al. described a related catalyst, in which both CO ligands have been replaced by a chiral *P*,*P*-chelate ligand providing evidence that structurally defined cationic Fe sandwich complexes are indeed efficient catalysts for Diels–Alder reactions [33].

The Lewis acidity is mainly caused by the positive total charge of complexes like **23** and is further amplified by the π-accepting CO ligands, while the Cp ligand buffers the acidity, overall resulting in mild Lewis acids. Replacement of CO by the less π-acidic PPh$_3$ resulted in a catalytically inactive species. To retain the catalytic activity, the neutral ligand thus has to mimic the binding characteristics of CO. Accordingly, with P(OMe)$_3$ catalytic activity was noticed [34, 35]. Kündig et al. utilized a C_2-symmetric chiral ligand exhibiting similar electronic binding characteristics as CO as a result of P-bound pentafluorophenyl moieties [36]. Complex **24** can be isolated as a solid by precipitation and can be used as the catalyst precursor, although it was found to be instable in solution.

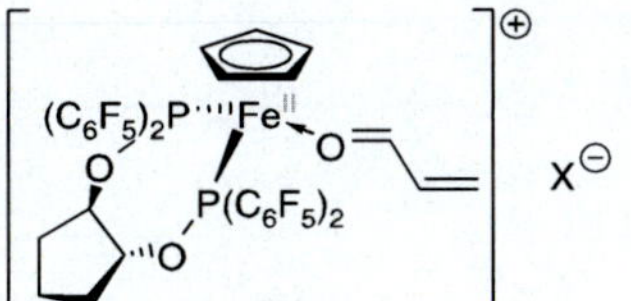

24

24 (5 mol%) catalyzes the cycloaddition of various enals **25**, α,β-substituted or -unsubstituted, and different acyclic and cyclic dienes such as cyclopentadiene with high enantioselectivity favoring in most cases the *exo*-products **26** (Fig. 13). An *endo* product **27** was preferentially formed with cyclohexadiene and a modest *endo* preference was found for the reaction of cyclopentadiene with acrolein. 2,6-Di-*tert*-butylpyridine was usually added to trap Brönsted acid impurities, which otherwise result in reduced *ee* values and varying reaction rates. Reaction temperatures higher than −20°C were not useful as the catalyst slowly decomposes.

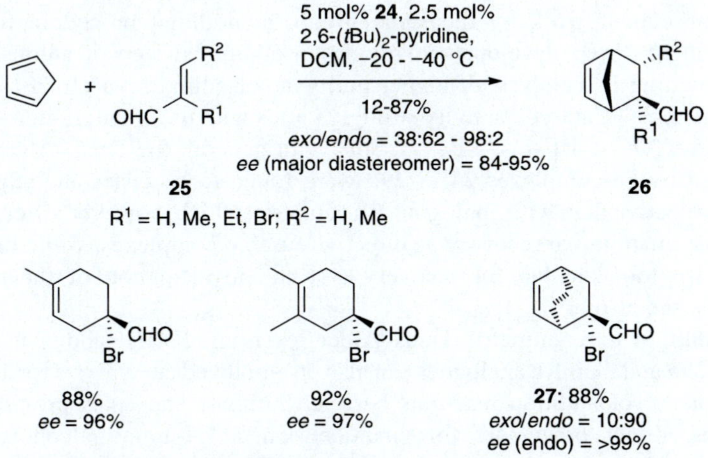

Fig. 13 Diels–Alder reactions with enals catalyzed by **24**

The enantioselectivity was explained by the preference for a transition state **28**, in which the diene approaches the C_α-*Si* face of the coordinated dienophile, which adopts an *s-trans* conformation (Fig. 14).

In 1999, Kündig and Bruin reported a closely related catalyst system **29a**, in which a more readily accessible ligand has been employed [37]. Catalytic activity and stability are strongly dependent upon the nature of the neutral ligand L. While the acetonitrile complex **29b** is stable, yet catalytically inactive, complex **29a** with L = acrolein is stable only in the solid state, but decomposes as a solution in DCM

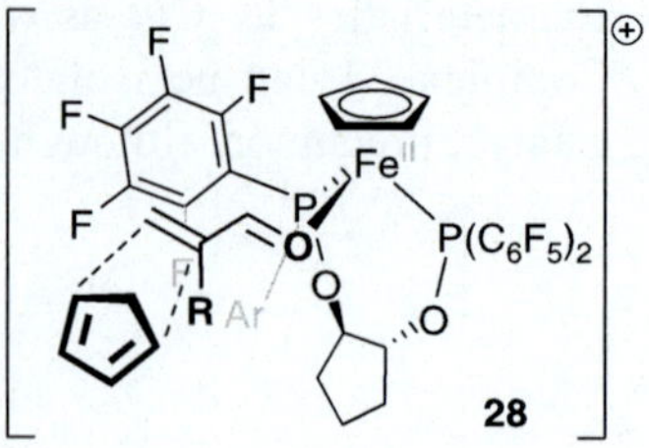

Fig. 14 Explanation of the enantioselectivity by the preferred transition state **28**

in the absence of an excess of free aldehyde at temperatures above −20°C. Stereoselectivities and yields obtained for the Diels–Alder reaction of enals either match or slightly exceed the results obtained with **24**.

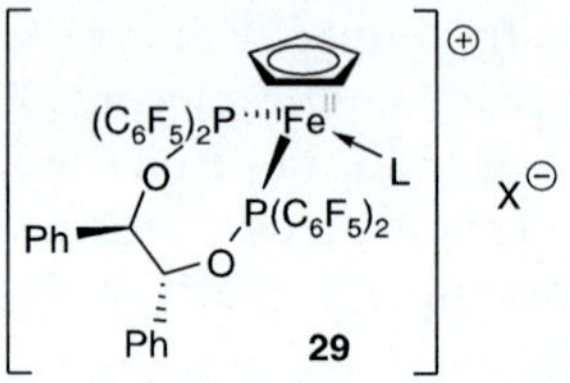

29a: L = acrolein
29b: L = MeCN

In subsequent work by the same group, a modified procedure for catalyst formation has been developed involving an additional step of anion metathesis thus providing Fe catalysts of higher purity superseding the use of Brönsted acid scavengers.[38] Catalytic activity strongly varies with the counterion, increasing in the order TfO^- < BF_4^- < PF_6^- < SbF_6^-. In general, the Diels–Alder reactions catalyzed by Fe-complexes **24** or **29a** were found to be faster and slightly more enantioselective than with analogous Ru catalysts, which, however, offer the benefit of almost quantitative recovery as directly reusable complexes, while the Fe complexes are too unstable for recovery (for the investigation of the related Ru catalysts, see also [39]).

In addition to asymmetric Diels–Alder reactions, Fe(II) sandwich complexes such as **29a** and their Ru analogues can also be employed as catalysts for asymmetric 1,3-dipolar cycloaddition reactions between nitrones and enals providing isoxazolidines, direct precursors for enantioenriched 1,3-amino alcohols. Usually, Lewis acids coordinate nitrones preferentially (and often irreversibly) over simple α,β-unsaturated aldehydes, explaining the often found necessity of two-point binding dipolarophiles. Any cycloaddition of a mono-coordinating enal with a nitrone would thus be due to a noncatalyzed background reaction. Nevertheless, in 2002, Kündig et al. could present the first examples of catalytic asymmetric 1,3-dipolar cycloadditions of nitrones with monocoordinating dipolarophiles based on a judicious choice of the metal ligand environment and Lewis acidity (Fig. 15) [40]. Using **29a** as catalyst (5 mol%), cycloadducts of diarylnitrones were obtained in excellent yield and with high enantiomeric excess. With methacrolein, only *endo*-products were obtained, yet with low regioselectivity. For the acyclic diarylnitrones it was found that the Ru analogous catalysts are superior [41–43].

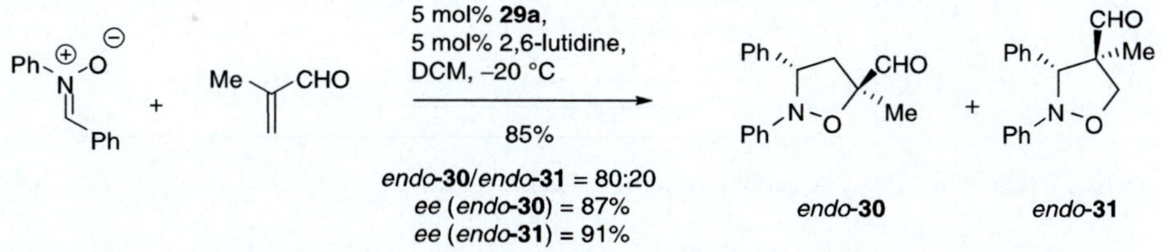

Fig. 15 [3 + 2] cycloaddition reactions of diarylnitrones with enals catalyzed by **29a**

Cyclic N-oxides demonstrated a higher reactivity (Fig. 16). To suppress a thermal background reaction, the nitrone concentration had to be maintained low. These conditions afforded the cycloaddition products **32–35** in excellent yields and high enantiomeric purities as a single regioisomer. For cyclic substrates, the Fe-catalyst **29a** performed significantly better than the corresponding Ru analogues. X-Ray crystal structure analysis of the latter as a complex with methacrolein shows that the enal adopts an *s*-trans conformation. The configuration of the cycloaddition products are consistent with an *endo*-approach of the nitrone to the C_α-*Si*-face of the coordinated enal at the (*R*,*R*)-catalyst site.

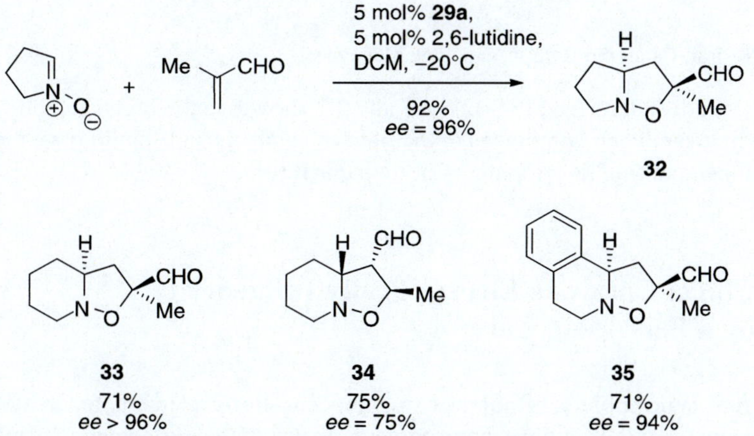

Fig. 16 [3 + 2] cycloaddition reactions of cyclic nitrones with enals catalyzed by **29a**

The cationic complex $[CpFe(CO)_2(THF)]BF_4$ (**23**) can also catalyze the proton reduction from trichloroacetic acid by formation of Fe-hydride species and may be considered as a bioinspired model of hydrogenases ("*Fe-H Complexes in Catalysis*") [44]. This catalyst shows a low overvoltage (350 mV) for H_2 evolution, but it is inactivated by dimerization to $[CpFe(CO)_2]_2$.

Hydride species were also formed in the dehydrogenative coupling of hydrosilanes with DMF [45]. The catalytic system is applicable to tertiary silanes, which are known to be difficult to be converted into disiloxanes (Fig. 17). The catalytic reaction pathway involves the intermediacy of a hydrido(disilyl)iron complex

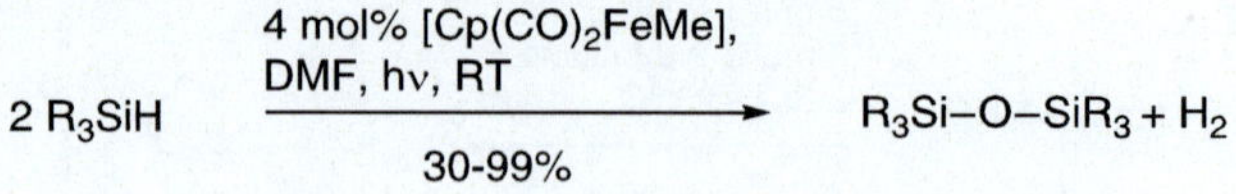

Fig. 17 Dehydrogenative coupling of tertiary silanes

(*"Fe-H Complexes in Catalysis"*). Similarly, also di-*tert*-butyltin dihydride could be dimerized [46].

Cyclopentadienone iron alcohol complexes like **37** were generated from the reactions of [2,5-$(SiMe_3)_2$-3,4-$(CH_2)_4(\eta^5$-$C_4COH)$]$Fe(CO)_2H$ (**36**) and aromatic aldehydes [47]. This process can be used for the iron-catalyzed hydrogenation of aldehydes (Fig. 18 and *"Fe-H Complexes in Catalysis"*).

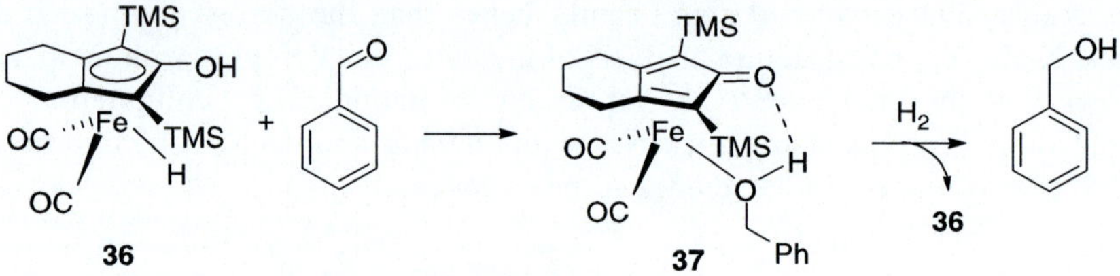

Fig. 18 Iron-Catalyzed Hydrogenation of Aldehydes

The studies described in Sects. 2.1 and 2.2 showcase the high synthetic potential of Fe half sandwich complexes in the field of catalysis and should pave the way for additional exciting developments in the near future.

3 Chiral Catalysts Electronically Influenced by a Ferrocene Core

The ferrocene moiety is not just an innocent steric element to create a three-dimensional chiral catalyst environment. Instead, the Fe center can influence a catalytic asymmetric process by electronic interaction with the catalytic site, if the latter is directly connected to the sandwich core. This interaction is often comparable to the stabilization of α-ferrocenylcarbocations **3** (see Sect. 1) making use of the electron-donating character of the Cp_2Fe moiety, but can also be reversed by the formation of ferrocenium systems thereby increasing the acidity of a directly attached Lewis acid. Alternative applications in asymmetric catalysis, for which the interaction of the Fe center and the catalytic center is less distinct, have recently been summarized in excellent extensive reviews and are outside the scope of this chapter [48, 49]. Moreover, related complexes in which one Cp ring has been replaced with an η^6-arene ligand, and which have, for example, been utilized as catalysts for nitrate or nitrite reduction in water [50], are not covered in this chapter.

3.1 *Planar Chiral Ferrocenyl Metallacycle Catalysts*

Palladacycles are defined as compounds with a Pd-C σ-bond with the Pd being stabilized by one or two neutral donor atoms, typically forming 5- or 6-membered rings [51]. Ferrocenyl palladacycles constitute a particularly attractive catalyst class partly due to the element of planar chirality. The first diastereoselective cyclopalladation of a chiral ferrocene derivative was reported in 1979 by Sokolov [52, 53].

There had been doubts about the utility of palladacycles in asymmetric catalysis, raised by the failure to achieve enantioselectivity as a result of a slow release of low ligated Pd(0) (naked Pd) [54]. However, recent success of several planar chiral palladacycles in highly enantioselective *aza*-Claisen reactions and in a number of other applications proves that the coordination shell of the Pd(II) species is not necessarily destroyed during the catalytic action.

3.1.1 The *Aza*-Claisen Rearrangement

In 1997 the first asymmetric *aza*-Claisen rearrangement was reported by Overman et al. [55], which made use of diamines as bidentate ligands for Pd(II), allowing for moderate enantioselectivities. In the same year, Hollis and Overman described the application of the planar chiral ferrocenyl palladacycle **38** as a catalyst for the enantioselective *aza*-Claisen rearrangement of benzimidates **39** (Fig. 19) [56]. A related ferrocenyl imine palladacycle provided slightly inferior results, while a benzylamine palladacycle lacking the element of planar chirality was not able to provide any enantioselectivity [57].

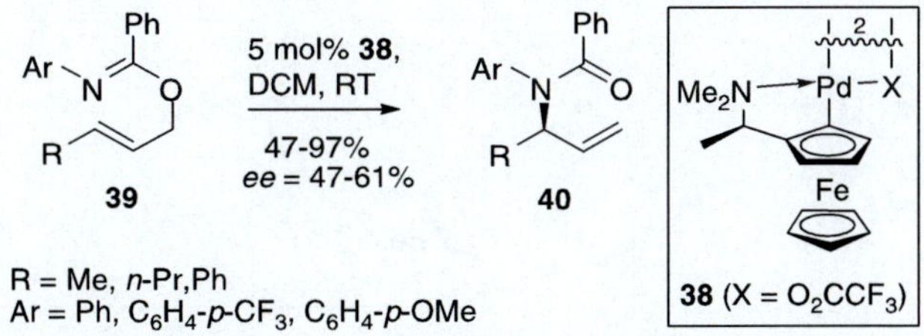

Fig. 19 Asymmetric *Aza*-Claisen Rearrangement of benzimidates **39** using amino palladacycle **38**

A significant breakthrough was achieved by Overman and Donde in 1999: they reported the first highly selective catalyst **41** for the *aza*-Claisen rearrangement of benzimidates **39** (Fig. 20) [58]. Enantioselectivities were in most cases good to very good.

Kang et al. described ferrocene bispalladacycle **42** (Fig. 21) [59]. Enantioselectivities and yields obtained were comparable to the results obtained by Overman with **41** [58].

In 2005, Moyano et al. [60] reported a new type of chiral dimeric ferrocene palladacycle **43** that lacked the element of planar chirality and involved three

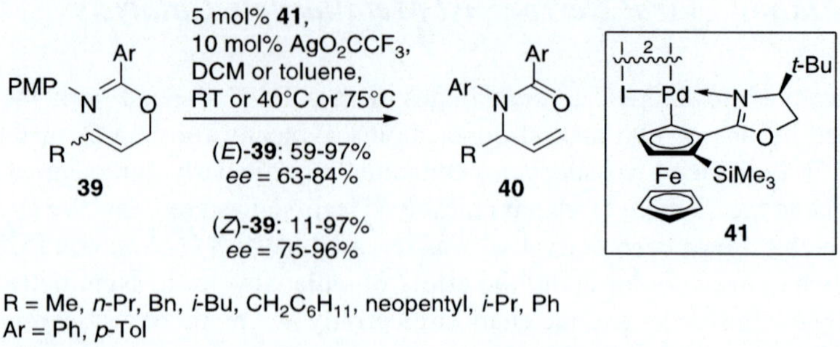

Fig. 20 Asymmetric *aza*-Claisen rearrangement of benzimidates using oxazoline palladacycle precatalyst **41**

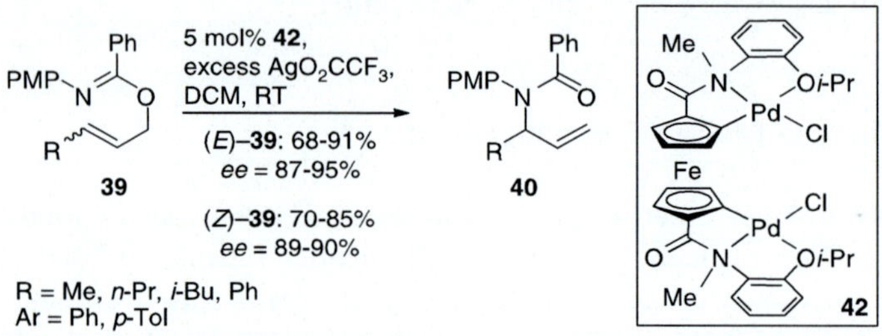

Fig. 21 Asymmetric *aza*-Claisen rearrangement of benzimidates using bispalladacycle precatalyst **42**

Pd-centers as catalysts for the rearrangement of benzimidate substrates. While the enantioselectivity reached a practical value, the yield was only moderate (Fig. 22).

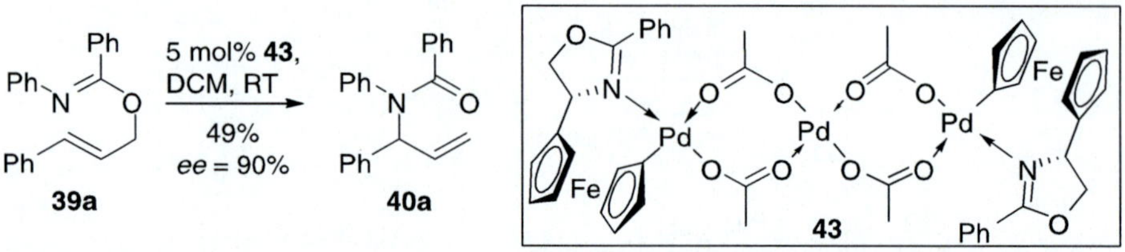

Fig. 22 Asymmetric *aza*-Claisen rearrangement of benzimidate **39a** using the trispalladium species **43**

In 2005 [61], Overman and coworkers described the application of their oxazoline system **41** to synthetically attractive trifluoroacetimidates **44** (Fig. 23) forming trifluoroacetamides **45**, which, in contrast to benzamides **40**, can be readily transferred into free primary allylic amines.

A related planar chiral Co-based oxazoline palladacycle COP-X (**46**) was later found to be of higher synthetic utility as it permitted the use of benzimidates, [62] as well as allylic trifluoro- [63] and trichloroacetimidates [64, 65]. **46** was found to be superior to its ferrocene analogue **41** [61] in a number of aspects such as ease of

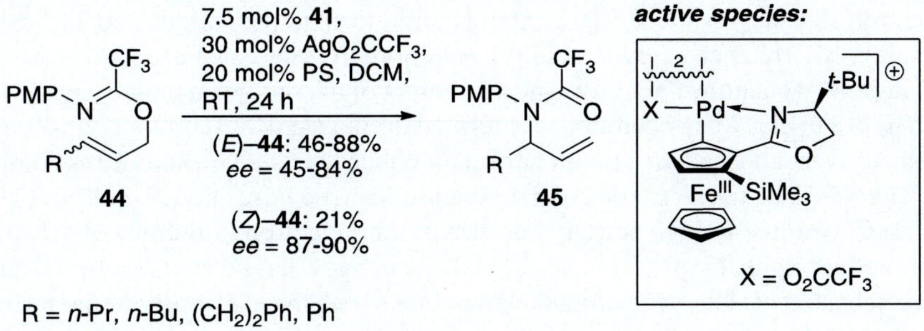

Fig. 23 Asymmetric *aza*-Claisen rearrangement of trifluoroacetimidates using precatalyst **41**

preparation [**41** was not accessible by direct cyclopalladation due to oxidative decomposition by Pd(II)], catalyst stability, activity, and substrate scope.

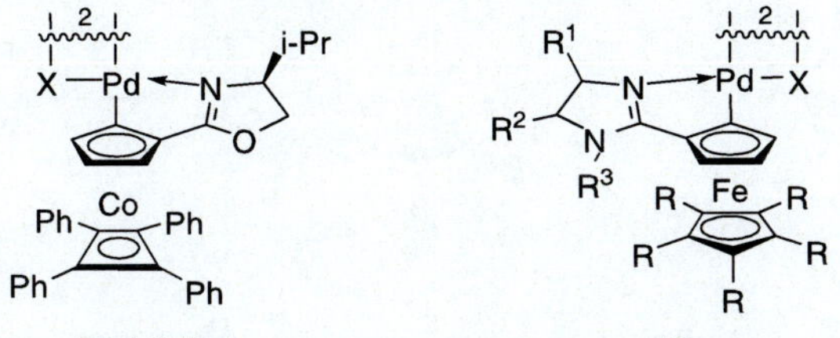

Unfortunately, in the case of trifluoroacetimidates COP-Cl (**46**) still required catalyst loadings, which are not useful for large-scale applications [10 mol% Pd (II)], while long reaction times were necessary for high conversion. Moreover, the scope was limited to substrates bearing α-unbranched alkyl substituents R at the 3-position of the allylic imidate.

Peters and coworkers have systematically addressed the issue of catalytic activity for the asymmetric rearrangement of trihaloacetimidates. These investigations utilized a modular design of a planar chiral carbophilic Lewis acid allowing for an adjustment of electronic and steric properties. Although all previous ferrocene-based catalysts had not been as efficient as COP-X (46), ferrocene moieties were chosen as backbone for planar chiral catalyst systems for two main reasons: (a) the cobalt-based sandwich complexes proved to be very sensitive with regard to the cyclopalladation outcome depending on the oxazoline substitution pattern in terms of reactivity and configuration, thus limiting the number of accessible catalyst structures [66]; (b) for the cobalt-based sandwich backbone, the electron density could not be extensively variegated, in contrast to ferrocenes for which, for example, the oxidation potentials are strongly dependent upon the substitution pattern [67].

Along these lines, imidazoline [68, 69] rather than oxazoline coordination sites better fitted into a modular concept in order to understand how the electronic properties should ideally be, since electron density can be readily adjusted by the choice of the N substituent [70]. The Cp′ ligand C_5R_5 in **47** permits steric and electronic tuning. With the electron-withdrawing effect of five Ph groups R, the electron density on the Pd(II)-center is significantly decreased, and the catalyst displays an enhanced Lewis acidity. Additionally, the enhanced bulk has led to an improved enantioselectivity. The iron center of the sandwich moiety can be either Fe(II) or Fe(III). A ferrocenium core formed during catalyst activation via oxidation with a silver salt is in fact a major reason for enhanced catalytic activity as compared to COP (46) due to electronic communication with the Pd-center. N-Sulfonyl groups R^3 in 47 resulted in high activity because of their electron-withdrawing effect. The sulfonyl group in **48** offers the additional advantages that Pd can be introduced by a direct diastereoselective cyclopalladation due to enhanced stability of the ferrocene core against oxidative decomposition and owing to a chirality transfer from R^1 to the neighboring N atom effecting a preferred conformation in which the sulfonyl moiety points away from the ferrocene moiety (Fig. 24).

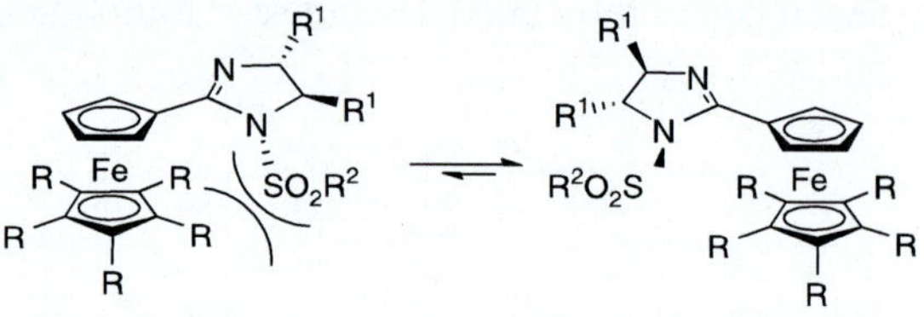

48

Fig. 24 Conformational equilibrium for **48** caused by chirality transfer to the sulfonylated *N*-atom

Different counteranions X^- coordinating to the Pd-center of **47a** were employed by breaking the chloro bridge of the catalytically inert dimeric precursor with different silver salts. AgO_2CCF_3 provided the best results. Enantioselectivities were good to excellent and the yields were in general high (Fig. 25). The catalyst loadings could be decreased by a factor of 100 as compared to the previous results, performing catalysis under almost solvent-free conditions.

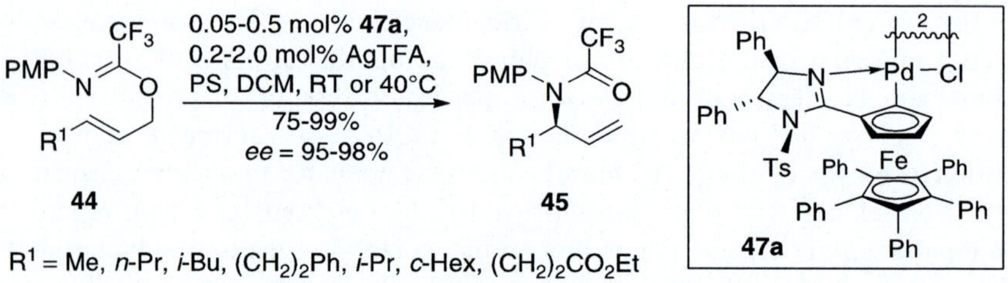

Fig. 25 Asymmetric *aza*-Claisen rearrangement of trifluoroacetimidates **44** using precatalyst **47a**

The high catalytic activity also enabled *aza*-Claisen rearrangements to form *N*-substituted quaternary stereocenters (Fig. 26) [71]. The catalyst does not need to distinguish between differently sized substituents on the double bond of **49** (e.g., $R = CD_3$, $R^1 = CH_3$, $ee = 96\%$), indicating that coordination of the olefin is the stereoselectivity predetermining step. The imidate-N-atom subsequently attacks intermediate **47-I** from the face remote to the Pd-center totally resulting in a

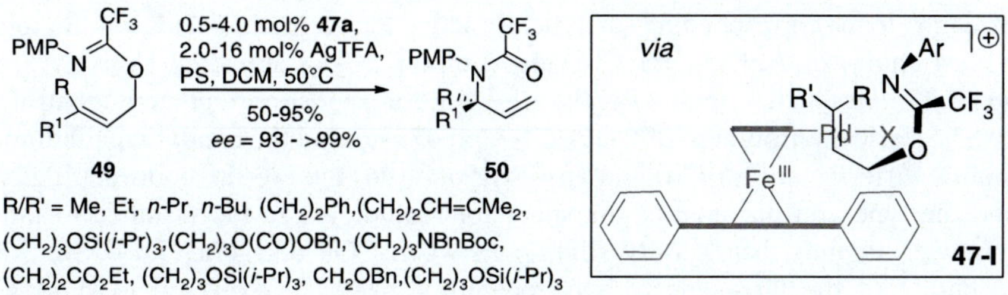

Fig. 26 Asymmetric *aza*-Claisen rearrangement of trifluoroacetimidates **49** generating *N*-substituted quaternary stereocenters

stereospecific reaction outcome. The allylic amides **50** can be transferred to almost enantiopure quaternary α- and β-amino acids. **47a** was also examined for the rearrangement of imidates carrying different *N*-substituents to give rise to various secondary allylic amines after reductive amide cleavage [72].

To examine if the higher catalytic activity and selectivity of **47a** as compared to the COP-X system **46** is mainly caused by the pentaphenyl ferrocenium or by the imidazoline moiety, oxazoline **53-Cl** was prepared in diastereomerically pure form starting from carboxylic acid **51** and (*S*)-valinol via oxazoline **52** (Fig. 27) [73].

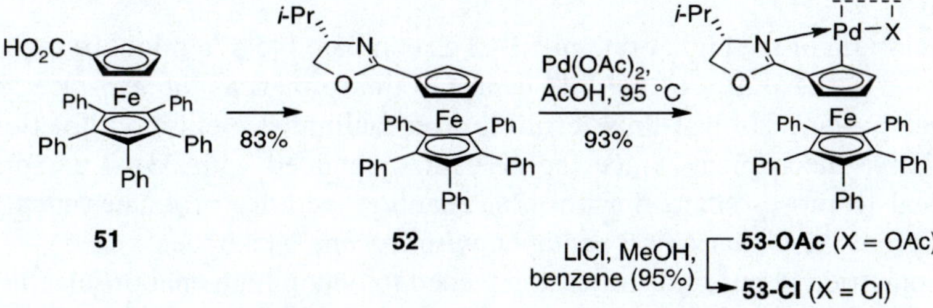

Fig. 27 Synthesis of pentaphenyl ferrocene oxazoline palladacycle precatalyst **53-Cl**

In difference to COP-X (**46**), the *i*-Pr group is pointing toward the sandwich core. $AgNO_3$ generates a highly active and selective catalyst (Fig. 28). Complete

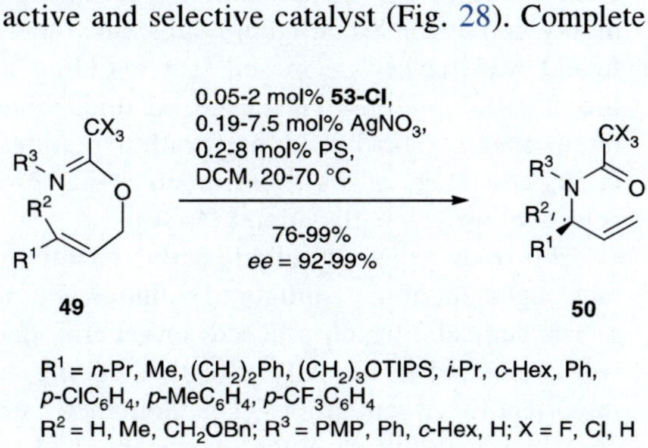

Fig. 28 Asymmetric *Aza*-Claisen rearrangement using precatalyst **53-Cl**

oxidation to the corresponding catalytically active ferrocenium species is accomplished with 4 equiv. of $AgNO_3$ per Cl-bridged dimer. **53-Cl** activated by $AgNO_3$ is in general more reactive than **47a** still allowing for almost perfect stereocontrol.

The steric environment of COP-X **46** and **47a** around the catalytic palladium site mainly differs in a Ph (**47a**) and an *i*-Pr group (**46**) next to the coordinating N-site and the type and distance of the spectator ligand. While the distance of the two sandwich ligands differs only slightly between COP and **47a** (3.4 Å vs. 3.3 Å), oxidation of the ferrocene to a ferrocenium species is expected to shorten this distance further. Overall, the steric hindrance to access the Pd-center is more distinct for **47a**. These steric effects are capable to explain the higher *ee* obtained with **47a**.

From an electronic point of view, the differences are more pronounced since the overall charge of the ligand (−1 for COP, 0 for **47a** after oxidation to the ferrocenium species) changes. As Pd(II) most likely acts as a carbophilic Lewis acid coordinating to an olefin, the lower electron density in **47a** is suitable to explain the higher reactivity found for this complex.

When comparing COP-X **46** to oxazoline **53-Cl**, there are two major differences: (1) the *i-Pr* residue on the oxazoline moiety in **53-Cl** is pointing toward the Cp′-spectator ligand, leaving the space above the Pd square plane unhindered, while in COP, the substituent is pointing away from the sandwich core. (2) **53-Cl** can be oxidized to a ferrocenium species, again providing a less electron-rich ligand.

The origin of the higher rate of **53-Cl** as compared to **47a** might be explained by the better accessibility of the Pd^{II}-center in the former, as the *exo*-face above the Pd-square plane is open thus resulting in a facilitated olefin coordination via an associative mechanism. Since the *ee*-values obtained with **53-Cl** are practically identical to those obtained with **47a**, enantioselectivity originates nearly exclusively from the planar chiral pentaphenyl ferrocene backbone.

Geometrically pure substrates are needed to obtain high enantioselectivities due to the stereospecific reaction outcome. **53**, however, is less active for *Z*-configured allylic imidates. As *Z*-olefins are, in general, more readily available in isomerically pure form, for example by semihydrogenation of triple bonds, this substrate class would be more interesting for technical applications. To develop a catalyst highly active for *Z*-configured substrates, the electron density of the second Cp ligand was further decreased, yet avoiding a complicated catalyst preparation for practical reasons. Ferrocene bisimidazoline bispalladacycle **55** fulfills these requirements. The ligand preparation requires only three steps from ferrocene taking advantage of the C_2 symmetry (Fig. 29). Ligand **54** undergoes a diastereoselective biscyclopalladation [74–76].

55 is to date the only highly active enantioselective catalyst for the *aza*-Claisen rearrangement of (*Z*)-configured trifluoroacetimidates (*Z*)-**44** (Fig. 30) [74, 75]. The rearrangements, which proceed, in general, under almost solvent-free conditions, were found to be equally effective on mg- and g-scale and tolerant to many important functional groups. The monomeric structure of the active catalyst species **56**, which is in this case not a ferrocenium species, was determined by NMR.

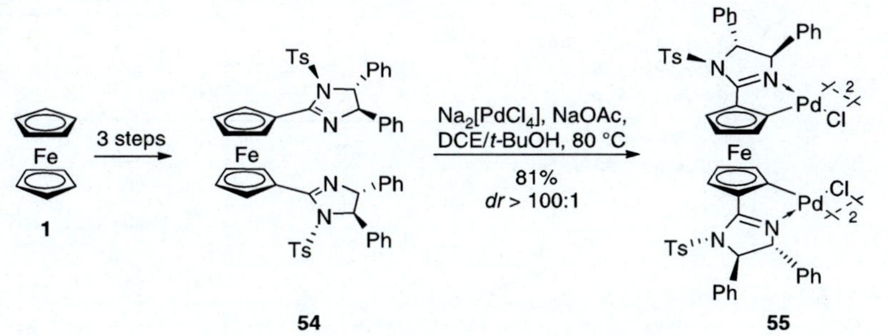

Fig. 29 Formation of **55** via direct diastereoselective biscyclopalladation

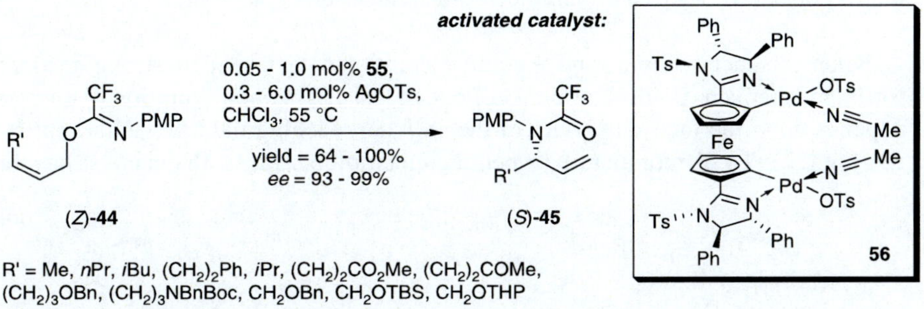

Fig. 30 Asymmetric *aza*-Claisen rearrangement of (*Z*)-configured trifluoroacetimidates **44**

3.1.2 Bispalladium-Catalyzed Michael-Addition of α-Cyanoacetates

Ferrocen-1,1′-diylbismetallacycles are conceptually attractive for the development of bimetal-catalyzed processes for one particular reason: the distance between the reactive centers in a coordinated electrophile and a coordinated nucleophile is self-adjustable for specific tasks, because the activation energy for Cp ligand rotation is very low. In 2008, Peters and Jautze reported the application of the bis-palladacycle complex **56a** to the enantioselective conjugate addition of α-cyanoacetates to enones (Fig. 31) [74–76] based on the idea that a soft bimetallic complex capable of simultaneously activating both Michael donor and acceptor would not only lead to superior catalytic activity, but also to an enhanced level of stereocontrol due to a highly organized transition state [77]. An α-cyanoacetate should be activated by enolization promoted by coordination of the nitrile moiety to one Pd(II)-center, while the enone should be activated as an electrophile by coordination of the olefinic double bond to the carbophilic Lewis acid [78].

The loading of the activated monomeric catalyst could be decreased to a level of 0.04 mol% for certain substrates [turn over number: 2,450]. The developed process does not require the use of inert gas techniques and in most cases chromatographic purification was not necessary to obtain analytically pure products as no side products were formed and the catalyst could be separated by filtration.

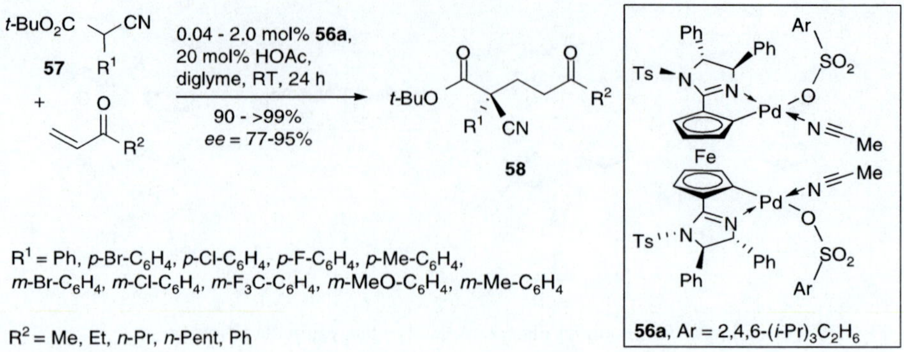

Fig. 31 Bis-Pd(II)-catalyzed asymmetric Michael addition of α-cyanoesters **57** to enones

Kinetic, spectroscopic, and enantioselectivity data provided strong evidence for a mechanism involving bimetallic catalysis. The configurational outcome depends upon the face selectivity of the enol approaching the Michael acceptor in **59** (Fig. 32). To differentiate between the enantiotopic faces, the catalyst has thus

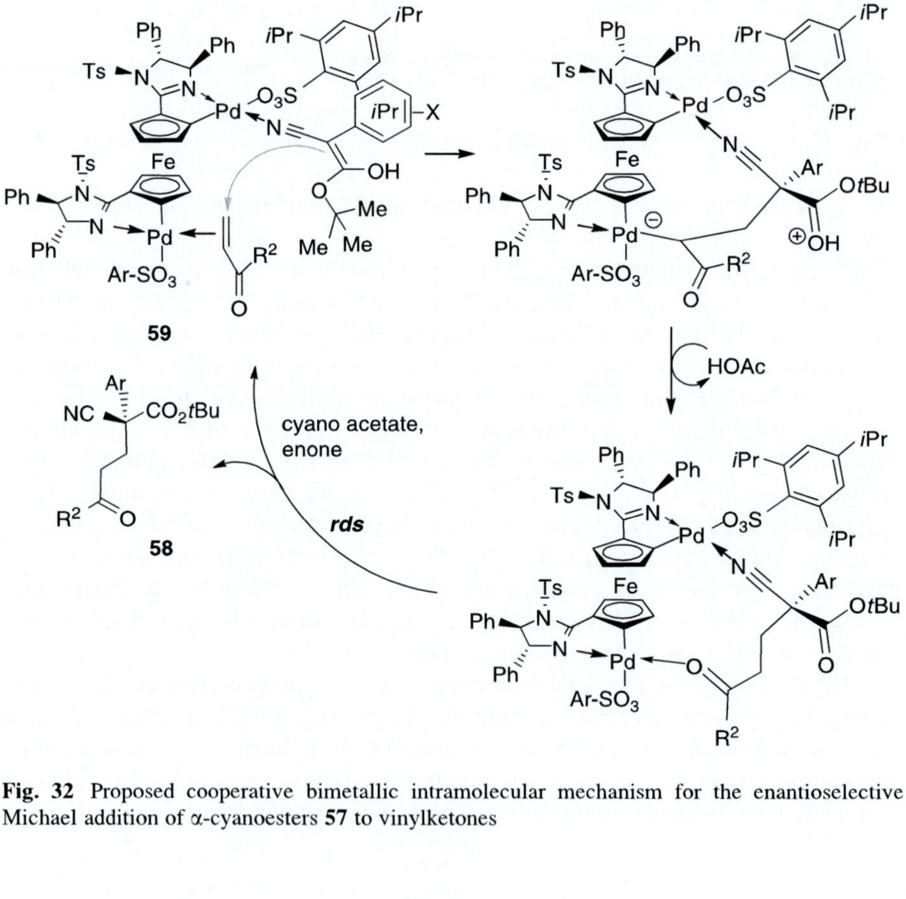

Fig. 32 Proposed cooperative bimetallic intramolecular mechanism for the enantioselective Michael addition of α-cyanoesters **57** to vinylketones

(a) to control the conformation of **59** with regard to the C–CN σ-bond and (b) to direct the enone. Control of the reactive conformation is achieved by the use of a bulky ester moiety and an especially large sulfonate counteranion. The direction of the enone attack is accomplished by the cooperative mechanism as control experiments have shown. Using the corresponding mono-palladacycle, the product was formed with low and inversed enantioface selection due to preferential attack of MVK on the *Re*-face of the enol.

The initial rate of the model reaction follows a first-order dependence for the activated catalyst, the Michael donor, and the Michael acceptor. The rate determining step is not the C–C bond formation or protonolysis but the decomplexation of the bidentate product. This was evidenced by the relationship between the initial conversion and the reaction time. Extrapolation to $t_0 = 0$ h provides a positive intercept. In other words, upon addition of the reagents, the C–C bond formation occurs almost instantaneously. The amount of product at t_0 correlates within the experimental error to the double precatalyst loading since the dimeric precatalyst forms two active monomeric catalyst species.

3.1.3 Ferrocene Bisimidazoline Platinacycle-Catalyzed Intramolecular Friedel–Crafts Alkylations

Ligand exchange processes are relatively slow with Pt [79, 80]. Pt-catalysts allowing for a more rapid ligand exchange could therefore lead to enhanced activity. A mono-platinacycle complex **60** was thus designed in which the Pt-center binds to two imidazoline units: one connected to the same Cp plane as the metal, the second one to the Cp′ ligand resulting in severe structural distortion [81]. Cycloplatination of bisimidazoline **54** (Fig. 33) occurs on treatment with $K[(H_2C{=}CH_2)PtCl_3]$. Both imidazoline units bind in a *trans*-mode to the same Pt entity resulting in a unique geometry in which the C-atom connected to the metal center is strongly pyramidalized (angle Cp-Pt: 159°, deviation of Pt from the upper Cp plane: 0.74 Å). The Cp rings are strongly tilted toward each other resulting in a close contact between Pt and Fe (3.19 Å). The conformational freedom of ligand **54** with regard to rotation

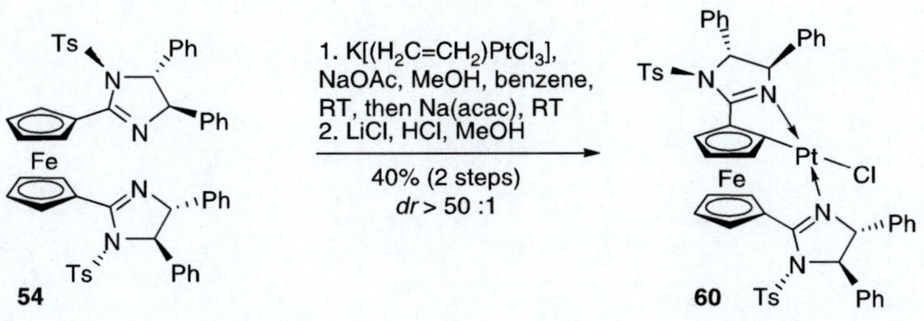

Fig. 33 Formation of platinacycle **60** by direct diastereoselective cycloplatination

along the Fe–Cp axes in combination with the possible deviation from the nonparallel Cp orientation are crucial factors to allow for such catalyst geometry.

To establish proof of principle, the intramolecular Friedel–Crafts alkylation with unactivated olefins was selected (Fig. 34). Platinacycle **60** is in fact sufficiently active to catalyze this process for the first time with indole systems **61** carrying a disubstituted olefin moiety. The highest reactivity was attained by activating **60** with $AgO_2CC_3F_7$ delivering the targeted products **62** in good yield and with high *ee* values. The corresponding nondistorted platinacycle with only one imidazoline unit provided product in poor yield under identical reaction conditions.

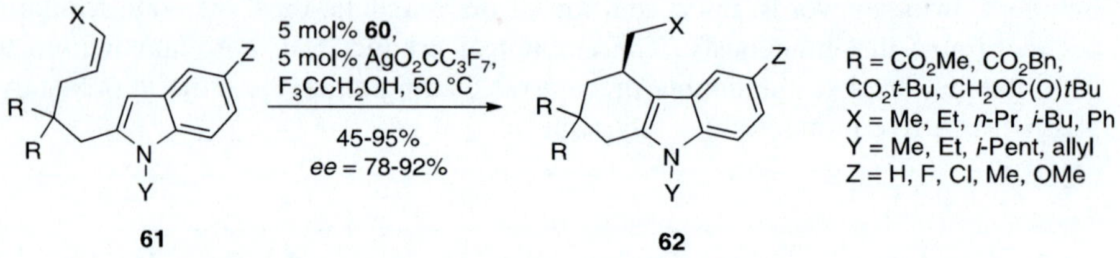

Fig. 34 Catalytic asymmetric Friedel–Crafts alkylation of indoles catalyzed by **60**

Both electron-withdrawing and -donating substituents Z on the aromatic nucleus afforded smooth conversion and similar enantioselectivity despite the projected difference in indole nucleophilicity.

The high enantioselectivity again can be rationalized by enantioface-selective alkene coordination in **63** (Fig. 35). The olefin moiety is expected to bind *trans* to the upper imidazoline moiety [70, 73] thereby releasing the catalyst strain. Coordination at this position may, in principal, afford four different isomers assuming the stereoelectronically preferred perpendicular orientation of the alkene and the Pt(II) square plane. In the coordination mode shown, steric repulsion between both olefin substituents and the ferrocene moiety is minimized. Outer-sphere attack of the indole core results in the formation of the product's stereocenter.

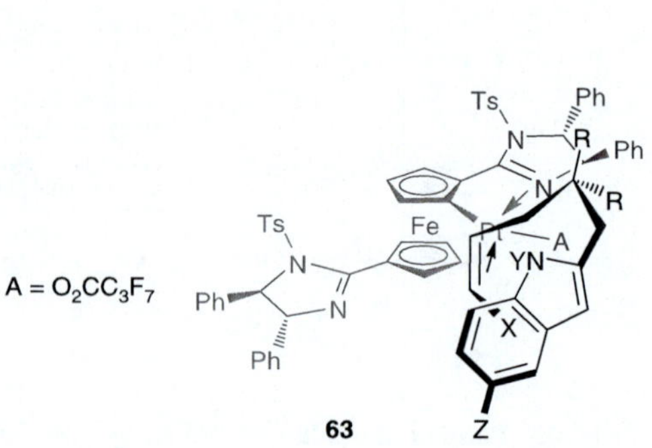

Fig. 35 Explanation of the enantioselectivity by face selective olefin coordination

3.2 *Planar Chiral Ferrocenes as Lewis- or Brønsted-Base Catalysts*

In 1996, Fu et al. reported the synthesis of the planar chiral heterocycles **64**, formally DMAP fused with a ferrocene core [82]. While the original synthesis provided racemic **64a** in only 2% overall yield requiring a subsequent resolution by preparative HPLC on a chiral stationary phase, a recently improved synthesis furnished the racemic complexes **64** in 32–40% yield over seven steps. A subsequent resolution with di-*p*-toluoyltartaric or dibenzoyltartaric acid gave access to the enantiomers with >99% *ee* (28–44% yield for each isomer in this step) [83].

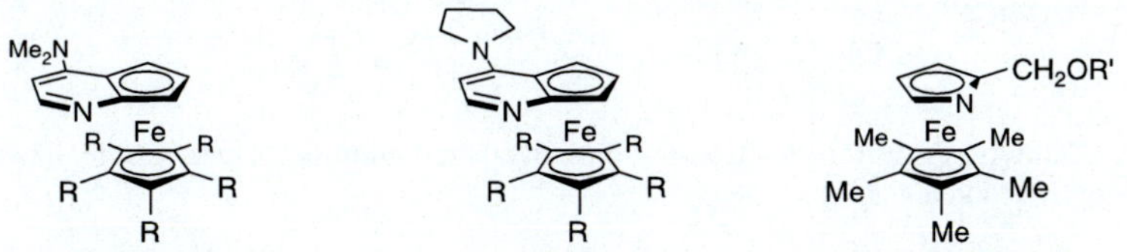

R = Me 64a
R = Ph 64b

R = Me 64c
R = Ph 64d

R' = TES 65a
R' = TBS 65b

The [Fe-Cp]-fragment does not only play the role of an additional steric element introducing planar chirality into the otherwise flat pyridine system. Substitution at the pyridine 2-position usually cuts the nucleophilicity of the nitrogen atom thus limiting the possibilities to achieve efficient chirality transfer using nucleophilic pyridine catalysts [84]. Ferrocene, however, functions as a strong electron donor (see Sect. 1) and thus restores the nucleophilicity impaired by substitution.

Extending the same concept of a planar chiral nucleophilic or basic heterocyclic Fe-sandwich complex, *aza*-ferrocenes **65** were prepared. The latter have also been successfully applied as bidentate ligands in transition metal catalysis [85].

Complexes like **64** and **65** can act by two general ways: either as a Brønsted-base or as a nucleophilic catalyst, depending on the type of reaction and substrate. However, the exact mechanistic pathway is in a few cases speculative to some extent as the distinction between the two mechanistic routes is sometimes rather difficult.

3.2.1 Brønsted-Base Catalysis

64a/c is, mainly due to the electron-donating nature of the pentamethyl ferrocene core, a comparatively strong base capable of deprotonating acidic compounds RX-H like phenols [86], hydrazoic acid [87], or 2-cyanopyrrole [88]. The generated corresponding anionic base RX^- can then undergo addition to a disubstituted ketene **66** (Fig. 36) to form a prochiral enolate **67** which is subsequently protonated by **64-H**$^+$ in an enantioselective manner.

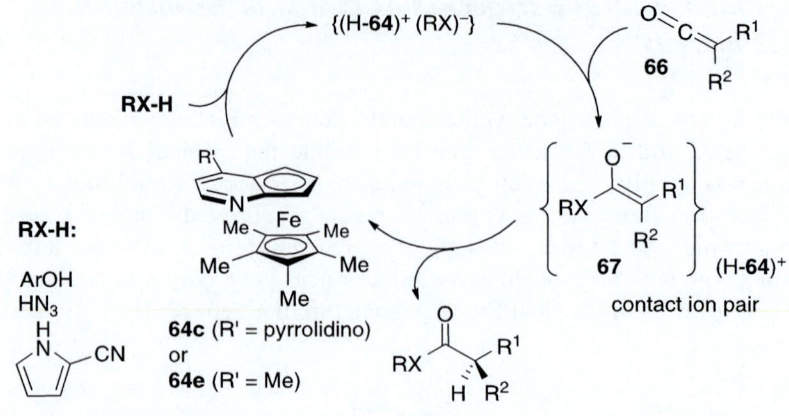

Fig. 36 Catalytic cycle for the Brønsted-base catalyzed addition of phenols, hydrazoic acid, or 2-cyanopyrrol to ketenes

To ensure that proton transfer takes place from the protonated catalyst **64-H**$^+$ and not from the acidic reagent itself, apolar solvents favoring contact rather than solvent separated ion pairs as well as a slow addition of the acidic substrate RX-H are required. In addition, it was sometimes found beneficial to lower the basicity of the catalyst, thus rendering the protonated species [catalyst-H$^+$] more acidic for the stereo-determining protonation of the enolate. This was accomplished by formally replacing NR_2 by Me (see **64e**, Fig. 36).

Employing this method, enantioenriched phenol esters **68**, amides **69**, and carbamates **70** (after Curtius rearrangement of the intermediate acyl azide) were prepared in yields often greater than 90% with *ee*-values reaching up to 97% (generally 80–95%, see Fig. 37).

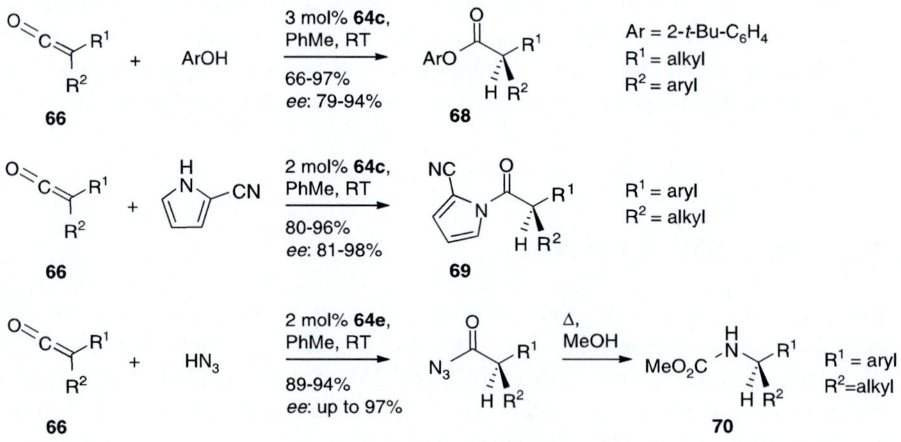

Fig. 37 Addition of phenols, 2-cyano-pyrrol, or hydrazoic acid to ketenes

3.2.2 Lewis-Base Catalysis via Intermediate Formation of a Chiral Zwitterionic Enolate

In an alternative mode of catalyst action, a disubstituted ketene **66** initially suffers a nucleophilic attack of **64**, leading to zwitterionic enolate **71** (Fig. 38).

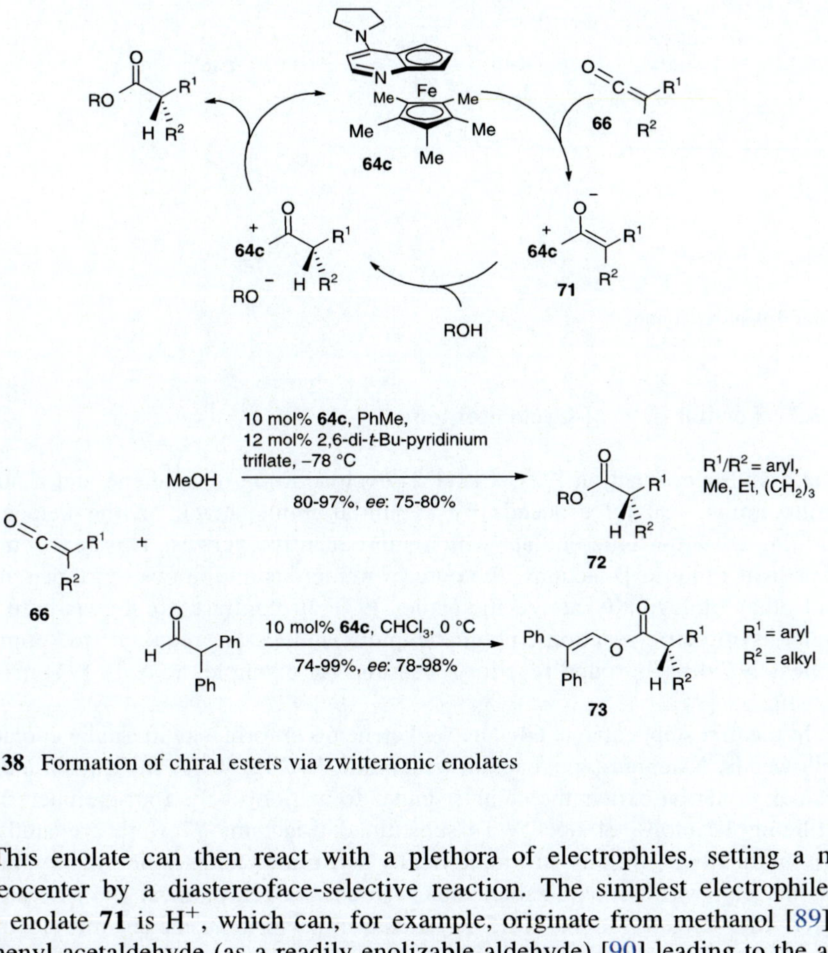

Fig. 38 Formation of chiral esters via zwitterionic enolates

This enolate can then react with a plethora of electrophiles, setting a new stereocenter by a diastereoface-selective reaction. The simplest electrophile to trap enolate **71** is H^+, which can, for example, originate from methanol [89] or diphenyl acetaldehyde (as a readily enolizable aldehyde) [90] leading to the acylated catalyst species (Fig. 38). The free catalyst is regenerated by acyl-group transfer to methanol(ate) or the aldehyde-derived enolate, producing methyl or enolesters **72/73** in good yields and enantioselectivities.

In a related reaction, enolate **71** is undergoing an electrophilic chlorination with 2,2,6,6-tetrachloro-cyclohexanone (**74**, Fig. 39), eventually leading to α-chlorinated enol esters **75** [91]. However, a different mechanism cannot be completely ruled out, where the catalyst is not acylated by the ketene, but chlorinated by the tetrachloro-ketone to form **[64c-Cl]**$^+$ as the reactive species.

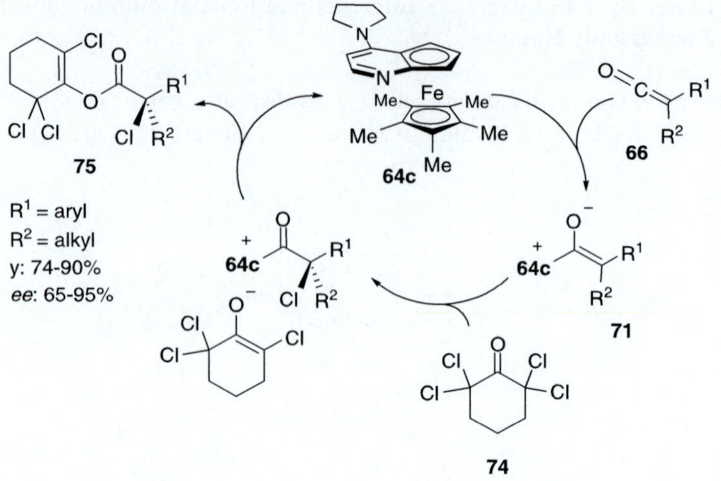

Fig. 39 Chlorination

3.2.3 Formal [2 + 2]-Cycloadditions

The Staudinger reaction [92], a [2 + 2]-cycloaddition of a ketene and a nucleophilic imine, usually proceeds by an initial imine attack on the ketene thus forming a zwitterionic enolate which subsequently cyclizes. This reaction is an expedient route to β-lactams, the core of numerous antibiotics (e.g., penicillins) and other biologically active molecules [93]. In contrast, for Lewis-base catalyzed asymmetric reactions, nonnucleophilic imines are required (to suppress a noncatalyzed background reaction), bearing, for example, an *N*-Ts [94] or -Boc-substituent [95].

In the first step, catalyst **64c** attacks ketene **66** to form a zwitterionic enolate **71**, followed by Mannich-type reaction with imine **76** (Fig. 40). A subsequent intramolecular acylation expels the catalyst under formation of the four-membered ring. Utilizing 10 mol% of **64c**, *N*-Ts substituted β-lactams **77** were prepared from symmetrically as well as unsymmetrically substituted ketenes **66**, mainly, but not exclusively, with nonenolizable imines **76** as reaction partners [96]. Diastereoselectivities ranged from 8:1 to 15:1, yields from 76 to 97%, and enantioselectivities from 81 to 94% *ee* in the case of aliphatic ketenes **66** or 89 to 98% *ee* for ketenes bearing an aromatic substituent. Applying complexes **65** or the more bulky and less electron-rich **64b**, *ee* values below 5% were obtained.

Interestingly, the mechanistic pathway seems to change if the imine is highly electrophilic (Fig. 41): *N*-trifluoromethyl sulfonyl imines **78** react quantitatively with catalyst **64c** to form zwitterion **79** which then attacks the ketene nucleophilically similar to a regular Staudinger reaction (a similar mechanism is also likely for the formation of β-sultams [97–99]). Depending on the imine, diastereoselectivities ranged from 80:20 to 98:2. Although yields (60–89%) and enantioselectivities

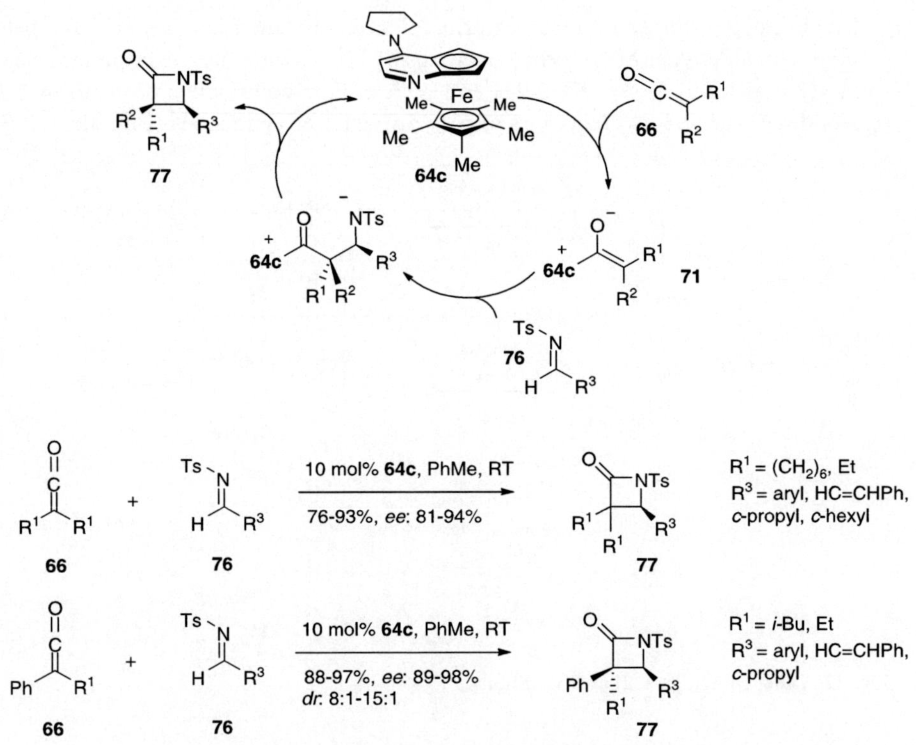

Fig. 40 Formation of β-lactams via asymmetric Staudinger-type reactions, catalyzed by **64c**

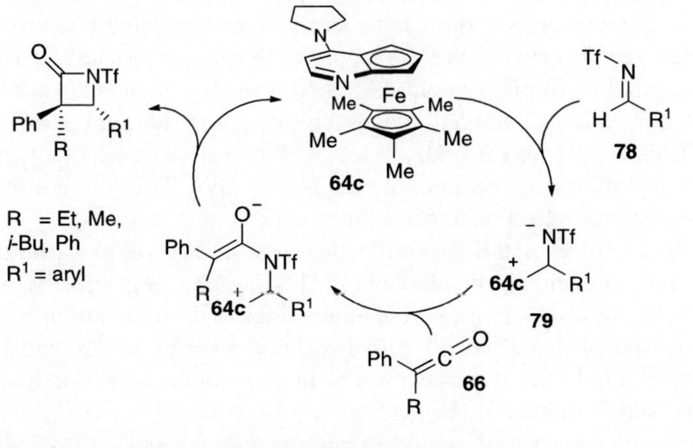

Fig. 41 Mechanistic proposal for highly electron-poor imines

(*ee* = 63–98%) were generally lower than for *N*-Ts imines **76**, both methodologies are complementary as *N*-Tf imines **78** provide, predominantly, the *trans*-diastereomers, while the *N*-Ts-counterparts **76** mainly form the *cis*-isomers.

Making use of the same reaction principle, disubstituted ketenes **66** have been reacted with aldehydes **80** to form β-lactones **81** [100], with diazo-compounds **82** to form 1,2-diazetidin-3-ones **83** [101] and with nitroso-compounds **84** to form 1,2-oxazetidin-3-ones **85** as precursors of α-hydroxy carboxylic acids (Fig. 42) [102].

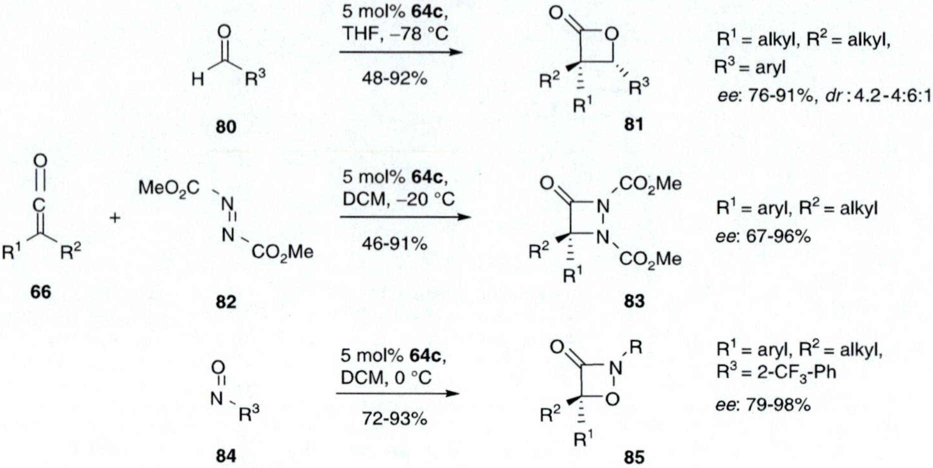

Fig. 42 Other formal [2 + 2] cycloadditions

3.2.4 Nucleophilic Catalysis: Acyl-Transfer Reactions

Enantioenriched alcohols and amines are valuable building blocks for the synthesis of bioactive compounds. While some of them are available from nature's "chiral pool", the large majority is accessible only by asymmetric synthesis or resolution of a racemic mixture. Similarly to DMAP, **64b** is readily acylated by acetic anhydride to form a positively charged planar chiral acylpyridinium species [**64b-Ac**]$^+$ (Fig. 43). The latter preferentially reacts with one enantiomer of a racemic alcohol by acyl-transfer thereby regenerating the free catalyst. For this type of reaction, the C_5Ph_5-derivatives **64b/d** have been found superior.

With 1–2 mol% of **64b**, racemic mixtures of aryl-alkyl carbinols **86** [103], propargylic [104] and allylic alcohol [105] **88** and **87**, respectively, were resolved (Fig. 43). The best selectivities were attained for aryl-alkyl-carbinols **86**, where the unreacted isomer was obtained with excellent *ee*s after ~55% conversion, while propargyl alcohols **88** required clearly higher conversions for high *ee*s in the remaining starting material [106].

Since amines react more readily than alcohols in noncatalyzed reactions with anhydrides, the reaction is more difficult and initially required stoichiometric catalyst loadings [107], but could be performed in a catalytic sense with an O-acylated azlactone as acylating agent, which does not react with a benzylic amine at $-50°C$, but is capable of acylating the catalyst [108, 109]. Depending on the bulkiness of the substrate, selectivities ranged from $s = 11$ to 27 ($s = [k_{\text{enantiomer 1}}]/[k_{\text{enantiomer 2}}]$).

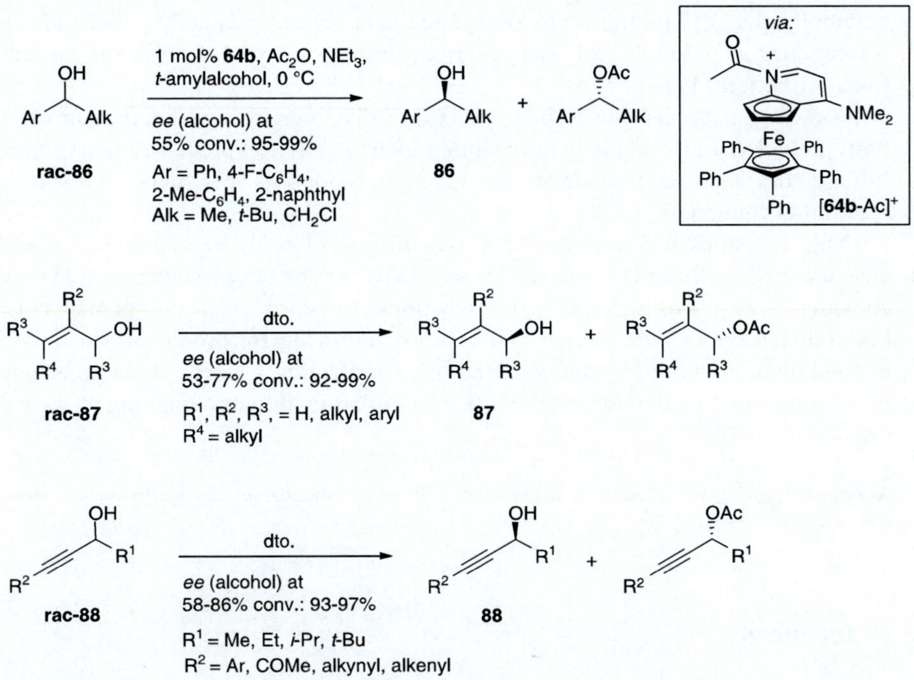

Fig. 43 Kinetic resolution via acylation of alcohols

Alternative catalytic asymmetric acylation reactions studied prochiral silyl iminoketenes **89** [110] (Fig. 44, top) and silyl ketene acetals **90** [111, 112] (Fig. 44, middle), leading to the formation of quaternary stereocenters. Furthermore, the

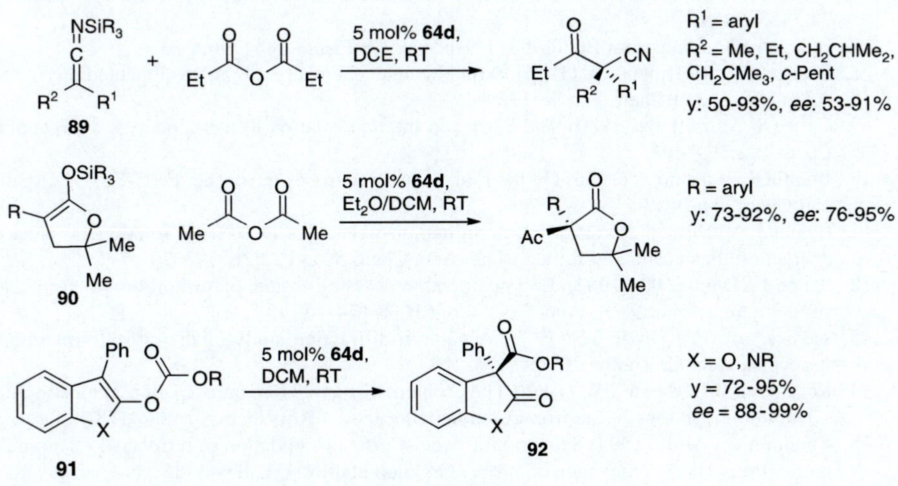

Fig. 44 Further acylation reactions catalyzed by **64d**

intramolecular acyl-migration in O-acylated azlactones [113] and O-acylated benzofuranones or oxindoles **91** was reported, leading to oxindoles **92** was reported (Fig. 44, bottom) [114].

Moreover, it is possible to open racemic azlactones by acyl bond cleavage to form protected amino acids in a dynamic kinetic resolution process. As azlactones suffer a fast racemization under the reaction conditions, eventually all starting material is converted [115].

Other reactions not described here are formal [3 + 2] cycloadditions of α,β-unsaturated acyl-fluorides with allylsilanes [116], or the desymmetrization of *meso* epoxides [117]. For many of the reactions shown above, the planar chiral Fe-sandwich complexes are the first catalysts allowing for broad substrate scope in combination with high enantioselectivities and yields. Clearly, these milestones in asymmetric Lewis-base catalysis are stimulating the still ongoing design of improved catalysts.

Acknowledgment We gratefully acknowledge financial support by F. Hoffmann-La Roche (Basel).

References

1. Miller SA, Tebboth JA, Tremaine JF (1952) Dicyclopentadienyliron. J Chem Soc 632–635
2. Kealy TJ, Pauson PL (1951) A new type of organo-iron compound. Nature 168:1039–1040
3. Fischer EO, Pfab W (1952) Cyclopentadiene-metallic complex, a new type of organo-metallic compound. Z Naturforsch B 7:377–379
4. Ruch E, Fischer EO (1952) Binding in dicyclopentadienyliron. Z Naturforsch B 7:676
5. Wilkinson G, Rosenblum M, Whiting MC, Woodward RB (1952) The structure of iron bis-cyclopentadienyl. J Am Chem Soc 74:2125–2126
6. Woodward RB, Rosenblum M, Whiting MC (1952) A new aromatic system. J Am Chem Soc 74:3458–3459
7. Barusch MR, Lindstrom EG (1958) Ferrocene. US Patent 2834796
8. Lauher JW, Hoffmann R (1976) Structure and chemistry of bis(cyclopentadienyl)-ML_n complexes. J Am Chem Soc 98:1729–1742
9. Carter S, Murrell JN (1980) The barrier to internal rotation in metallocenes. J Organomet Chem 192:399–408
10. Bohn RK, Haaland A (1966) On the molecular structure of ferrocene, $Fe(C_5H_5)_2$. J Organomet Chem 5:470–476
11. Haaland A, Nilsson JE (1968) Determination of barriers to internal rotation by electron diffraction. Ferrocene and ruthenocene. Acta Chem Scand 22:2653–2670
12. Seiler P, Dunitz JD (1982) Low-temperature crystallization of orthorhombic ferrocene: structure analysis at 98 K. Acta Crystallogr B 38:1741–1745
13. Takusagawa F, Koetzle TF (1979) A neutron diffraction study of the crystal structure of ferrocene. Acta Crystallogr B 35:1074–1081
14. Rosenblum M, Abbate FW (1966) The problem of metal atom participation in electrophilic substitution reactions of the iron group metallocenes. J Am Chem Soc 88:4178–4184
15. Cunningham AF Jr (1994) Sequential Friedel-Crafts diacetylation of ferrocene: interannular proton transfers as a mechanistic probe. Organometallics 13:2480–2485
16. Watts WE (1979) Ferrocenylcarbocations and related species. J Organomet Chem Library 7:399–459

17. Behrens U (1979) Übergangsmetall-Fulven-Komplexe: XIV. Kristall- und Molekülstruktur von Ferrocenyldiphenylcarbenium-tetrafluoroborat, $[C_5H_5FeC_5H_4C(C_6H_5)_2]^+$ BF_4^-. Ein Fulven-Eisen-Komplex? J Organomet Chem 182:89–98
18. Bourke SC, MacLachlan MJ, Lough AJ, Manners I (2005) Ring-opening protonolysis of Sila [1]ferrocenophanes as a route to stabilized silylium ions. Chem Eur J 11:1989–2000
19. Klare HFT, Bergander K, Oestreich M (2009) Taming the silylium ion for low-temperature Diels–Alder reactions. Angew Chem Int Ed 48:9077–9079
20. Schlögl K (1967) Stereochemistry of metallocenes. Top Stereochem 1:39–89
21. Cahn RS, Ingold CK, Prelog V (1966) Specification of molecular chirality. Angew Chem Int Ed Engl 5:385–415
22. Whitesides TH, Shelly J (1975) Thermolysis and photolysis of (cyclopentadiene)iron tricarbonyl. Evidence for a radical mechanism involving iron(I). J Organomet Chem 92:215–226
23. Shackleton TA, Mackie SC, Fergusson SB, Johnston LJ, Baird MC (1990) The chemistry of $(\eta^5\text{-}C_5H_5)Fe(CO)_2H$ revisited. Organometallics 9:2248–2253
24. Yasuda S, Yorimitsu H, Oshima K (2008) Synthesis of aryliron complexes by palladium-catalyzed transmetalation between $[CpFe(CO)_2I]$ and aryl Grignard reagents and their chemistry directed toward organic synthesis. Organometallics 27:4025–4027
25. Jonas K, Schieferstein L (1979) Simple route to Li- or Zn-metalated η^5-cyclopentadienyliron-olefin complexes. Angew Chem Int Ed Engl 18:549–550
26. Jonas K, Schieferstein L, Krüger C, Tsay YH (1979) Tetrakis(ethylene)irondilithium and Bis (η^4–1,5-cyc1ooctadiene) irondilithium. Angew Chem Int Ed Engl 18:550–551
27. Jonas K (1985) Reactive organometallic compounds obtained from metallocenes and related compounds and their synthetic applications. Angew Chem Int Ed Engl 24:295–311
28. Jonas K, Krüger C (1980) Alkali metal-transition metal π-complexes. Angew Chem Int Ed Engl 19:520–537
29. Jonas K (1990) New findings in the arene chemistry of the 3d transition metals. Pure Appl Chem 62:1169–1174
30. Fürstner A, Martin R, Majima K (2005) Cycloisomerization of enynes catalyzed by iron (0)−ate complexes. J Am Chem Soc 127:12236–12237
31. Fürstner A, Majima K, Martin R, Krause H, Kattnig E, Goddard R, Lehmann CW (2008) A cheap metal for a "Noble" task: preparative and mechanistic aspects of cycloisomerization and cycloaddition reactions catalyzed by low-valent iron complexes. J Am Chem Soc 130:1992–2004
32. Bonnesen PV, Puckett CL, Honeychuck RV, Hersh WH (1989) Catalysis of Diels–Alder reactions by low oxidation state transition-metal Lewis acids: fact and fiction. J Am Chem Soc 111:6070–6081
33. Kündig EP, Bourdin B, Bernardinelli G (1994) Asymmetric Diels–Alder reactions catalyzed by a chiral iron Lewis acid. Angew Chem Int Ed 33:1856–1858
34. Saha AK, Hossain MM (1993) Synthesis of a polymer-bound iron Lewis acid and its utilization in Diels–Alder reactions. Tetrahedron Lett 34:3833–3836
35. Olson AS, Seitz WJ, Hossain MM (1991) Transition metal catalysis of the Diels–Alder reaction. Tetrahedron Lett 32:5299–5302
36. Kündig EP, Dupré C, Bourdin B, Cunningham A Jr, Pons D (1994) New C_2-chiral bidentate ligands bridging the gap between donor phosphine and acceptor carbonyl ligands. Helv Chim Acta 77:421–428
37. Bruin ME, Kündig EP (1998) A new chiral ligand for the Fe-Lewis acid catalysed asymmetric Diels–Alder reaction. Chem Commun 2635–2636
38. Kündig EP, Saudan CM, Viton F (2001) Chiral cyclopentadienyl-iron and -ruthenium Lewis acids containing the electron-poor BIPHOP-F ligand: a comparison as catalysts in an asymmetric Diels–Alder reaction. Adv Synth Catal 343:51–56
39. Kündig EP, Saudan CM, Bernardinelli G (1999) A stable and recoverable chiral Ru Lewis acid: synthesis, asymmetric Diels–Alder catalysis and structure of the Lewis acid methacrolein complex. Angew Chem Int Ed 38:1220–1223

40. Viton F, Bernardinelli G, Kündig EP (2002) Iron and ruthenium Lewis acid catalyzed asymmetric 1,3-dipolar cycloaddition reactions between nitrones and enals. J Am Chem Soc 124:4968–4969
41. Bădoiu A, Brinkmann Y, Viton F, Kündig EP (2008) Asymmetric Lewis acid-catalyzed 1,3-dipolar cycloadditions. Pure Appl Chem 80:1013–1018
42. Bădoiu A, Bernardinelli G, Mareda J, Kündig EP, Viton F (2008) Iron- and ruthenium-Lewis acid catalyzed asymmetric 1,3-dipolar cycloaddition reactions between enals and diaryl nitrones. Chem Asian J 3:1298–1311
43. Bădoiu A, Bernardinelli G, Mareda J, Kündig EP, Viton F (2009) Iron- and ruthenium-Lewis acid catalyzed asymmetric 1,3-dipolar cycloaddition reactions between enals and diaryl nitrones. Chem Asian J 4:1021–1022
44. Artero V, Fontecave M (2008) Hydrogen evolution catalyzed by {$CpFe(CO)_2$}-based complexes. C R Acad Sci II 11:926–931
45. Itazaki M, Ueda K, Nakazawa H (2009) Iron-catalyzed dehydrogenative coupling of tertiary silanes. Angew Chem Int Ed 48:3313–3316
46. Sharma HK, Arias-Ugarte R, Metta-Magana AJ, Pannell KH (2009) Dehydrogenative dimerization of di-tert-butyltin dihydride photochemically and thermally catalyzed by iron and molybdenum complexes. Angew Chem Int Ed 48:6309–6312
47. Casey CP, Guan H (2009) Cyclopentadienone iron alcohol complexes: synthesis, reactivity, and implications for the mechanism of iron-catalyzed hydrogenation of aldehydes. J Am Chem Soc 131:2499–2507
48. Gómez Arrayás R, Adrio J, Carretero JC (2006) Recent applications of chiral ferrocene ligands in asymmetric catalysis. Angew Chem Int Ed 45:7674–7715
49. Dai LX, Hou XL (2010) Chiral ferrocenes in asymmetric catalysis. Wiley-VCH, Weinheim
50. Rigaut S, Delville MH, Losada J, Astruc D (2002) Water-soluble mono- and star-shaped hexanuclear functional organoiron catalysts for nitrate and nitrite reduction in water: syntheses and electroanalytical study. Inorg Chim Acta 334:225–242
51. Dupont J, Consorti CS, Spencer J (2005) The potential of palladacycles: more than just precatalysts. Chem Rev 105:2527–2572
52. Sokolov VI, Troitskaya LL, Reutov OA (1977) Asymmetric induction in the course of internal palladation of enantiomeric 1-dimethylaminoethylferrocene. J Organomet Chem 133:C28–C30
53. Kuz'mina LG, Struchkov YT, Troitskaya LL, Sokolov VI, Reutov OA (1979) Absolute configuration of the $(-)$-R,S_p enantiomer of cyclo-1-(1'-dimethylaminoethylferrocene)-2-(acetylacetonato)palladium. Izv Akad Nauk SSSR Ser Khim 7:1528–1534
54. Beletskaya IP, Cheprakov AV (2004) Palladacycles in catalysis – a critical survey. J Organomet Chem 689:4055–4082
55. Calter M, Hollis TK, Overman LE, Ziller J, Zipp GG (1997) First enantioselective catalyst for the rearrangement of allylic imidates to allylic amides. J Org Chem 62:1449–1456
56. Hollis TK, Overman LE (1997) Cyclopalladated ferrocenyl amines as enantioselective catalysts for the rearrangement of allylic imidates to allylic amides. Tetrahedron Lett 38:8837–8840
57. Cohen F, Overman LE (1998) Planar-chiral cyclopalladated ferrocenyl amines and imines as enantioselective catalysts for allylic imidate rearrangements. Tetrahedron Asymmetry 9:3213–3222
58. Donde Y, Overman LE (1999) High enantioselection in the rearrangement of allylic imidates with ferrocenyl oxazoline catalysts. J Am Chem Soc 121:2933–2934
59. Kang J, Yew KH, Kim TH, Choi DH (2002) Preparation of bis[palladacycles] and application to asymmetric *aza*-Claisen rearrangement of allylic imidates. Tetrahedron Lett 43:9509–9512
60. Moyano A, Rosol M, Moreno RM, López C, Maestro MA (2005) Oxazoline-mediated interannular cyclopalladation of ferrocene: chiral palladium(II) catalysts for the enantioselective *Aza*-Claisen rearrangement. Angew Chem Int Ed 44:1865–1869

61. Anderson CE, Donde Y, Douglas CJ, Overman LE (2005) Catalytic asymmetric synthesis of chiral allylic amines. Evaluation of ferrocenyloxazoline palladacycle catalysts and imidate motifs. J Org Chem 70:648–657
62. Kang J, Kim TH, Yew KH, Lee WK (2003) The effect of face-blocking in the enantioselective *aza*-Claisen rearrangement of allylic imidates. Tetrahedron Asymmetry 14:415–418
63. Overman LE, Owen CE, Pavan MM, Richards CJ (2003) Catalytic asymmetric rearrangement of allylic N-aryl trifluoroacetimidates. A useful method for transforming prochiral allylic alcohols to chiral allylic amines. Org Lett 5:1809–1812
64. Anderson CE, Overman LE (2003) Catalytic asymmetric rearrangement of allylic trichloroacetimidates. A practical method for preparing allylic amines and congeners of high enantiomeric purity. J Am Chem Soc 125:12412–12413
65. Nomura H, Richards CJ (2007) An investigation into the allylic imidate rearrangement of trichloroacetimidates catalyzed by cobalt oxazoline palladacycles. Chem Eur J 13:10216–10224
66. Prasad RS, Anderson CE, Richards CJ, Overman LE (2005) Synthesis of *tert*-leucine-derived cobalt oxazoline palladacycles. Reversal of palladation diastereoselectivity and application to the asymmetric rearrangement of N-aryl trifluoroacetimidates. Organometallics 24:77–81
67. Noviandri I, Brown KN, Fleming DS, Gulyas PT, Lay PA, Masters AF, Phillips L (1999) The decamethylferrocenium/decamethylferrocene redox couple: a superior redox standard to the ferrocenium/ferrocene redox couple for studying solvent effects on the thermodynamics of electron transfer. J Phys Chem B 103:6713–6722
68. Peters R, Fischer DF (2005) Preparation and diastereoselective *ortho*-metalation of chiral ferrocenyl imidazolines: remarkable influence of LDA as metalation additive. Org Lett 7:4137–4140
69. Peters R, Xin ZQ, Fischer DF, Schweizer WB (2006) Synthesis and Diastereoselective Ortho-Lithiation/Cyclopalladation of Enantiopure [2-Imidazolyl]-1′,2′,3′,4′,5′-pentamethylferrocenes and -1′, 2′, 3′, 4′, 5′-pentaphenylferrocenes. Organometallics 25:2917–2920
70. Weiss ME, Fischer DF, Xin ZQ, Jautze S, Schweizer WB, Peters R (2006) Practical, highly active, and enantioselective ferrocenyl-imidazoline palladacycle catalysts (FIPs) for the *Aza*-Claisen rearrangement of *N-para*-methoxyphenyl trifluoroacetimidates. Angew Chem Int Ed 45:5694–5698
71. Fischer DF, Xin ZQ, Peters R (2007) Asymmetric formation of allylic amines with *N*-substituted quaternary stereocenters by Pd^{II}-catalyzed *Aza*-Claisen rearrangements. Angew Chem Int Ed 46:7704–7707
72. Xin ZQ, Fischer DF, Peters R (2008) Catalytic asymmetric formation of secondary allylic amines by *Aza*-Claisen rearrangement of trifluoroacetimidates. Synlett 1495–1499
73. Fischer DF, Barakat A, Xin ZQ, Weiss ME, Peters R (2009) The asymmetric *Aza*-Claisen rearrangement: development of widely applicable pentaphenylferrocenyl palladacycle catalysts. Chem Eur J 15:8722–8741
74. Jautze S, Seiler P, Peters R (2007) Macrocyclic ferrocenyl-bisimidazoline palladacycle dimers as highly active and enantioselective catalysts for the *Aza*-Claisen rearrangement of *Z*-configured *N-para*-methoxyphenyl trifluoroacetimidates. Angew Chem Int Ed 46: 1260–1264
75. Jautze S, Seiler P, Peters R (2008) Synthesis of nearly enantiopure allylic amines by *Aza*-Claisen rearrangement of *Z*-configured allylic trifluoroacetimidates catalyzed by highly active ferrocenylbispalladacycles. Chem Eur J 14:1430–1444
76. Jautze S, Diethelm S, Frey W, Peters R (2009) Diastereoselective bis-cyclopalladation of ferrocene-1,1′-diyl bis-imidazolines: translation of central via axial into planar chirality. Organometallics 28:2001–2004
77. Jautze S, Peters R (2008) Enantioselective bimetallic catalysis of Michael additions forming quaternary stereocenters. Angew Chem Int Ed 47:9284–9288
78. Jautze S, Peters R (2010) Catalytic asymmetric Michael additions of α-cyano acetates. Synthesis 365–388

79. Fürstner A, Davies PW (2007) Catalytic carbophilic activation: catalysis by platinum and gold π-acids. Angew Chem Int Ed 46:3410–3449
80. Chianese AR, Lee SJ, Gagné MR (2007) Electrophilic activation of alkenes by platinum(II): so much more than a slow version of palladium(II). Angew Chem Int Ed 46:4042–4059
81. Huang H, Peters R (2009) A highly strained planar chiral platinacycle for catalytic activation of internal olefins in the Friedel-Crafts alkylation of indoles. Angew Chem Int Ed 48:604–606
82. Ruble JG, Fu GC (1996) Chiral π-complexes of heterocycles with transition metals: a versatile new family of nucleophilic catalysts. J Org Chem 61:7230–7231
83. Wurz RP, Lee EC, Ruble JC, Fu GC (2007) Synthesis and resolution of planar-chiral derivatives of 4-(dimethylamino)pyridine. Adv Synth Catal 349:2345–2352
84. Nguyen HV, Butler DCD, Richards CJ (2006) A metallocene-pyrrolidinopyridine nucleophilic catalyst for asymmetric synthesis. Org Lett 8:769–772
85. Fu GC (2006) Application of planar-chiral heterocycles as ligands in asymmetric catalysis. Acc Chem Res 39:853–860
86. Wiskur SL, Fu GC (2005) Catalytic asymmetric synthesis of esters from ketenes. J Am Chem Soc 127:6176–6177
87. Dai X, Nikkei T, Romero JAC, Fu GC (2007) Enantioselective synthesis of protected amines by the catalytic asymmetric addition of hydrologic acid to ketenes. Angew Chem Int Ed 46:4367–4369
88. Hodous BL, Fu GC (2002) Enantioselective addition of amines to ketenes catalyzed by a planar-chiral derivative of PPY: possible intervention of chiral Brönsted-acid catalysis. J Am Chem Soc 124:10006–10007
89. Hodous BL, Ruble JC, Fu GC (1999) Enantioselective addition of alcohols to ketenes catalyzed by a planar-chiral azaferrocene: catalytic asymmetric synthesis of arylpropionic acids. J Am Chem Soc 121:2637–2638
90. Schäfer C, Fu GC (2005) Catalytic asymmetric couplings of ketenes with aldehydes to generate enol esters. Angew Chem Int Ed 44:4606–4608
91. Lee EC, McCauley KM, Fu GC (2007) Catalytic asymmetric synthesis of tertiary alkyl chlorides. Angew Chem Int Ed 46:977–979
92. Staudinger H (1907) Zur Kenntnis der Ketene. Diphenylketen. Liebigs Ann Chem 356:51–123
93. Page MI (ed) (1997) The chemistry of β-lactams. Chapman and Hall, London
94. Taggi AE, Hafez AM, Wack H, Young B, Ferraris D, Lectka T (2002) The development of the first catalyzed reaction of ketenes and imines: catalytic, asymmetric synthesis of β-lactams. J Am Chem Soc 124:6626–6635
95. Zhang YR, He L, Wu X, Shao PL, Ye S (2008) Chiral N-heterocyclic carbene catalyzed Staudinger reaction of ketenes with imines: highly enantioselective synthesis of *N*-Boc β-lactams. Org Lett 10:277–280
96. Hodous BL, Fu GC (2002) Enantioselective Staudinger synthesis of β-lactams catalyzed by a planar-chiral nucleophile. J Am Chem Soc 124:1578–1579
97. Lee EC, Hodous BL, Bergin E, Shih C, Fu GC (2005) Catalytic asymmetric Staudinger reactions to form β-lactams: an unanticipated dependence of diastereoselectivity on the choice of the nitrogen substituent. J Am Chem Soc 127:11586–11587
98. Zajac M, Peters R (2009) Catalytic asymmetric synthesis of β-Sultams as precursors for taurine derivatives. Chem Eur J 15:8204–8222
99. Zajac M, Peters R (2007) Catalytic asymmetric formation of β-Sultams. Org Lett 9:2007–2010
100. Wilson JE, Fu GC (2004) Asymmetric synthesis of highly substituted β-lactones by nucleophile-catalyzed [2 + 2] cycloadditions of disubstituted ketenes with aldehydes. Angew Chem Int Ed 43:6358–6360
101. Berlin JM, Fu GC (2008) Enantioselective nucleophilic catalysis: the synthesis of Aza-β-lactams through [2 + 2] cycloadditions of ketenes with azo compounds. Angew Chem Int Ed 120:7156–7158

102. Dochnahl M, Fu GC (2009) Catalytic asymmetric cycloaddition of ketenes and nitroso compounds: enantioselective synthesis of α-hydroxycarboxylic acid derivatives. Angew Chem Int Ed 48:2391–2393
103. Ruble JC, Tweddell J, Fu GC (1998) Kinetic resolution of arylalkylcarbinols catalyzed by a planar-chiral derivative of DMAP: a new benchmark for nonenzymatic acylation. J Org Chem 63:2794–2795
104. Tao B, Ruble JC, Hoic DA, Fu GC (1999) Nonenzymatic kinetic resolution of propargylic alcohols by a planar-chiral DMAP Derivative: crystallographic characterization of the acylated catalyst. J Am Chem Soc 121:5091–5092
105. Bellemin-Laponnaz S, Twedel J, Ruble JC, Breitling FM, Fu GC (2000) The kinetic resolution of allylic alcohols by a non-enzymatic acylation catalyst; application to natural product synthesis. Chem Commun 1009–1010
106. Birman VB, Li X (2008) Homobenzotetramisole: an effective catalyst for kinetic resolution of aryl-cycloalkanols. Org Lett 10:1115–1118
107. Ie Y, Fu GC (2000) A new benchmark for the non-enzymatic enantioselective acylation of amines: use of a planar-chiral drivative of 4-pyrrolidinopyridine as the acylating agent. Chem Commun 119–120
108. Arai S, Bellemin-Laponaz S, Fu GC (2001) Kinetic resolution of amines by a nonenzymatic acylation catalyst. Angew Chem Int Ed 40:234–236
109. Arp FO, Fu GC (2006) Kinetic resolutions of indolines by a nonenzymatic acylation catalyst. J Am Chem Soc 128:14264–14265
110. Mermerian AH, Fu GC (2005) Nucleophile-catalyzed asymmetric acylations of silyl ketene imines: application to the enantioselective synthesis of verapamil. Angew Chem Int Ed 44:949–952
111. Mermerian AH, Fu GC (2005) Catalytic enantioselective construction of all-carbon quaternary stereocenters: synthesis and mechanistic studies of the C-acylation of silyl ketene acetals. J Am Chem Soc 127:5604–5607
112. Mermerian AH, Fu GC (2003) Catalytic enantioselective synthesis of quaternary stereocenters via intermolecular C-acylation of silyl ketene acetals: dual activation of the electrophile and the nucleophile. J Am Chem Soc 125:4050–4051
113. Ruble JC, Fu GC (1998) Enantioselective construction of quaternary stereocenters: rearrangements of O-acylated azlactones catalyzed by a planar-chiral derivative of 4-(pyrrolidino) pyridine. J Am Chem Soc 120:11532–11533
114. Hills ID, Fu GC (2003) Catalytic enantioselective synthesis of oxindoles and benzofuranones that bear a quaternary stereocenter. Angew Chem Int Ed 42:3921–3924
115. Liang J, Ruble JC, Fu GC (1998) Dynamic kinetic resolutions catalyzed by a planar-chiral derivative of DMAP: enantioselective synthesis of protected α-amino acids from racemic azlactones. J Org Chem 63:3154–3155
116. Bappert E, Mueller P, Fu GC (2006) Asymmetric [3 + 2] annulations catalyzed by a planar-chiral derivative of DMAP. Chem Commun 2604–2606
117. Tao B, Lo MMC, Fu GC (2001) Planar-chiral pyridine N-oxides, a new family of asymmetric catalysts: exploiting an η^5-C_5Ar_5 ligand to achieve high enantioselectivity. J Am Chem Soc 123:353–354

Top Organomet Chem (2011) 33: 177–213
DOI: 10.1007/978-3-642-14670-1_6

Catalysis by Means of Complex Ferrates

Markus Jegelka and Bernd Plietker

Abstract Ferrates represent a class of complex iron-based anions that has a long-standing tradition in chemistry. A variety of different ferrates are accessible with their metal centers in oxidation states ranging from +III up to −II. Although they are structurally well characterized, it was only very recently that their potential as catalysts in organic reaction was explored. This chapter provides a first insight into the organometallic basics of this class of compounds, and subsequently highlights the most interesting applications of different ferrates in catalysis.

Keywords Allylic substitution · Catalysis · Cross-coupling · Cycloaddition · Cycloisomerisation · DNIC · Ferrate · Hydrogenase · Iron

Contents

M. Jegelka and B. Plietker (✉)
Institute of Organic Chemistry, University of Stuttgart, Pfaffenwaldring 55, 70569 Stuttgart, Germany
e-mail: Markus.Jegelka@oc.uni-stuttgart.de, Bernd.Plietker@oc.uni-stuttgart.de

Abbreviations

Ac	Acetyl
acac	Acetylacetonate
act	Activating group
Alk	Alkyl
Ar	Aryl
bmim	Butylmethylimidazolium
cat	Catalytic
cod	Cyclooctadiene
Cp	Cyclopentadienyl
dbm	Dibenzylmethide
DCM	Dichloromethane
DME	1,2-Dimethoxyethane
DMF	*N,N*-dimethylformamide
DMM	Dimethylmalonate
DMPU	1,3-Dimethyl-3,4,5,6-tetrahydro-2(1H)-pyrimidinone
DNIC	Dinitrosyliron complex
dppe	Bis(diphenylphosphino)ethane
dppv	*cis*-1,2-Bis(diphenylphosphino)ethene
dr	Diastereomeric ratio
ECCE	Electrochemical-chemical-chemical-electrochemical
EECC	Electrochemical-electrochemical-chemical-chemical
en	1,2-Ethylenediamine
equiv	Equivalent(s)
Et	Ethyl
ET	Electron transfer
h	Hour(s)
HMPA	Hexamethylphosphoric triamide
i-Bu	*iso*-Butyl
i-Pr	Isopropyl
LG	Leaving group
LMW-DNIC	Low molecular weight dinitrosyliron complex

Me	Methyl
Mes	Mesityl, 2,4,6-trimethylphenyl
min	Minute(s)
mol	Mole(s)
MTBE	*tert*-Butylmethylether
NHC	*N*-heterocyclic carbene
Nu	Nucleophile
Ph	Phenyl
Pin	Pinacol
pip	Piperidinium
PTA	Phosphatriazaadamantane
rt	Room temperature
TBAF	Tetrabutylammonium fluoride
THF	Tetrahydrofuran
tmeda	*N,N,N′,N′*-tetramethyl-1,2-ethylenediamine
TMS	Trimethylsilyl
TsOH	Toluenesulfonic acid
Unk	Unknown ligand

1 Introduction

The chemistry of iron is closely connected to the chemistry of its corresponding complex ions, the ferrates. As Lewis acids, iron salts tend to form complex anions, for example, by the reaction of NaCl and $FeCl_3$ to form the corresponding ferrate salt $NaFeCl_4$. These complexes are known to act as single-electron acceptors. A variety of complexes are accessible by simple anion exchange. Depending on the Fe-source employed the metal center occupies oxidation states of +3 or +2. The spin ground states of the complexes however are strongly dependent on the ligand field (Fig. 1). Hence, the ligand field energy 10 Dq for different coordination modes correlates with the specific binding properties of a ligand (expressed through the f-factor) and the specific perturbation factor of the metal center (expressed through the g-factor).

Hence, ferrates with weak ligand field binders (e.g., halides, pseudohalides) occupy the spin ground state of 5/2 [for Fe(+3)], 4/2 [for Fe(+2)], 2/2 [for Fe(0)], and 0/2 [for Fe(−2)] and represent paramagnetic high-spin complexes. The configuration of these iron-complexes is – depending on the size of the corresponding anion – either octahedral or – for sterically more demanding ligands – tetrahedral, for example, $[FeCl_4]^-$ **1** versus $[Fe(CN)_6]^{3-}$. Changing the nature of the ligand to stronger ligands, for example, alkyl, aryl, thiolate, and amines, leads to the formation of diamagnetic low-spin complexes (Fig. 1).

While Fe(+3)- and Fe(+2)-ate complexes are formed by the coordination of four anionic ligands, more electron-rich metal centers tend to bind neutral or even

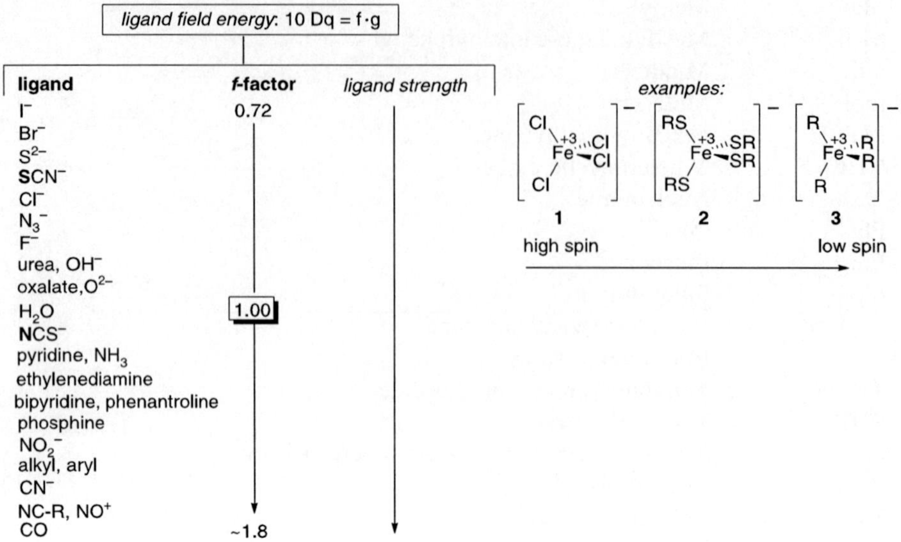

Fig. 1 Influence of the ligand field on the spin ground state of Fe(+3)-ate complexes (R = alkyl, aryl) [1]

Table 1 Complex tetrahedral ferrates used in catalysis

Complex type[a]	Fe oxidation state[b, c]			
	+3	+2	0	−2
FeA_4	+	+	+	+
FeA_3N	−	+	+	+
FeA_3C	−	−	+	+
FeA_2N_2	−	−	+	+
$FeA_2N_1C_1$	−	−	+	+
FeA_2C_2	−	−	−	−
FeA_1N_3	−	−	+	+
$FeA_1N_2C_1$	−	−	−	+
$FeA_1N_1C_2$	−	−	−	+
FeA_1C_3	−	−	−	−
FeN_4	−	−	−	+
FeN_3C_1	−	−	−	+
FeN_2C_2	−	−	−	−
FeN_1C_3	−	−	−	−
FeC_4	−	−	−	−

[a]Ligand A: anionic (e.g., halides, pseudohalides, alkyl, aryl, thiolate, alkoxide); ligand N: neutral (e.g., amines, imines, phosphines, carbenes, nitriles, isonitriles, NO, CO); ligand C: cationic (e.g., NO^+)
[b]Ferrates that have been used in catalysis are highlighted in grey
[c]Ferrates that have not been used in catalysis so far are highlighted in black

cationic ligands that are able to decrease the electron density of the iron due to a strong back bonding (e.g., NO and CO). A variety of different mono nuclear ferrates are referred to in the literature (Table 1). However, this chapter focuses only on catalytic applications involving mononuclear complex anions of the general composition listed in Table 1. Furthermore, electron-rich, low-valent but neutral Fe-complexes, for example, $(Ph_3P)_2Fe(NO)_2$ or $(RS)_2Fe(NO)_2$, are beyond the scope of this chapter. Despite the recent interest in these complexes as NO-delivery compounds for pharmaceutical application, no catalytic transformation of these complexes has been published so far, to the best of our knowledge.

As already pointed out above, the interaction of ligand field and electronic properties of the ligand are key issues for the overall chemical reactivity of the resulting ferrates. In certain cases, the strong metal–ligand bond allows for an efficient electron transfer from the ligand to the metal and vice versa. The use of noninnocent ligands allows for an intermediate stabilization of the metal center in a formally unusual oxidation state which complicates the use of the simplified oxidation state models for the explanation of the chemical properties of the metal complexes. Non-innocent redox-active ligands are found in nature, for example, in [FeFe]-hydrogenases **4** of the general structure shown in Scheme 1. Although the exact mechanism of electron transfer and spin ground state are still under debate, there is strong evidence that in these complexes the Fe-center proximal to the cysteine residue acts as an electron acceptor (eq. 1 in Scheme 1). After reduction of the Fe(+2) to Fe(0) the μ-sulfur ligands act as electron shuffles to give the reactive species in which the distal Fe-center is reduced to Fe(0). All further chemical transformations are performed at this metal center. The concept of using noninnocent ligands in order to improve the catalytic activity by stabilization of the reactive metal center was successfully applied by the group of Chirik. Mössbauer spectroscopic investigations indicated the active species to be an Fe(+2)-diradical complex **5b** rather than the isoelectronic Fe(0)-diimine species **5a**. Hence, the redox

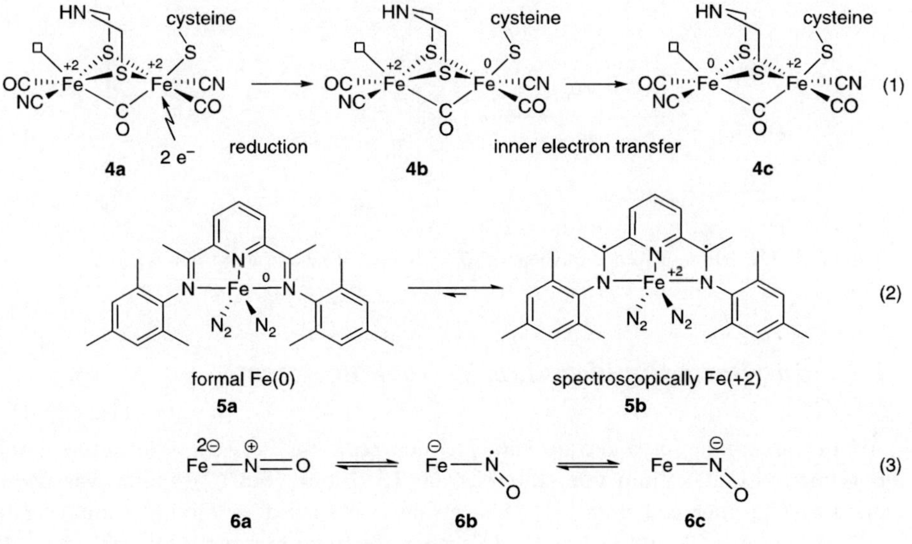

Scheme 1 Non-innocent ligands in Fe-complexes

chemistry of the complexes does not necessarily happen at the metal center; moreover, the metal center acts only as the point of electron transfer between ligand and substrate (eq. 2 in Scheme 1). The group of non-innocent ligands (thiolates as anionic ligand, imines as neutral ligands) is complemented by the cationic nitrosyl ligand. Depending on the character of this ligand a cationic, linear bound species **6a** as well as a neutral, bend **6b** or an anionic bend **6c** conformation was observed. Hence, this ligand can accept or donate up to two electrons (eq. 3 in Scheme 1).

As already pointed out above, the following chapter will concentrate on the most recent advances in catalysis using complex ferrates. It is divided by the type of ferrate employed in order to demonstrate common reaction profiles.

2 Catalysis with Fe(+3)-Ate Complexes

2.1 *Potassium Hexacyanoferrate*

Until now examples for catalytic reactions involving ferrates with iron in the oxidation state of +3 are very rare. One example is the hexacyanoferrate **8**-catalyzed oxidation of trimethoxybenzenes **7** to dimethoxy-*p*-benzoquinones **9**/**10** by means of hydrogen peroxide which was published by Matsumoto and Kobayashi in 1985 [2]. Using hexacyanoferrate **8** product **9** was favored while other catalysts like $Fe(acac)_3$ or $Fe_2(SO_4)_3$ favored product **10** (Scheme 2). The oxidation is supposed to proceed via the corresponding phenols which are formed by the attack of OH radicals generated in the Fe/H_2O_2 system.

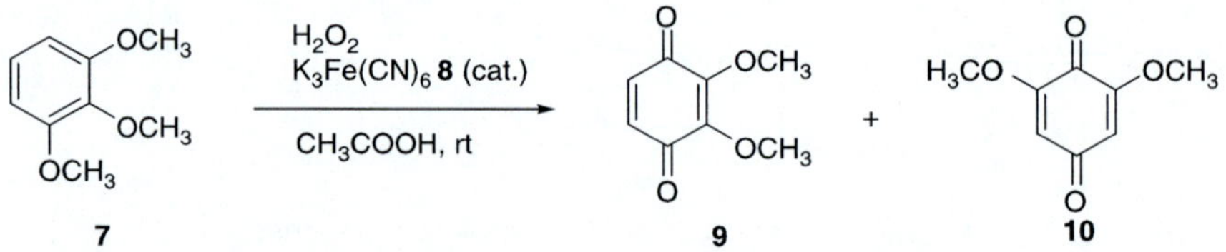

Scheme 2 Oxidation of trimethoxybenzene **7** by means of hexacyanoferrate **8** [2]

2.2 *Butylmethylimidazolium Tetrachloroferrate*

Another example for a ferrate catalyst incorporating Fe(+3) is the ionic liquid butylmethylimidazolium tetrachloroferrate **13** (bmim-[$FeCl_4$]) which was developed by Gaertner and Bica [3]. This ferrate is prepared by mixing commercially available bmim-Cl and $FeCl_3 \cdot 6H_2O$ with subsequent removal of water. The ionic liquid was shown to be capable of catalyzing the cross-coupling of aryl Grignard

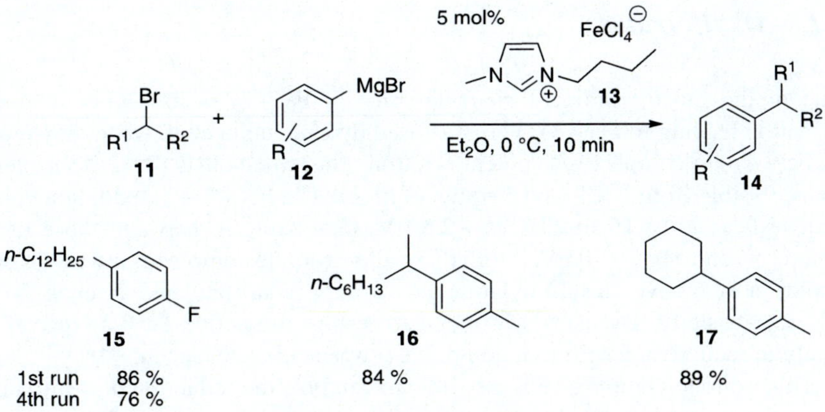

Scheme 3 Cross-coupling with Gaertner's ionic liquid ferrate catalyst **13** [3]

reagents **12** with alkyl halides **11** while applications as catalyst for Friedel–Crafts acylation were known before [4]. Various primary and secondary alkyl halides **11** bearing β-hydrogens were coupled with aryl magnesium reagents **12** in good yields (Scheme 3). The catalyst itself proved to be air and moisture stable and could be recycled several times. The reaction showed a preference for electron-poor aromatic Grignard reagents, while attempts to run the reaction with tertiary alkyl halides or phenyllithium either failed or led to drastically reduced yields. Considering the oxidation state of iron in the ionic liquid, it should be noted that the ionic liquid presumably serves only as a precatalyst, while the catalytically active species have to be generated through in situ reduction, leading to a low-valent iron species.

2.3 Alkylferrates

Although the formation of tetraalkylferrates, for example, in the first step of a cross-coupling reaction starting from Fe(+3)-salts and alkyl Grignard reagents, seems possible, until now there is no analytical or spectroscopic proof for such a species. In fact, it was shown that Fe(+3) is reduced by MeLi or MeMgBr to Fe(+2) under release of ethane [5, 6].

3 Catalysis with Fe(+2)-Ate Complexes

In contrast to Fe(+3)-ate complexes a variety of examples for catalytic reactions using Fe(+2)-ate complexes are known in literature.

3.1 Alkylferrates

One of the most prominent characteristics of Fe(+2) is its ability to undergo oxidation leading to Fe(+3). This was used by Uchiyama et al. when they reported on Fe(+2)-ate complexes as potent electron transfer catalysts [7, 8]. These ferrates are accessible from $FeCl_2$ and 3 equiv. of MeLi. The Fe(+2/+3) oxidation potential of $[Me_3Fe(+2)]Li$ **19** in THF is −2.50 V, thus being in between those of SmI_2 (−2.33 V) and Mg (−3.05 V). With these alkyliron-ate complexes it was possible to realize a reductive desulfonylation of various *N*-sulfonylated amines **20** with different basicity. By using Mg metal to restore the active Fe(+2) species **19** a catalytic reductive desulfonylation process was achieved (Scheme 4).

This iron-ate complex **19** is also able to catalyze the reduction of 4-nitroanisole to 4-methoxyaniline or Ullmann-type biaryl couplings of bis(2-bromophenyl) methylamines **31** at room temperature. In contrast, the corresponding bis(2-chlorophenyl)methylamines proved to be unreactive under these conditions. A shift to the dianion-type electron transfer(ET)-reagent $[Me_4Fe]Li_2$ afforded the biaryl as well with the dichloro substrates at room temperature, while the dibromo substrates proved to be reactive even at −78°C under these reaction conditions. This effect is attributed to the more negative oxidation potential of dianion-type $[Me_4Fe]Li_2$.

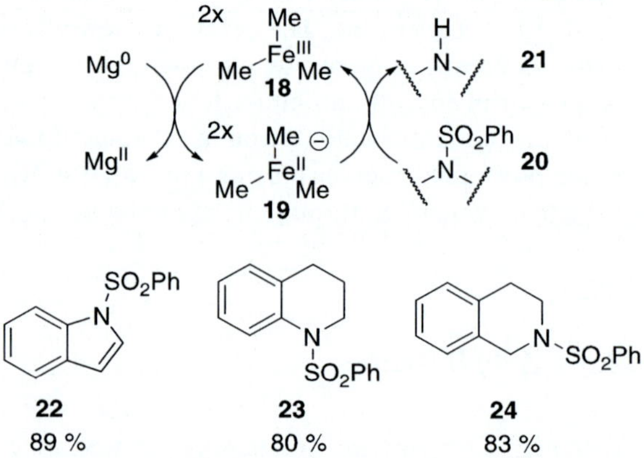

Scheme 4 Catalytic deprotection of *N*-phenylsulfonyl amines **20** using alkyliron-ate complexes [7]

Another possibility of fine-tuning the oxidation/reduction potential of the catalysts is the variation of ligands. The authors showed that the pinacol coupling of benzaldehyde using $[Me_3Fe]Li$ **19** was not possible due to side reactions like overreduction or ligand transfer, which is attributed to the too high (−2.50 V) oxidation potential compared to the reduction potential of benzaldehyde (−1.73 V). Thus, the pinacol coupling of various carbonyl compounds **25** could be achieved by lowering the electron-releasing ability of the complex, for example, by using $[(t\text{-BuO})_3Fe]K$ **26** as catalyst (−1.86 V) (Scheme 5).

As shown above, iron salts are able to catalyze cross-coupling reactions. While most of the known procedures nowadays start with commercially available iron

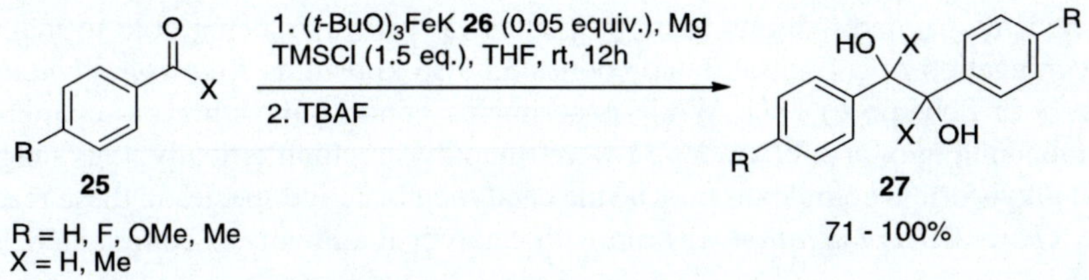

Scheme 5 Catalytic pinacol coupling reaction using [(*t*-BuO)$_3$Fe]K **26** [7]

salts like $FeCl_3$ or Fe(acac)$_3$, little has been known about the catalytically active species. While studying the mechanism of cross-coupling reactions, Fürstner et al. managed to isolate the tetraalkylferrate complex [(Me$_4$Fe)(MeLi)][Li(OEt$_2$)]$_2$ **28** from the reaction of $FeCl_3$ with excess MeLi at low temperatures (Scheme 6) [9, 10]. Thus, MeLi is able to reduce Fe(+3) to only Fe(+2), but no further reduction is possible because the methyl group is devoid of any β-hydrogen, which could otherwise be eliminated. This complex **28** is an exceptionally sensitive red solid, which vigorously ignites in air and rapidly decomposes above 0°C. Despite this sensitivity, an X-ray structure of this complex was obtained. While **28** is not able to induce methylation of chlorobenzoate or iodobenzoate to any appreciable extent, it reacts with more activated substrates such as acid chlorides **29** or enol triflates in good yields (eq. 1 in Scheme 6). Besides its nucleophilic properties, complex **28** can also act as a single-electron-transfer reagent, which was shown by the Ullmann-type coupling of aryl-bromides **31** (cf. reaction of [Me$_4$Fe]Li$_2$ by Uchiyama et al., eq. 2

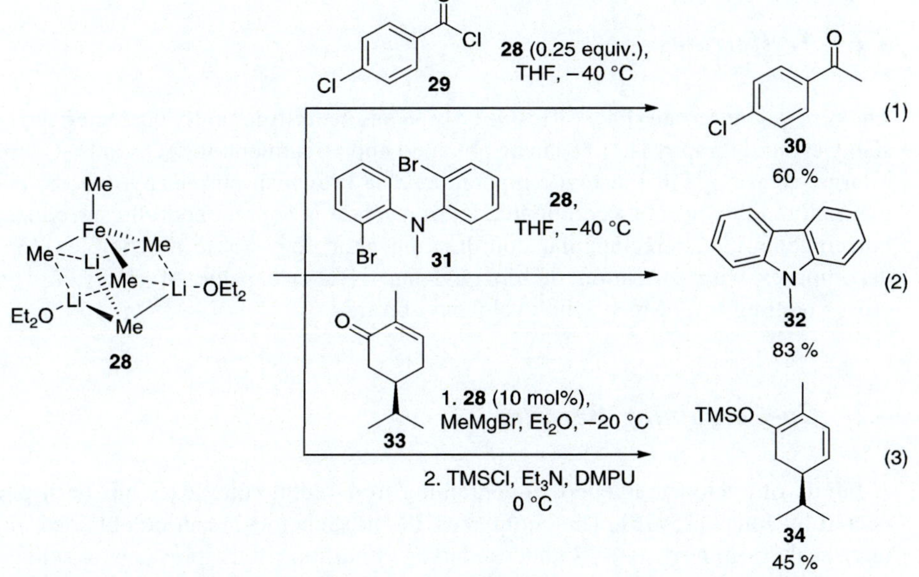

Scheme 6 Structure and reactivity of tetramethylferrate-complex **28** [9]

in Scheme 6), and exhibits basic properties as well, thus being able to induce the deconjugation of α,β-unsaturated enones **33** also known as *Kharash deconjugation* (eq. 3 in Scheme 6) [11]. While experiments concerned with cross-coupling and homocoupling of aryl-bromides **31** were run only stoichiometrically, thus suggesting that alkyl-ferrate complexes may be the catalytically active species in these reactions, the *Kharash deconjugation* was run with catalytical amounts of complex **28**.

3.2 Arylferrates

Similar to the tetramethylferrate **28**, Fürstner et al. were successful in isolating a tetraphenylferrate(+2), which was synthesized from $FeCl_2$ and 4 equiv. of PhLi and analyzed as $[Ph_4Fe][Li(Et_2O)_2][Li(1,4\text{-dioxane})]$ **35** [9]. This ferrate is a mustard-colored, highly pyrophoric solid, which decomposes with the formation of biphenyl. Biphenyl can also be generated by treating **35** with an oxidant such as iodine, 1,2-dichloroethane, or dry air. This observation, which is also a known problem in the cross-coupling chemistry of PhMgX with less activated substrates, and the fact that ferrate **35** catalyzes the cross-coupling of PhMgBr with cyclohexyl bromide appear to hint that such a ferrate **35** may be passed through in cross-coupling chemistry.

4 Catalysis with Fe(0)-Ate Complexes

4.1 Arylferrates

The zero valent ferrate $[Ph_4Fe][Li(OEt_2)]_4$ **36** has been structurally characterized by X-ray crystallography [12]. It can be obtained upon treatment of $FeCl_3$ in Et_2O with a large excess of PhLi at low temperature as a brownish-purple, pyrophoric and thermally highly unstable compound. Its iron center is not tetrahedrally surrounded but contains planar rectangular coordination geometry. Same as complex **35** it decomposes with formation of biphenyl and is also capable of catalyzing the cross-coupling of PhMgBr with cyclohexyl bromide.

4.2 Cp-Containing Ferrates

A family of cyclopentadiene(Cp)-containing iron–olefin complexes has been pioneered by Jonas [13–15]. The complexes **38**–**40** (Scheme 7) can be obtained in a large scale from ferrocene **37** under reducing conditions in the presence of suitable coordinating olefins. Complex **38** is a highly air-sensitive, crystalline material, whereas complexes **39** and **40** are more robust due to their cyclooctadiene (cod)

Scheme 7 Cp-iron(0)-ate complexes (**a**) Li, ethylene (1 bar), THF, −50°C → 0°C, then tmeda; 45% (23 g scale), (**b**) (1) Li, cod, THF, −50°C, (2) tmeda, Et_2O, rt, 75%, (**c**) Li, cod, DME, −50°C → RT, 97% (85 g scale) [9]

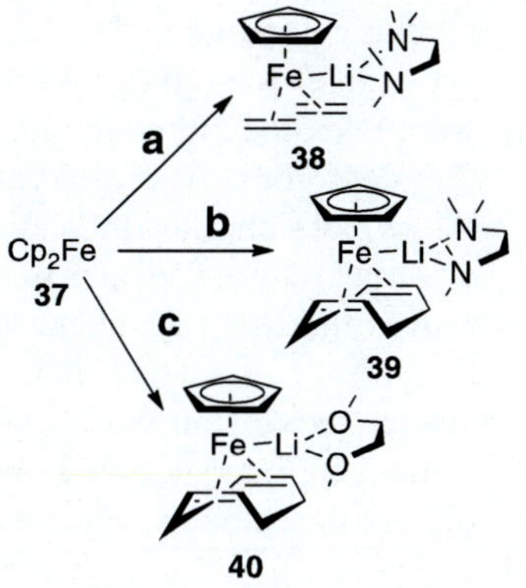

ligand. Nevertheless, the electron-rich iron centers are stabilized through substantial back-bonding from the metal into the π^* orbital of the alkenes, which can be seen from X-ray structures in a considerable elongation of the double bonds of the alkenes.

Fürstner et al. showed that complex **38** can be used as an effective catalyst in cross-coupling reactions, even in the more challenging coupling of cyclohexyl bromide with PhMgBr [9]. **38** exhibits a nucleophilic character, as it was shown to insert into allylic halides, coordinating the new allyl ligand in a π-coordinating fashion. Furthermore, it is even capable of undergoing oxidative addition of chlorobenzene, which Pd(0)-catalysts can only achieve by the use of special ligands with high donor capacity and steric demand. In spite of these interesting characteristics of complex **38**, which indicate that Fe(0)-ate complexes may be intermediates in cross-coupling reactions, it should be noted that these reactions proceed remarkably slower than those catalyzed by Fe(−2)-ate complexes (cf. Sect. 1).

Complex **38** also turned out to be an efficient catalyst for cycloisomerization reactions of enynes **41** (Scheme 8) [16, 17]. This seems reasonable if one considers the fact that Fe(0) is isoelectronic to Rh(+1), which is also a catalyst for Alder-ene cycloisomerizations [18, 19].

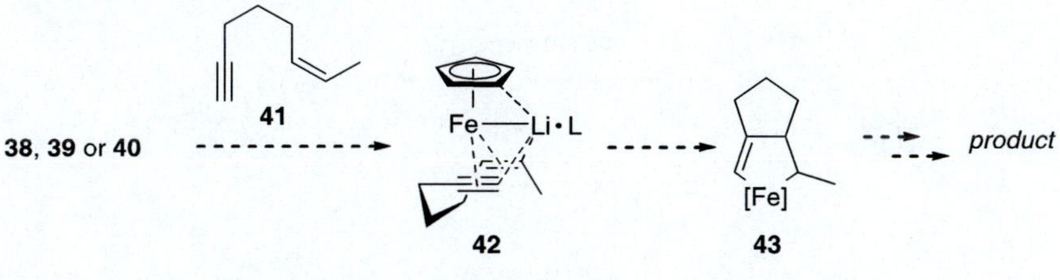

Scheme 8 Catalytic role of iron in cycloisomerization reactions [17]

The reaction in the presence of *bis*-ethylene complex **38** proceeds in 6 h, while complexes **39** and **40** need a somewhat longer reaction time (Scheme 9). This may be explained by the more facile and irreversible substitution of the volatile ethylene ligands compared to the chelating cod ligand. The reaction was shown to be broadly applicable, tolerating various substituents on the alkyne. Even terminal alkynes were tolerated although one might assume the Fe(0)-ate complexes to be basic and thus deprotonating the alkyne which is not the case. Tertiary amines do not interfere in the reaction, thus indicating that the iron center only exhibits a moderate Lewis acidity, if any. The fact that even aryl halides remain untouched in this reaction proves that the rate of the oxidative cyclization must be remarkably higher than that of the oxidative insertion into C–X bonds. The products formed almost always exhibit an exocyclic double bond in (*E*)-configuration, while the newly formed endocyclic double bond is (*Z*)-configurated in 5- to 8-membered rings and (*E*)-configurated in ≥10-membered rings.

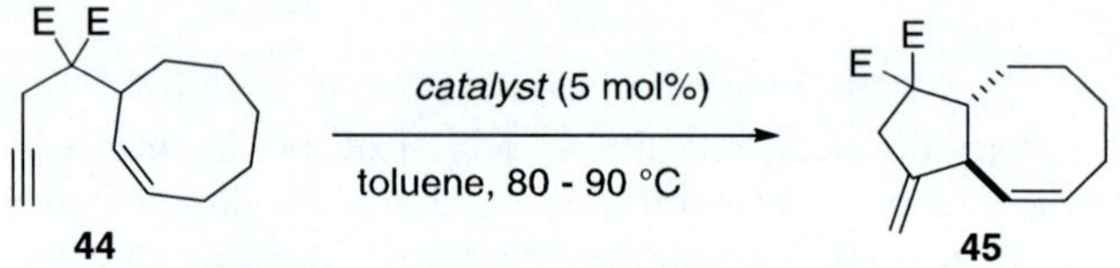

catalyst	yield (%)
$[CpFe(C_2H_4)_2][Li(tmeda)]$ **38**	83
[CpFe(cod)][Li(dme)] **40**	80
[CpFe(cod)][Li(tmeda)] **39**	82

Scheme 9 Enyne cycloisomerization catalyzed by different Fe(0)-ate complexes **38–40**; E = COOEt [17]

With regard to the mechanism of the cycloisomerization, Fürstner et al. found strong evidence of a metallacyclic intermediate. By labeling the allylic position of enynes **46** and **48,** they showed that reactions yielding *trans*-annulated rings **47** transferred the deuterium atom to the exocyclic double bond (eq. 1 in Scheme 10), whereas *cis*-annulated rings **49** formed with complete preservation of the position of the deuterium atom (eq. 2 in Scheme 10). This corresponds well to a metallacyclic

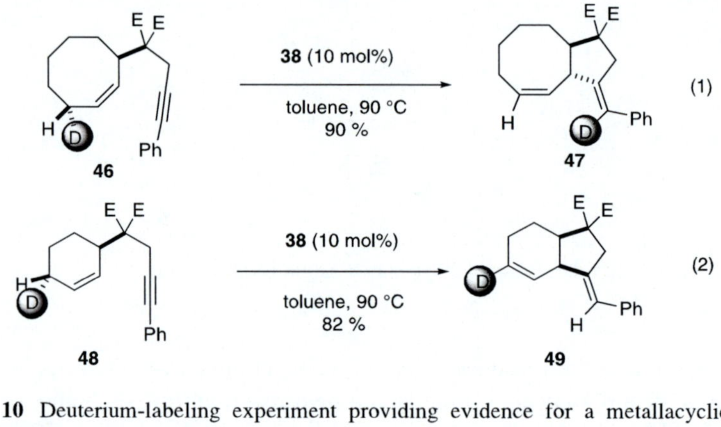

Scheme 10 Deuterium-labeling experiment providing evidence for a metallacyclic reaction mechanism [17]

intermediate pathway, whereas a mechanism based on C–H-activation involving an iron–allyl intermediate should proceed with some degree of scrambling of the deuterium in the product.

Ferrate **38** also turned out to catalyze [5 + 2]-cycloadditions. In this context, various vinylcyclopropane derivatives **52** were converted into the corresponding cycloheptadiene derivatives **53**. The products were obtained in good to excellent diastereoselectivites favoring the 1,2-*trans*-disubstituted isomer **53a**.

Equally to ferrate **38** ferrates **39** and **40** also catalyze Alder-ene cycloisomerizations [17]. Compared to **38**, they require somewhat longer reaction times, possibly due to the chelating cod-ligand, which is more difficult to substitute by the enyne. The presence of cod in the reaction turns out to be of advantage with more demanding substrates like acyclic enynes with a terminal alkene moiety where ferrate **38** is not reactive. Cod is assumed to stabilize the catalyst in its resting state as ancillary ligand.

Treating diene-yne derivatives **50** with ferrate **40** does not lead to the expected ene-allenes, instead the [4 + 2]-cycloaddition products **51** are obtained in moderate yields (eq. 1 in Scheme 11). As metal-catalyzed Diels–Alder-reactions of unactivated alkynes and dienophiles are assumed to proceed via metallacyclic intermediates, this supports the mechanism for the Alder-ene-reaction discussed before.

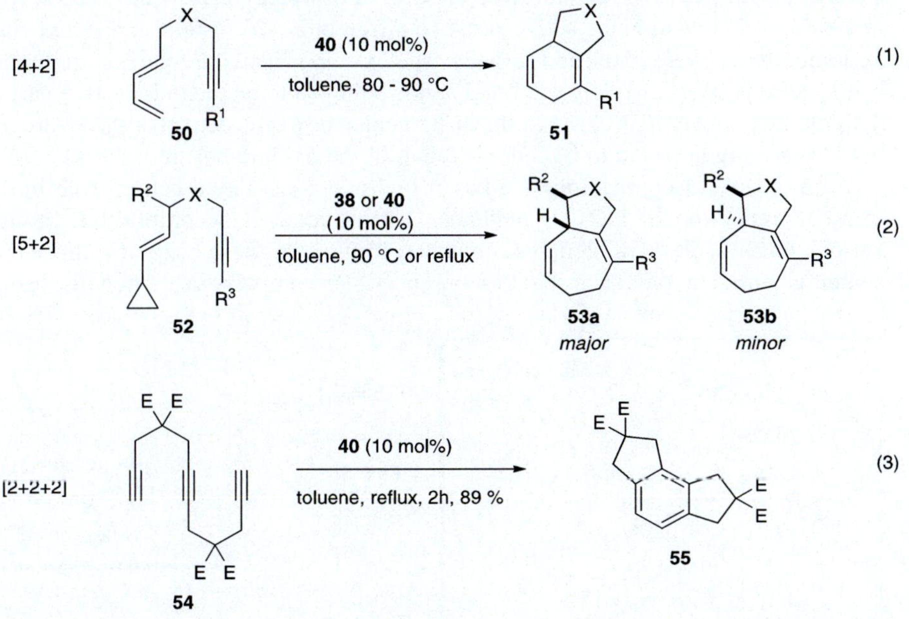

Scheme 11 Iron(0)-ate catalyzed cycloadditions; X = CE_2, NTs; E = COOEt [17]

Similar to this cycloaddition, ferrate **40** also proved to be catalytically active in [5 + 2]-cycloadditions, as discussed for ferrate **38** (eq. 2 in Scheme 11). As for the cycloisomerization reactions, ferrate **40** also turned out to be reactive toward conformationally less biased substrates (R^2 = H, eq. 2 in Scheme 11), whereas ferrate **38** was not catalytically active.

As complex **40** proved to be active in cycloaddition reactions and is isoelectronic to Rh(+1), which is a potent catalyst for [2 + 2 + 2] cycloadditions [20, 21], it was expected that **40** might also be active in those reactions, which is indeed the case. Triyne **54** could be converted to the [2 + 2 + 2]-cycloaddition product **55** in good yield (eq. 3 in Scheme 11). Mechanistically, this reaction is also assumed to proceed via a metallacyclic intermediate.

4.3 Fe-Hydrogenases and Model Complexes Thereof

In nature three different types of hydrogenases have been discovered. Besides binuclear hydrogenases such as [NiFe] and [FeFe]-hydrogenases **4**, a mononuclear hydrogenase, which was recently shown to contain also an iron center, is known under the name Hmd (H_2-forming methylenetetrahydromethanopterin dehydrogenase) or [Fe]-hydrogenase **56** [22]. All of those three enzymes were discovered to contain iron carbonyl sites, while [NiFe] and [FeFe]-hydrogenases **4** also contain CN ligands [23–25]. The crystal structure of [FeFe]-hydrogenase **4** has been solved in 1998 [26, 27]. Within the active site, two different Fe–S clusters, connected via a cysteine ligand, were identified (Fig. 2). The cysteine-ligated Fe-part of the binuclear species is coordinatively saturated and is believed to be the redoxactive part of the complex (Fe(+2)/Fe(0)). The distal Fe-center however activates the hydrogen but is generally believed to be redoxinactive in the oxidation state +2 [28].

Among several applications, Fe-based hydrogenases play a central role in the stepwise reduction of CO_2 to methane. This process is accomplished through various types of Fe-hydrogenases; however, in most of these enzymes, the active center is either a binuclear Fe–Fe- or an Ni–Fe-complex. Although the exact

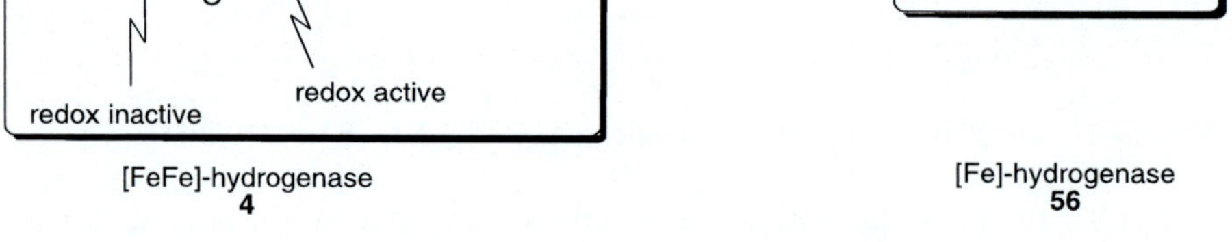

[FeFe]-hydrogenase
4

[Fe]-hydrogenase
56

Fig. 2 Active center of [FeFe]-hydrogenase **4** (*left*) and [Fe]-hydrogenase **56** (*right*); Unk = unknown ligand [22]

mechanism for H_2-activation by [FeFe]-hydrogenases **4** still remains unclear, some evidence points in the direction of an inner-electron transfer from the cysteine-ligated Fe-center in the binuclear complex. Because only the [FeFe]-hydrogenase **4** is able to directly activate H_2, a cooperative effect of the [Fe_4S_4]-cluster appears to be likely; however, the exact mechanism of H_2-activation is still a matter of debate. At the current stage of research, a two-electron transfer between two Fe-centers appears to proceed via a fast double single-electron transfer event with participation of the μ-S-bridge, rather than via a concerted two-electron shift. The non-innocence of ligands in Fe-chemistry seems to be one of the main prerequisites for catalytic reactions in which the Fe-atom undergoes electron transfers (cf. Sect. 1).

A different mechanism for reduction processes by [Fe]-hydrogenase **56** is assumed. The hydride generated by splitting dihydrogen is directly transferred to an electrophilic organic center in methenyltetrahydrocyanopterin. As no electrons need to be transferred this reaction requires only one metal center. Due to its structure the center of [Fe]-hydrogenase **56** does not count to the class of ferrates.

Concerning [FeFe]- **4** and [NiFe]-hydrogenases, several complexes mimicking the active sites have been synthesized and examined. Rauchfuss et al. showed that complexes of the type $[Fe(CN)_2(CO)_3]^{2-}$ have an electron-rich character, can easily be oxidized, but remain stable upon protonation in contrast to its monocyanide analogues [29, 30]. This complex exhibits trigonal bipyramidal geometry with the CN^- ligands in axial position. The observed Fe–CO and Fe–CN bond lengths indicate strong π-back-bonding in the Fe–CO-bond and predominantly σ-bond character in the Fe–CN-bond. This complex could be protonated by *p*-TsOH or NH_4PF_6 to yield $[Et_4N][Fe(CN)_2(CO)_3H]$ **57a**, which has an octahedral geometry with CO predominantly *trans* to the strong donor CN^-. It exhibits a pK_a value of 17.53 (in MeCN) and undergoes rapid H/D exchange in D_2O. Analytical data point out that the protonation occurs at the Fe-site and not at a CN^- ligand. Compared to its deprotonated form, it shows increased electrophilicity, thus reacting with phosphines or phosphites under decarbonylation to the corresponding ligand-exchanged complexes **59b** (Scheme 12).

When $[Fe(CN)_2(CO)_3H]^-$ **57a** is treated with excess *p*-TsOH the complex is completely consumed and evolution of H_2 and an equimolar amount of CO

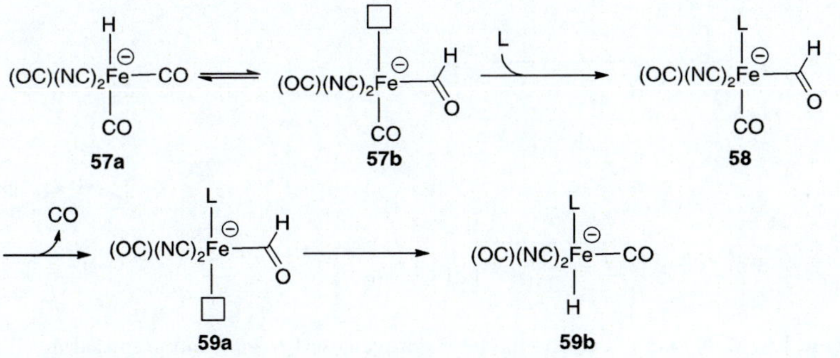

Scheme 12 Mechanism for decarbonylative ligand exchange [30]

detected. Equally, the protonation of the phosphine-containing analogue $[Fe(CN)_2(CO)H(dppv)]^-$ (dppv = *cis*-$Ph_2PCH{=}CHPPh_2$) resulted in the release of H_2. The authors hypothesize that protonation generates a labile H_2 ligand, which is displaced by FeC*N* on another molecule of the complex to give a polymer which depolymerizes in MeCN to give $[Fe(CN)_2(CO)(NCMe)(dppv)]$. The fact that the charge-neutral dihydrogen complexes were not isolable corresponds well to the high efficiency of the negatively charged active site of [NiFe]-hydrogenase, while most stable H_2 complexes are cations [31]. When $[Fe(CN)_2(CO)_3]^{2-}$ was treated with methylating reagents, a methylation at Fe was observed, which was proved by IR-spectroscopy. When treated with dppv, this complex underwent CO-migration to form an Fe-acyl species, also analyzed by IR-spectroscopy. In contrast, reaction with acid chlorides forms *N*-acyl isocyanate complexes without acylation at Fe. Thus, the function of a hydrogenase could be mimicked, which implicates a Fe–CN–CO–H_2 intermediate.

Darensbourg et al. could show that binuclear Fe(+1)Fe(+1) model complexes **60–62** of the [FeFe]-hydrogenase **4** are capable of electrocatalytical generation of hydrogen in the presence of acetic acid as a proton source [32, 33]. For the all-CO-complex **60** electrochemical measurements pointed to an EECC mechanism (electrochemical–electrochemical–chemical–chemical), where protonation occurs only at the stage of a Fe(0)Fe(0)-complex (mechanism a in Scheme 13). In the complexes containing PMe_3 or PTA ligands, **61/62** protonation occurs even at the level

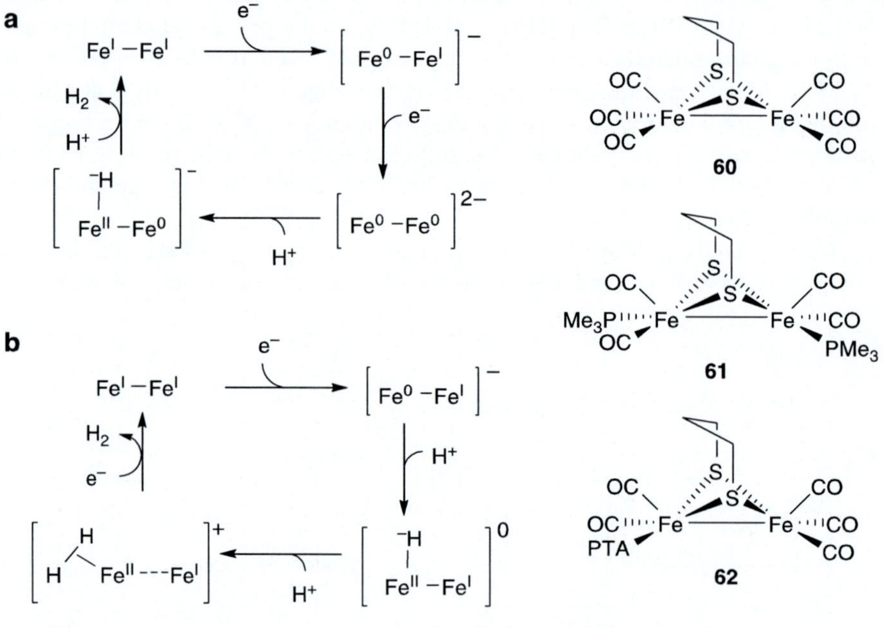

Scheme 13 Model complexes for [FeFe]-hydrogenase **4** (*right*) and proposed mechanisms for electrocatalytic hydrogen generation (*left*) (**a**) EECC mechanism, (**b**) ECCE mechanism; PTA = phosphatriazaadamantane [33]

of a Fe(0)Fe(+1) complex, that is, after one-electron reduction. This mechanism is called the ECCE mechanism (electrochemical–chemical–chemical–electrochemical mechanism b in Scheme 13) and seems to be favored in the presence of strong donor ligands like PMe_3 or PTA, which mimic the role of CN^- ligands in the [FeFe]-hydrogenase **4**. Although these model complexes **60–62** are neutral nonferrate complexes, they mimic the activity of a naturally occurring complex, which can be seen as a ferrate, containing as it does CN^- instead of the phosphorous ligands of the model complexes, thus resulting in a negatively charged iron complex.

5 Catalysis with Fe(−1)-Ate Complexes

5.1 Dinitrosyl-Ferrates

Ferrates that exhibit the rare oxidation state of −1 have been shown to catalyze oxygen activation and transfer to electrophilic substrates like phosphines or alkenes. When Postel et al. treated dinitrosyliron complexes (DNIC) of the type $[Fe(NO)_2Cl]_2$ with oxygen in the presence of mono- or bidentate [34, 35] phosphines or phosphites [36, 37], the nitrosyl ligands were oxidized to nitrato ligands, while iron was oxidized from the oxidation state of −1 to +3 to yield the complexes $Fe(NO_3)_2ClL_2$ (L = O_2dppe [34], $OPPh_3$, HMPA [38] ...), in which the phosphine ligands were also oxidized. Under anaerobic conditions, these complexes proved to be reluctant toward olefin oxidation, while the nitrato complexes reacted with phosphines to yield the corresponding phosphine oxides. Nitrosyl complexes were only detected as reaction products in the case of monodentate phosphine ligands. Nevertheless, under aerobic conditions, these nitrato complexes were capable of transferring oxygen catalytically to olefins, thus generating 2-cyclohexene-1-one **65** and 2-cyclohexene-1-ol **64** from cyclohexene **63**, forming

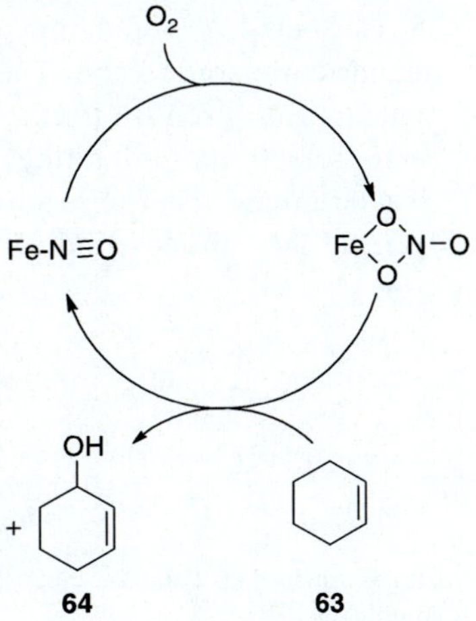

Scheme 14 Oxygen activation and transfer by dinitrosyliron complexes [35]

cyclohexene oxide only in trace amounts (Scheme 14). As the reaction products are the products of allylic oxidation which is a main characteristic for autoxidation processes, the authors assume the reaction mechanism to be a radical pathway with formation of a hydroperoxide and subsequent decomposition of the latter by the metal center.

6 Catalysis with Fe(−2)-Ate Complexes

6.1 Ferrates with Olefin Ligands

The Fe(−2)-ate complex [Fe(cod)$_2$]·[Li(dme)]$_2$ **66** described by Jonas [13–15] is available similar to the Cp-containing Fe(0)-ate complexes **38–40** from ferrocene **37** under reductive conditions in the presence of the coordinating olefin ligands. It exhibits activity in the cross-coupling of aryl chlorides with Grignard reagents as well as in the coupling of cyclohexyl bromide with ArMgX reagents [9].

Tetraethyleneferrate(−2) **67**, also described by Jonas [39, 40], can be obtained from ferrocene **37** as well through reduction with lithium under an ethylene atmosphere (5–8 bar) and subsequent addition of tmeda (Fig. 3). The X-ray structure of **67** shows a remarkably short Li–Fe distance and elongated ethylene C=C bonds compared to free ethylene (1.43 Å and 1.45 Å vs. 1.34 Å), which shows a strong back-donation of electrons from the iron into the π^* orbitals of the olefins. These occurrences of intermetallic bonds, the low formal oxidation state with d^{10} configuration, and the free coordination sites created upon release of the labile ethylene ligands suggest that this complex is a model complex for the clusters [Fe(MgX$_2$)$_n$], proposed by Bogdanovic [41] to be intermediates in cross-coupling chemistry.

This becomes clearer, as ferrate **67** was shown to exhibit high activity in cross-coupling reactions [9, 42], being somewhat more reactive than complexes **66** and **38** (cf. Sect. 4.2), which may partly be attributed to the higher lability of ethylene ligands compared to cod. The cross-coupling of alkyl halides with aryl Grignard reagents catalyzed by ferrate **67** is broadly applicable and exhibits an excellent chemoselectivity, tolerating functional groups that are normally attacked by Grignard reagents. The reaction is also compatible with lots of functional groups, even tertiary amines, indicating that complex **67** is not or only weakly Lewis

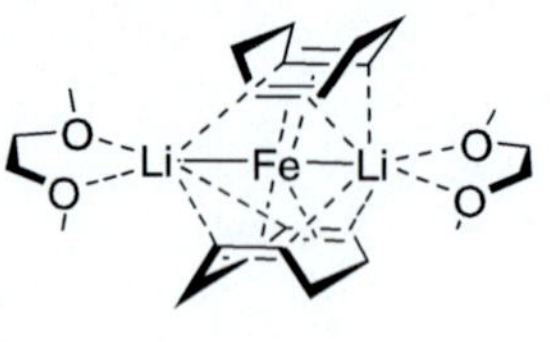

66

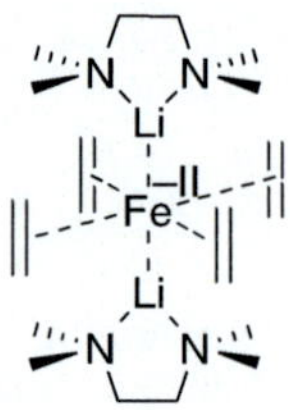

67

Fig. 3 Jonas' Fe(-2)-olefin complexes [9]

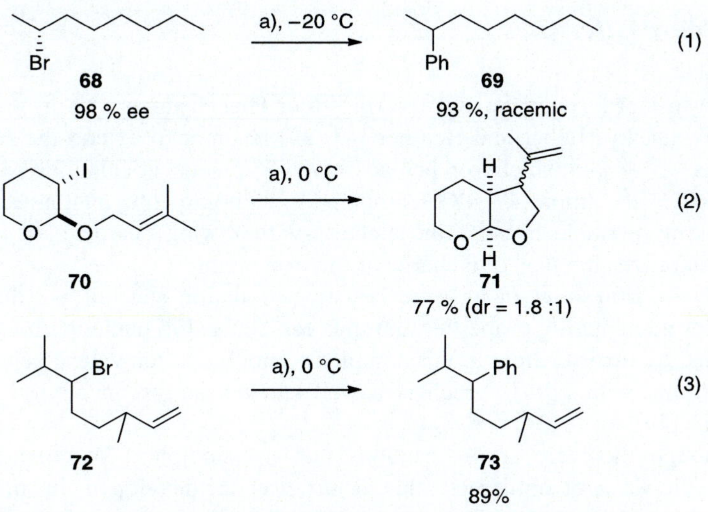

Scheme 15 Cross-coupling experiments to obtain mechanistical evidence; (**a**) complex **67** (5 mol%), PhMgBr, THF [9]

acidic. Furthermore, allylic, benzylic, and propargylic halides could be converted in cross-coupling reactions, the latter being converted without the formation of allenic by-products. Mechanistically, it is not clear whether the cross-coupling reactions of ferrate complexes proceed via an oxidative addition-transmetallation-reductive elimination pathway or via radical pathways. Some evidence was found that single-electron transfer may intervene in such reactions as optically pure (*R*)-2-bromooctane **68** (98% ee) was racemized when reacted with PhMgBr in the presence of catalytic amounts of complex **67** (eq. 1 in Scheme 15). Furthermore, some olefinic iodoacetals **70** underwent 5-*exo*-trig(dig)-cyclizations without cross-coupling (eq. 2 in Scheme 15). Despite these hints at a radical mechanism, substrate **72** leads to the corresponding cross-coupling product **73** instead of the cyclization product, which shows that both routes may be involved (eq. 3 in Scheme 15).

As complex **67** outperforms the Fe(0)-ate complexes in rate and yield, shown in the reaction of cyclooctenyl bromide with PhMgBr (full conversion: complex **67** <20 min, 81% yield, **38** 18 h, 39% yield), it was shown that both Fe(0)-ate and Fe (−2)-ate complexes should be intermediates in cross-coupling reactions, but the major contribution should be made by the route emanating from Fe(−2)-complexes. The superiority of Fe(−2)-ate complexes was also shown in the stoichiometric insertion of **67** into allylic halides, which proceeded much faster (<5 min) than with any higher valent iron complex (hours or days).

However, in contrast to Fe(0)-ate complexes **38–40** (cf. Sect. 4.2), complex **67** failed to catalyze Alder-ene cycloisomerizations, which may be attributed to the thermal lability of this complex [17].

6.2 Nitrosylferrates

The low-valent ferrate $[Fe(CO)_3(NO)]^-$ **76** or Hieber anion was discovered some 50 years ago by Hieber and Beutner [43, 44] in order to extend the *Hieber base reaction* [45, 46], in which iron pentacarbonyl **78** reacts with alkaline bases to form the $[Fe(CO)_4]^{2-}$ anion [47, 48]. Compared to its homoleptic analogue, the Hieber anion is more stable because the electron-withdrawing character of the nitrosyl ligand stabilizes the negative charge at the iron atom.

With the iron atom in its most negative oxidation state of −2 this complex possesses nucleophilic properties and thus can be used in nucleophilic substitution reactions. As the iron atom in this complex formally has ten valence electrons, it is isoelectronic with Pd(0), which is a well-known catalyst in allylic substitution reactions [49].

Although there are some reactions that use complex **76** stoichiometrically [50–58], it was not until 1979 that Roustan et al. developed the first catalytic application of complex **76-Na** (Scheme 16) [59, 60]. In his publication, he could show that catalytic amounts of complex **76-Na** react with an allylic chloride or acetate to form an allyl–iron-complex, which, in a second step, is substituted with a malonate to yield **77**. Most importantly, they observed a preference for the *ipso*-substitution-product **77a**, that is, the new C–Nu-bond was formed preferentially at the carbon atom that was substituted with the leaving group before.

With Roustan et al. using the sodium salt of the Hieber anion **76-Na**, the procedure was improved by Xu and Zhou in 1987 when they introduced the corresponding shelf-stable tetrabutylammonium salt **76-[Bu$_4$N]** which is available from $Fe(CO)_5$ **78**, $NaNO_2$ and Bu_4NBr (Scheme 17) [61, 62]. As well as discovered by Roustan they obtained the substitution products with an *ipso*-preference (Scheme 16) albeit in a significantly lower yield. In order to maintain the catalytic activity of the product, the reactions were performed under CO-gas atmosphere.

After these encouraging results, it is surprising that no further investigations have been performed on this reaction, which might be attributed to problems regarding catalyst stability, reproducibility, or the use of a problematic CO gas atmosphere. The *comeback* of the Hieber anion dates back to 2006 when our group

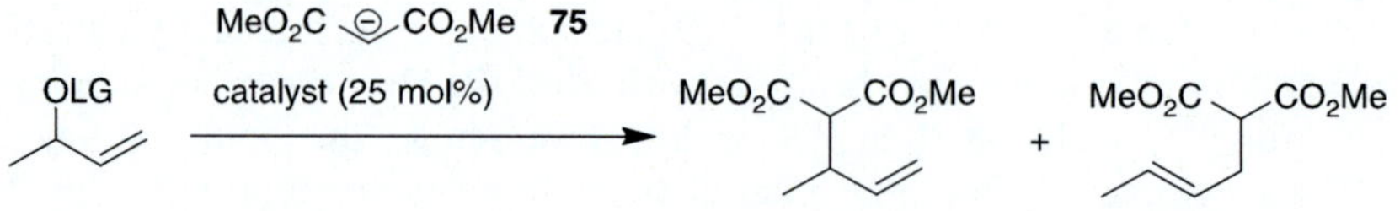

74 | 77a | 77b

80 : 20

		yield
Roustan (1979):	$Na[Fe(CO)_3(NO)]$ **76-Na**, THF, reflux	85 %
Xu (1987):	$[Bu_4N][Fe(CO)_3(NO)]$ **76-[Bu$_4$N]**, CO(g), THF, reflux	35 %

Scheme 16 Fe-catalyzed allylic substitution

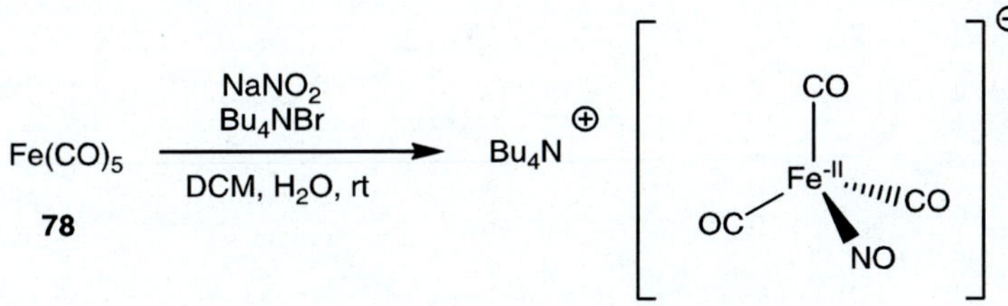

Scheme 17 Synthesis of $[Bu_4N][Fe(CO)_3NO]$ **76-[Bu₄N]** from $Fe(CO)_5$ **78**

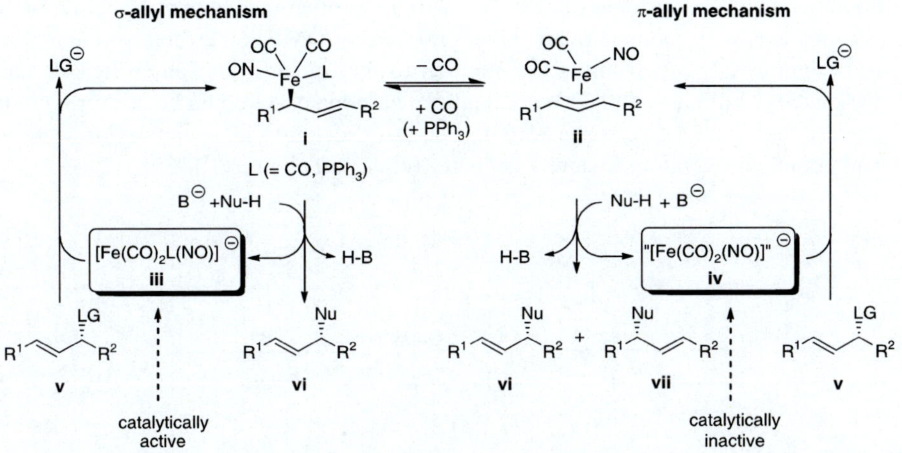

Scheme 18 σ- vs. π-allyl mechanism in Fe-catalyzed allylic substitutions

envisioned the use of a monodentate σ−donor-ligand like PPh_3 for the stabilization of the intermediate allyl-Fe complex as a CO-surrogate. The replacement of one CO-ligand by the phosphine was thought to prevent the formation of a catalytically inactive π-allyl-Fe-complex (Scheme 18) [63].

Indeed, the use of PPh_3 led to an increased stability of the complex, additionally using DMF as a coordinating solvent increased its nucleophilicity [63]. Thus, various allylic carbonates could be transferred to the substitution products in good to excellent yields and high regioselectivities in favor of the *ipso* substitution products (Scheme 19). When using a carbonate as both leaving group and in situ base, no preformation of the nucleophile was necessary.

Subsequently, the scope of the reaction was extended to N-nucleophiles **82**. Because the inherent basicity of the substitution products **83** imposed some problems concerning catalyst decomposition, the addition of catalytic amounts of piperidine hydrochloride (pip·HCl) proved to be necessary. Under optimized reaction conditions different aromatic amines **82** were allylated with almost exclusive regioselectivites in favor of the *ipso* substitution products **83** (eq. 1 in Scheme 20) [64].

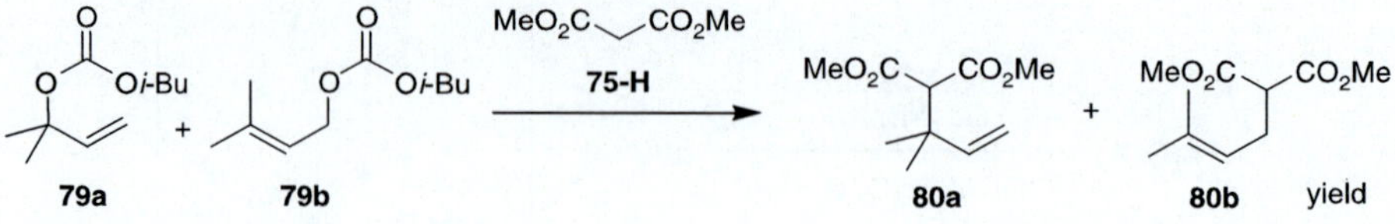

79a	:	79b	80a	:	80b	yield
100	:	0	98	:	2	81 %
0	:	100	7	:	93	61 %

Scheme 19 The Fe-catalyzed allylic alkylation (reagents and conditions: 2.5 mol% $[Bu_4N][Fe(CO)_3(NO)]$ **76-[Bu$_4$N]**, 3 mol% PPh_3, DMF, 80°C, 24 h) [63]

Recently, the scope of the allylic substitution has been extended to sulfinate salts **84** to obtain allylic sulfones **85**. Due to solubility problems of both nucleophile **84** and carbonate leaving group, a polar solvent mixture of DMF and 2-methoxyethanol had to be employed, which limits the reaction to the use of a phosphine ligand. Thus, various aryl sulfinates **84** and functionalized carbonates **81** could be converted to the corresponding allylic sulfones **85** with good to excellent yields and regioselectivites and complete retention of stereochemistry (eq. 2 in Scheme 20) [65].

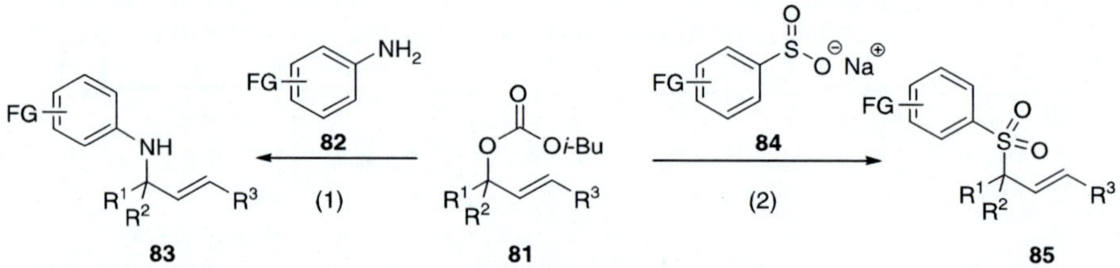

Scheme 20 Fe-catalyzed allylic amination and sulfonation reagents and conditions: (**a**) 5 mol% $[Bu_4N][Fe(CO)_3(NO)]$ **76-[Bu$_4$N]**, 5 mol% PPh_3, 30 mol% pip·HCl, DMF, 80°C; (**b**) 5 mol% $[Bu_4N][Fe(CO)_3(NO)]$ **76-[Bu$_4$N]**, 6 mol% $P(C_6H_5OMe)_3$, DMF/2-methoxyethanol 3:1, 80°C [64, 65]

The protocol of the allylic alkylation, which proceeds most likely via a σ-allyl-Fe-intermediate, could be further improved by replacing the phosphine ligand with an *N*-heterocyclic carbene (NHC) (Scheme 21) [66]. The addition of a *tert*-butyl-substituted NHC ligand **86** allowed for full conversion in the exact stoichiometric reaction between allyl carbonate and pronucleophile. Various C-nucleophiles were allylated in good to excellent regioselectivities conserving the π bond geometry of enantiomerically enriched (*E*)- and (*Z*)-carbonates **87**. Even chirality and prochirality transfer was observed (Scheme 21) [67].

However, by changing the ligand's topology, a significant change in the regioselective course of the reaction is observed. Whereas the *tert*-butyl-substituted ligand **86** allows for a regio- and stereoselective allylic substitution, an aryl-substituted ligand **92** forces the reaction to follow a π-allyl mechanism (eq. 1 in Scheme 22). This ligand-dependent mechanistic dichotomy resembles a promising starting point for the development of an asymmetric Fe-catalyzed allylic substitution. Later on it was shown that even isolated π-allyl Fe-complexes, which were

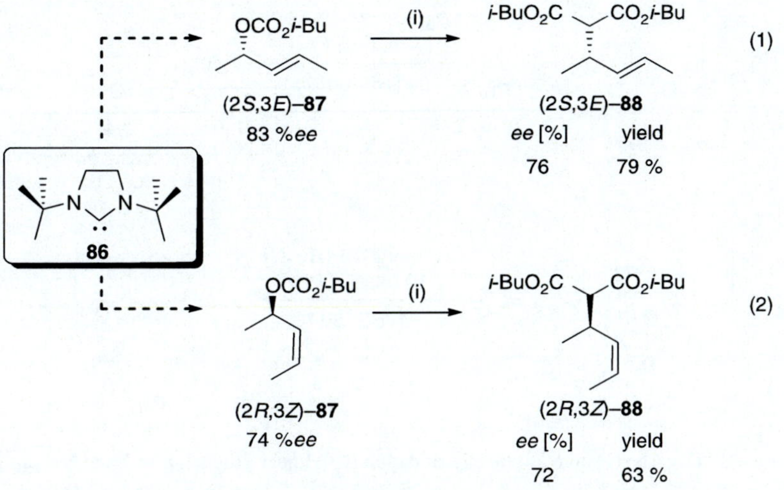

Scheme 21 Fe-catalyzed regio- and stereoselective allylic substitution in the presence of NHC-ligand **86** {reagent and conditions: (1) $[Bu_4N][Fe(CO)_3(NO)]$ **76-[Bu₄N]** (cat.), MTBE, 80°C} [67]

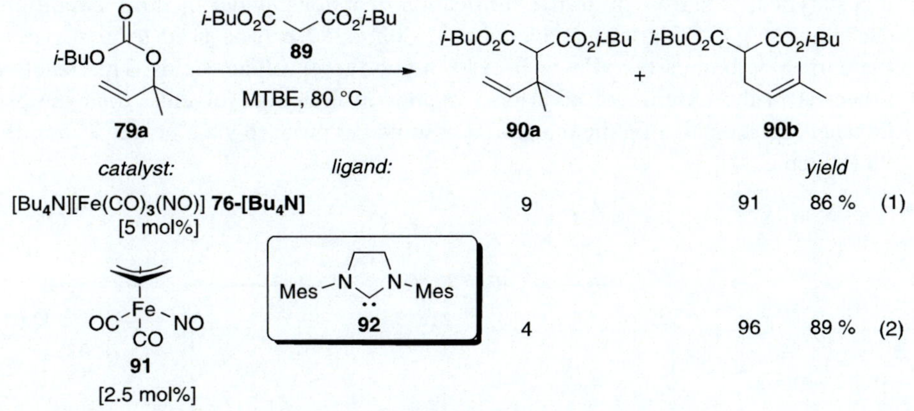

Scheme 22 Fe-catalyzed allylic substitution via π-allyl mechanism [67, 68]

previously reported to be catalytically inactive, are potent precatalysts in the presence of ligand **92** (eq. 2 in Scheme 22) [68].

The most recent application of catalyst **76-[Bu₄N]** was reported by Trivedi and Tunge [69]. Various aromatic allylic carbonates were converted in a decarboxylative allylic etherification into their corresponding allylic ethers (Scheme 23). The allylated products were obtained in high yields independently of the substitution pattern of the phenol or the electronic properties of the substituent. Regioselectivity is strongly dependent on the starting allyl ester, as cinnamyl carbonates **93** favor the more stable linear products **94** while crotyl carbonate **95** favors the branched

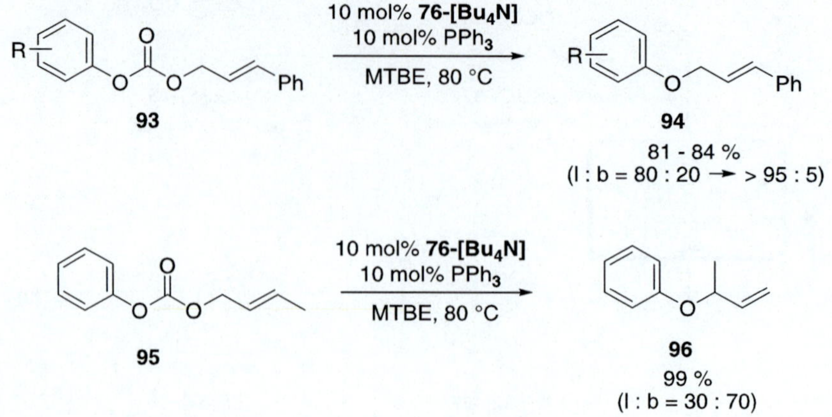

Scheme 23 Decarboxylative allylic etherification (l = linear product, b = branched product) [69]

product **96**. Thus, the observed regioselectivity is a kinetic selectivity which the authors attribute to the intermediacy of a π-allyl iron complex.

Besides allylic substitution reactions it was also shown that $[Fe(CO)_3(NO)]^-$ **76** is catalytically active in transesterification reactions under neutral conditions (Scheme 24) [70]. Various activated acyl donors **97** can be used to give rise to the corresponding carboxylic esters **100** in good to excellent yields. This reaction proceeds in the absence of additional ligands in nonpolar solvents, for example, hexane. Mechanistically, the reaction is assumed to proceed via a Fe–acyl-complex **98** (Scheme 24).

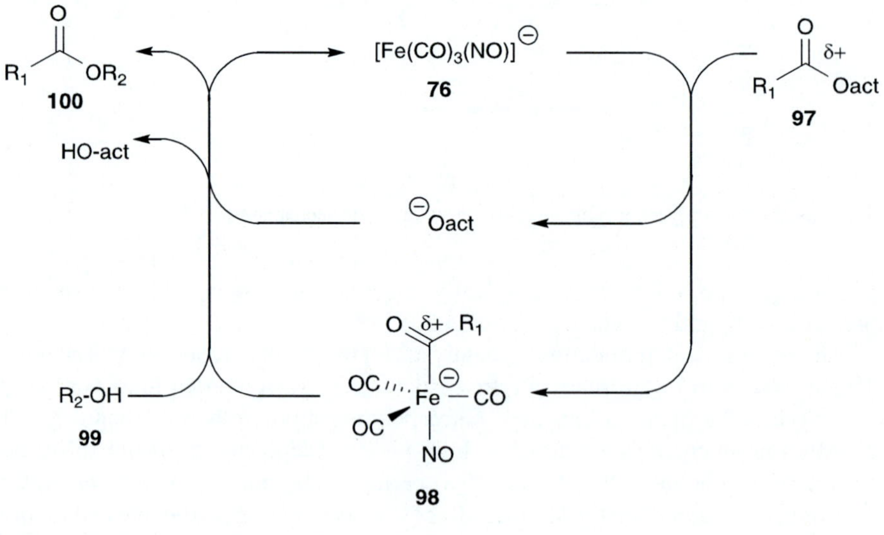

Scheme 24 Mechanistic hypothesis for the transesterification (act = activating group) [70]

7 Miscellaneous

Apart from catalysis with well-defined iron complexes a variety of efficient catalytic transformations using cheap and easily available Fe(+2) or Fe(+3) salts or Fe(0)-carbonyls as precatalysts have been published. These reactions may on first sight not be catalyzed by ferrate complexes (cf. Sect. 1), but as they are performed under reducing conditions ferrate intermediates as catalytically active species cannot be excluded. Although the exact nature of the low-valent catalytic species remains unclear, some of these interesting transformations are discussed in this section.

7.1 Cross-coupling Chemistry

The origins of Fe-catalyzed cross-coupling reactions date back to the 1970s when Kochi studied reactions of alkenyl halides with Grignard reagents in the presence of catalytic amounts of $FeCl_3$ [71, 72] or even to the 1940s when Kharasch and Fields studied the influence of transition metal salts on the reactivity of Grignard reagents [73]. First mechanistic proposals assumed the generation of a low-valent iron catalyst by reduction of iron salts with Grignard reagents [74, 75]. Various iron salts such as $FeCl_3$, also in the presence of a ligand [76], $Fe(acac)_3$, $Fe(dbm)_3$, FeF_3 in the presence of an NHC ligand [77], Fe(salen)Cl-complexes [76, 78], or even Fe powder [79] proved to be applicable in cross-coupling reactions. Besides the commonly used Grignard reagents, organometallic reagents derived from manganese [80], copper [81], or zinc [82] were also shown to be potent transmetallation reagents. Even a Grignard reagent on solid phase was used for cross-coupling [83]. With regard to the mechanism, some studies point to the involvement of a single-electron transfer or radical intermediate, as postulated by Bedford et al. [84–86] (Scheme 25), while Fürstner et al. provided evidence that these reactions may proceed via two different nonradical pathways (Scheme 26) [9]. Species unable to undergo β-hydride elimination, such as MeLi or PhMgBr, are thought to react via iron-ate complexes (cf. Sect. 3), while species with β-hydrogens, such as EtMgX, are assumed to form the so-called inorganic Grignard reagents $[Fe(MgX)_2]_n$ **101**, which were postulated earlier by Bogdanovic and Schwickardi [41]. These dark brown bimetallic clusters could not be isolated, but were used to transfer inactive organo chlorides into their corresponding Grignard reagents. The cluster is formed through reduction by small amounts of an easily accessible Grignard reagent (alkylation, then β-hydride elimination, and reductive elimination), subsequently forming an iron species **101** with a d^{10}-configuration and a formal oxidation state of −2. The aryl iron intermediate is then formed from the cluster and an aryl chloride. Subsequent Fe/Mg transmetallation releases the Grignard compound and regenerates the catalyst **101** (Scheme 27).

As catalytic shuttling between lots of oxidation states such as Fe(−2)/Fe(0) [87], Fe(0)/Fe(+2) [88], and Fe(+1)/Fe(+3) [89] were predicted to be involved in

Scheme 25 Postulated radical mechanism by Bedford et al. [86]

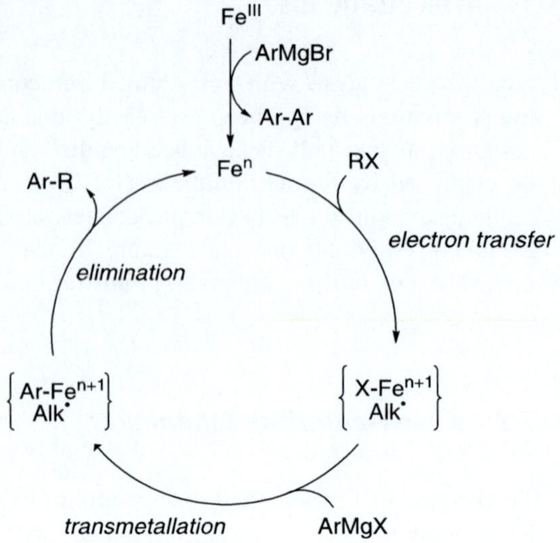

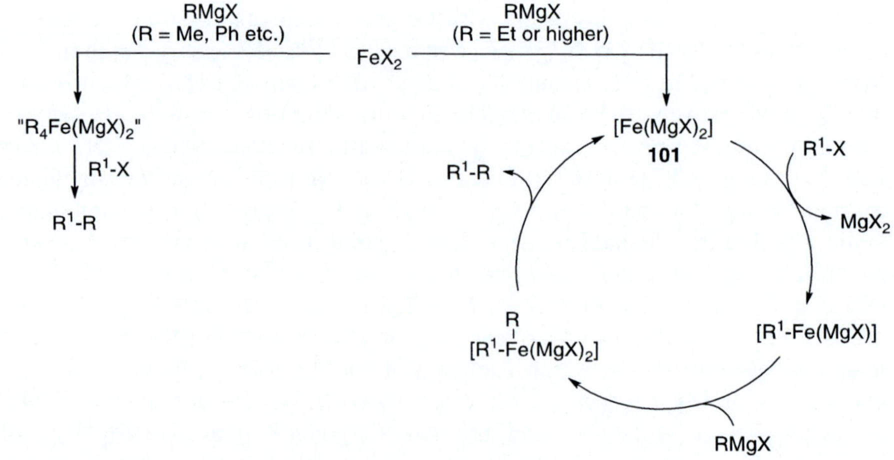

Scheme 26 Two different mechanisms for cross-coupling reactions [9]

cross-coupling reactions, it was indeed demonstrated by Fürstner et al. that all these oxidation states ranging from −2 to +3 may be traversed in these reactions, and the pair Fe(−2)/Fe(0) was shown to be the dominating one [9].

A novel strategy avoiding hazardous Grignard reagents was reported by Jacobi von Wangelin et al. [90]. Based upon the seminal contributions of Bogdanovic and Schwickardi [41], they developed a domino-iron-catalysis in which both the Grignard reagent and the active iron-species are formed in situ upon reaction between an alkyl halide, an Fe-salt, and magnesium (Scheme 28). Due to the fast

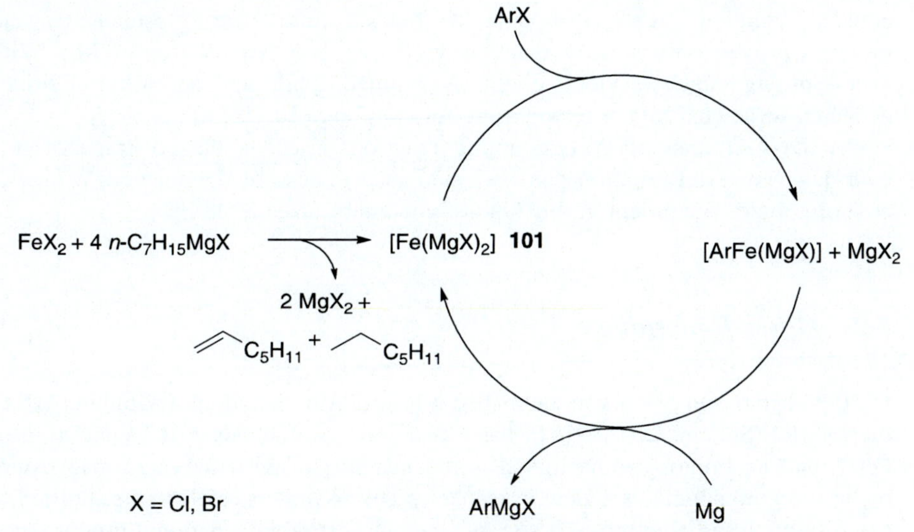

Scheme 27 Mechanism for the iron-catalyzed preparation of Grignard compounds [41]

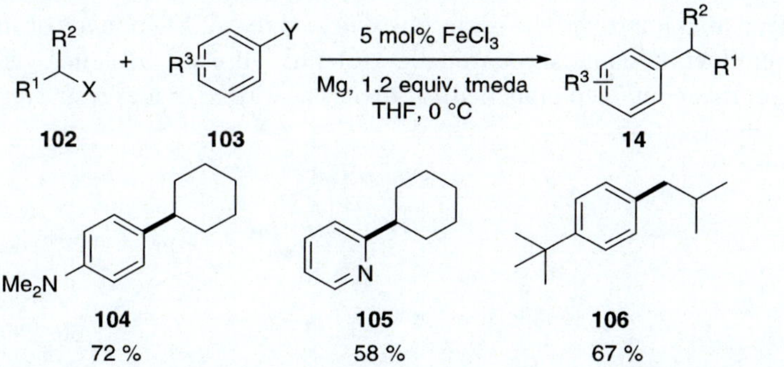

Scheme 28 Domino iron catalysis for cross-coupling of alkyl and aryl halides [90]

cross-coupling, the concentration of free Grignard reagents remained at a quasi-stationary level, thus avoiding problems encountered with larger amounts of active Grignard reagents (homo coupling, elimination, addition, etc.).

Various aryl **103** and alkyl halides **102** were cross-coupled to yield the corresponding alkyl benzenes **14** by using a slight excess of the alkyl halide **102**. $FeCl_3$ possesses a similar activity as $FeCl_2$ which gives further evidence for the reduction of Fe(+3)-salts by Grignard reagents. A reduction of the aryl halide **103** to toluene could be suppressed by the addition of tmeda, which is thought to stabilize both complexes of iron and magnesium (Scheme 28). Significant in this reaction is the high selectivity in favor of the cross-coupling product **14** leaving the amount of biaryl products in all cases below <9%. Various alkyl **102** and aryl halides **103** could be coupled at 0°C in good yields, but substrates bearing ester and cyano moieties proved only weakly reactive due to surface deactivation of Mg. With excess of Mg moderate yields could be obtained, while aryl and alkyl chlorides needed a somewhat higher temperature for coupling (20°C) [90].

As the vast area of cross-coupling reactions exceeds the coverage of this overview, more information can be found in a review by Jacobi von Wangelin covering the development in this field of chemistry up to 2008 [91].

7.2 Other Reactions

In 2009, Ritter and coworkers published a 1,4-addition reaction of α-olefins **110** to dienes **108** (Scheme 29) [92]. In these reactions, a precatalyst **107** consisting of $FeCl_2$ and an iminopyridine ligand is reduced in situ with activated magnesium metal to an unknown low-valent iron species **109**. With this catalyst, the addition of electron-poor and electron-rich styrenes bearing different functional groups, such as esters, ethers, or halides to different 1,3-dienes, could be achieved. Upon addition to unsymmetrical 1,3-dienes, the formation of two regioisomers occurred. The regioselectivity turned out to be ligand dependent; larger substituents on the 1,3-diene also improved the regioselectivity up to >20:1 in favor of the linear 1,4-diene product. In contrast to aromatic α-olefins, aliphatic α-olefins yielded product mixtures that contained branched 1,4-dienes as well as dienes resulting from double bond isomerization.

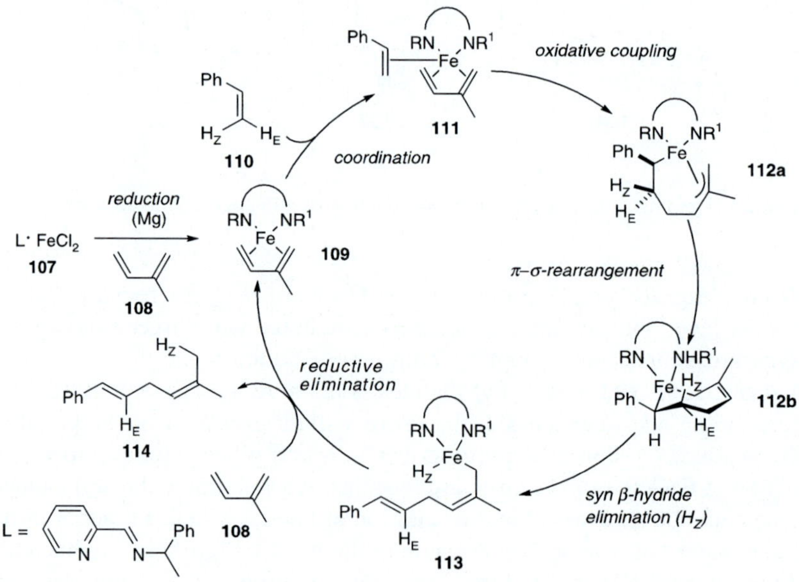

Scheme 29 Mechanistic hypothesis of Ritter's 1,4-addition to 1,3-dienes **108** [92]

As a mechanistic hypothesis, the authors assumed a reduction of the Fe(+2) by magnesium and subsequent coordination of the substrates, followed by oxidative coupling to form alkyl allyl complex **112a**. A $\pi-\sigma$ rearrangement, followed by a *syn* β-hydride elimination and reductive elimination, yields the linear product **114** with the 1,2-disubstituted (*E*)-double bond (Scheme 29). This hypothesis has been supported by deuterium labeling experiments, whereas the influence of the ligand on the regioselectivity still remains unclear.

Similar to the latter reaction Ritter also reported on an iron-catalyzed 1,4-hydroboration of 1,3-dienes (Scheme 30) [93] which is carried out under similar conditions with $FeCl_2$ and iminopyridine ligands as precatalysts. $FeCl_2$ is reduced in situ with magnesium and hydroboration proceeds at room temperature. By varying the substituent on the imine nitrogen of the ligand, regioselectivity could be controlled. Furthermore, the regioselectivity of 2-substituted 1,3-dienes increased with the size of the 2-substituent. The reaction tolerates electrophilic functionalities such as esters, exhibits a high chemoselectivity of diene versus olefin hydroboration, which the authors attribute to the affinity of 1,3-dienes to low-valent iron, and yields the products with solely (*E*)-configured double bond geometry. The authors assume a mechanistic scenario starting from the reduction of $FeCl_2$ with magnesium to a low-valent iron species, which is coordinated by the 1,3-diene. Oxidative addition of the borane is followed by migratory insertion of the diene into either the Fe–B or Fe–H bond. Reductive elimination finally yields the allylboranes. This proposed mechanism was supported by deuterium-labeling experiments, whereas the reason for the ligand-dependent regioselectivity has not been able to be explained until now (Scheme 30).

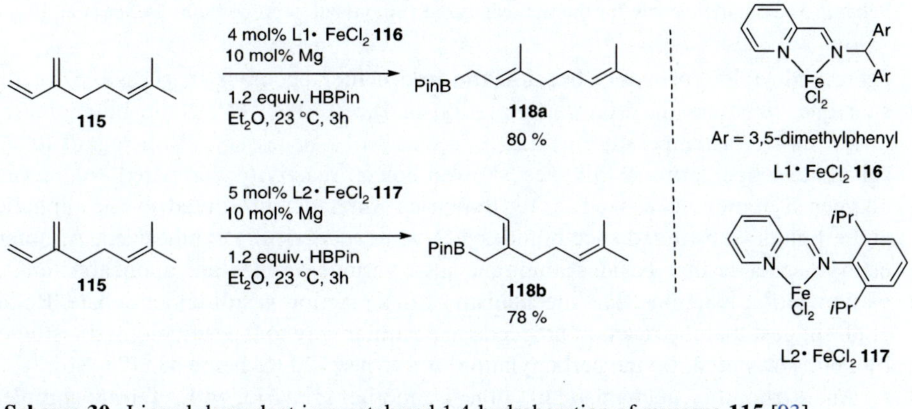

Scheme 30 Ligand-dependent iron-catalyzed 1,4-hydroboration of myrcene **115** [93]

An iron-catalyzed carbonylation reaction of alkynes **120** forming succinimides **121** by the aid of $Fe(CO)_5$ **78** or $[Fe_3(CO)_{12}]$ **119** has been reported by Beller et al. (Scheme 31) [94]. This reaction seems interesting as iron–carbonyl complexes are kinetically relatively inert. As a model system 3-hexyne was reacted with excess ammonia under 20 bar CO pressure. Employing a higher pressure leads to

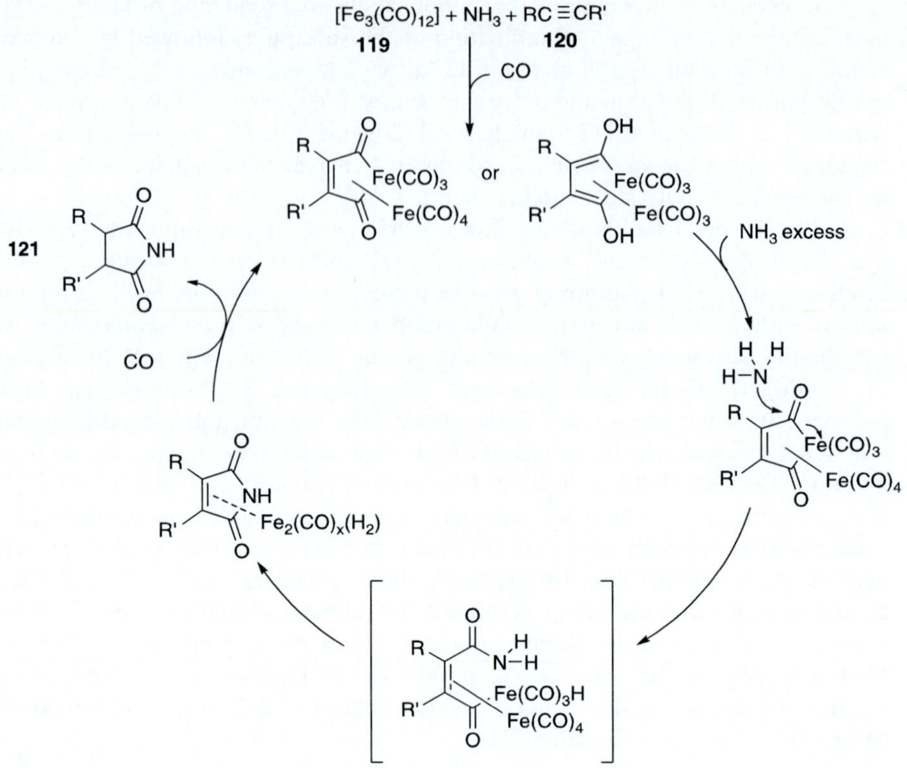

Scheme 31 Catalytic cycle for the iron-catalyzed carbonylation proposed by Beller et al. [94]

decreased yields, probably because the iron center becomes inaccessible for the substrate. Whereas the iron concentration of [$Fe_3(CO)_{12}$] **119** could be decreased to 0.5 mol%, a temperature of 120°C seems to be necessary. With regard to the substrate scope terminal alkynes showed lower reactivity compared to internal alkynes. Symmetrical as well as unsymmetrical aliphatic-substituted, cyclic aliphatic, and aromatic-substituted succinimides **121** were successfully synthesized. An interesting aspect is that besides ammonia also various amines are applicable in the carbonylation reactions. The mechanism of this reaction is still under debate. Beller et al. suggest that the reaction proceeds in a similar way to the stoichiometric studies by Periasamy et al. on the carbonylation of alkynes **120** (Scheme 31) [95, 96].

Apart from this mechanistic hypothesis, another scenario, with a ferrate complex as intermediate, may be possible. In 1928, Hieber discovered that $Fe(CO)_5$ **78** underwent a disproportionation in the presence of ethylenediamine **122** [97–101]. Depending on the reaction temperature, different ferrate complexes were formed that incorporated a $[Fe(en)_3]^{2+}$ cation (en = ethylenediamine) and mono-, di- or trinuclear ferrate anions (Scheme 32) [102–107]. As the reaction discussed above is also performed with amines at high temperatures, these ferrates may well be involved in the catalytic cycle of the carbonylation discussed above.

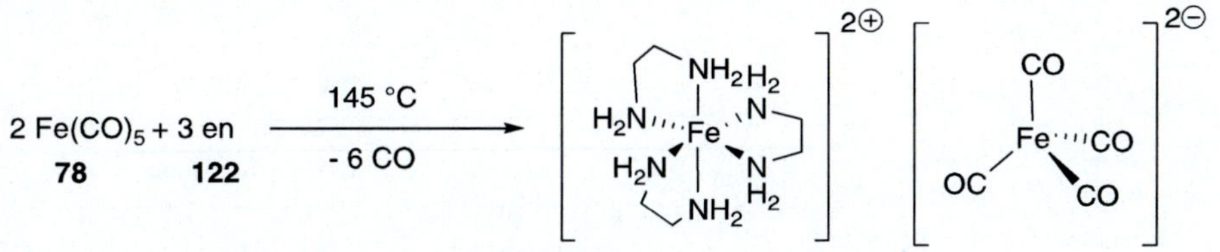

123

Scheme 32 Disproportionation of $Fe(CO)_5$ **78** reported by Hieber

Scheme 33 Iron-catalyzed reductive carbonylation reported by Reppe [108]

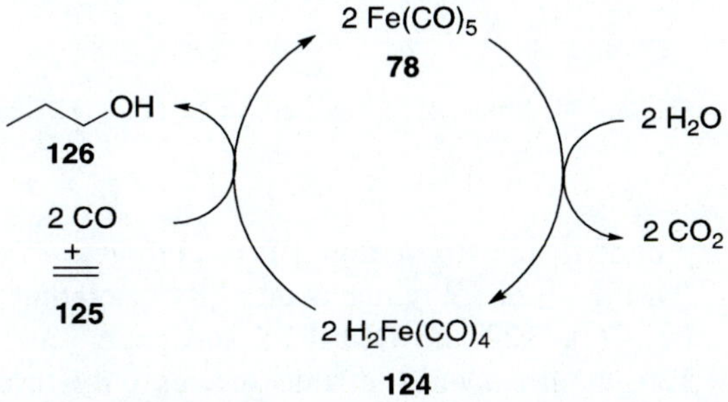

Another catalytic application emanating from the *Hieber base reaction* was developed by Reppe and Vetter [108]. They showed that 1-propanol **126** could be generated by treatment of ethylene **125** with catalytic amounts of $Fe(CO)_5$ **78** under CO-pressure and basic reaction conditions (Scheme 33). Thereby, trimethylamine and *N*-alkylated amino acid derivatives turned out to be optimal bases for this reaction. Like ethylene **125,** propylene could be transferred mainly to 1-butanol; diolefins like butadiene only reacted to monoalcohols. By employing these reaction conditions to olefins in the presence of ammonia, primary or secondary amines, mono-, di-, and trialkylamines were obtained whose alkyl chains were elongated with one carbon atom, compared to the olefins.

Years earlier, Nicholas and Ladoulis had found another example of reactions catalyzed by $Fe_2(CO)_9$ **127**. They had shown that $Fe_2(CO)_9$ **127** can be used as a catalyst for allylic alkylation of allylic acetates **129** by various malonate nucleophiles [109]. Although the regioselectivites were only moderately temperature-, solvent-, and substrate-dependent, further investigations concerned with the reaction mechanism and the catalytic species were undertaken [110]. Comparing stoichiometric reactions of cationic (η^3-allyl)$Fe(CO)_4^+$ and neutral (η^2-crotyl acetate)$Fe(CO)_4$ with different types of sodium malonates and the results of the $Fe_2(CO)_9$ **127**-catalyzed allylation they could show that these complexes are likely no reaction intermediates, because regioselectivites between stoichiometric and catalytic reactions differed. Examining the interaction of sodium dimethylmalonate **75** and $Fe_2(CO)_9$ **127** they found some evidence for the involvement of a coordinated malonate species in the catalytic reactions. With an excess of malonate they

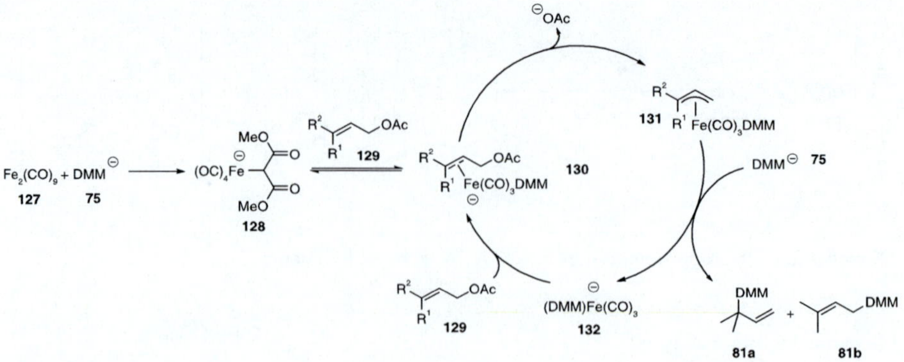

Scheme 34 Proposed mechanism for $Fe_2(CO)_9$ **127**-catalyzed allylic alkylation [110]

proposed the formation of a complex of the type $NaFe(CO)_4(DMM)$ **128** (DMM = dimethylmalonate). By isolating the product of the reaction of $Fe_2(CO)_9$ **127** and DMM **75** and examining its activity toward allylic acetates **129**, it turned out that this species only forms the substitution products in the presence of DMM **75**. This led to the proposed mechanism displayed in Scheme 34.

Dimethylmalonate **75** coordinates to a $Fe(CO)_4$ species, yielding a ferrate species **128**. This coordinates the allylic substrate under decarbonylation and by nucleophilic attack at the double bond an allyliron-species **131** is generated which undergoes substitution of the ferrate **132** by a dimethylmalonte molecule **75**. Although there is some evidence of this catalytically active ferrate **128**, until now it could not be fully analytically characterized and therefore the structure presented above still remains a hypothesis.

7.3 Dinitrosyliron Complexes

In vivo, nitric oxide can be stored and stabilized in the form of dinitrosyliron complexes (DNICs) and is probably released from cells in the form of low-molecular-weight DNICs (LMW-DNICs) [111–113]. The coordination of NO by [Fe–S] clusters is thought to result in impairment of metabolic functions, while the interplay between NO and [Fe–S] clusters at catalytical sites is thought to be crucial for the response to environmental signals within cells [114]. Thus, NO was discovered to strongly inhibit mammalian ferrochelatase by the degradation of the [2Fe–2S]-cluster to form monomeric cysteinyl-coordinated dinitrosyl complexes [115]. Furthermore, SoxR, a redox-sensitive transcription activator in *E. coli*, is suggested to be activated by nitrosylation of [2Fe–2S]-clusters to form protein-bound dinitrosyl adducts [116], while DNIC can be directly transformed back to ferredoxin [2Fe–2S]-cluster by the enzyme cysteine desulfurase [117, 118].

In recent years, several model complexes have been synthesized and studied to understand the properties of these complexes, for example, the influence of S- or N-ligands or NO-releasing abilities [119]. It is not always easy to determine the electronic character of the NO-ligands in nitrosyliron complexes; thus, forms of NO^- [120], neutral NO, or NO^+ [121] have been postulated depending on each complex. Similarly, it is difficult to determine the oxidation state of Fe; therefore, these complexes are categorized in the Enemark–Feltham notation [122], where the number of *d*-electrons of Fe is indicated. In studies on the nitrosylation pathway of thiolate complexes, Liaw et al. could show that the nitrosylation of complexes $[Fe(SR)_4]^{2-/1-}$ (R = Ph, Et) led to the formation of air- and light-sensitive mononitrosyl complexes $[Fe(NO)(SR)_3]^-$ in which tetrathiolate iron(+3) complexes were reduced to Fe(+2) under formation of $(SR)_2$. Further nitrosylation by NO yields the dinitrosyl complexes $[(SR)_2Fe(NO)_2]^-$, while nitrosylation by NO^+ forms the neutral complex $[Fe(NO)_2(SR)_2]$ and subsequently Roussin's red ester $[Fe_2(\mu\text{-}SR)_2(NO)_4]$ under reductive elimination forming $(SR)_2$. Thus, nitrosylation of biomimetic oxidized- and reduced-form rubredoxin was mimicked [121]. Lippard et al. showed that dinuclear Fe-clusters are susceptible to disassembly in the presence of NO [123].

Liaw et al. reported that conversions between the neutral sparteine $\{Fe(NO)_2\}^{10}$-complex **133** and the anionic $\{Fe(NO)_2\}^9$ $[Fe(NO)_2(S_2C_3H_6)]^-$ **137** proceed via the cationic sparteine $\{Fe(NO)_2\}^9$-complex **135** through oxidation by NO^+ and transfer of the $[Fe(NO)_2]^+$-unit to the chelating ligand $^-S\text{–}(CH_2)_3\text{–}S^-$ **136** (Scheme 35). The resulting anionic complex **137** is stable in contrast to the cationic complex **135**. The cationic complex **135** also acts as a $[Fe(NO)_2]^+$ donor in the presence of the DNIC $[(PhS)_2Fe(NO)_2]^-$ **138** to yield Roussin's red ester **139**. The bidentate thiol ligand $^-S\text{–}(CH_2)_3\text{–}S^-$ **136** promotes the stability of the anionic DNIC $\{Fe(NO)_2\}^9$

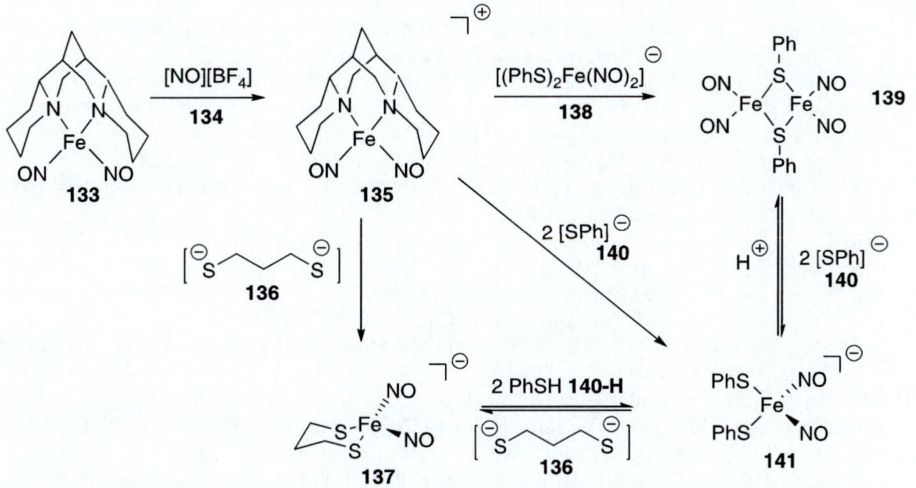

Scheme 35 Transformation between anionic, neutral, and cationic DNICs [124]

137 which is shown by the fact that the cationic sparteine complex **135** reacts with 2 equiv. of EtS^- to form the corresponding Roussin's red ester $[Fe_2(\mu\text{-}SEt)_2(NO)_4]$. This study implicates that the chelating effect of thiobiomolecules may play a role in stabilizing protein-bound DNICs, because the ligand $^-S\text{–}(CH_2)_3\text{–}S^-$ **136** is mimicking Cys–$(X)_n$–Cys binding motifs in proteins [124].

Catalysis has not been established with these DNICs until now; nevertheless, these complexes are interesting low-valent iron complexes that fit into the class of ferrates from which one can learn a lot about reactivity, stability, and electronic properties.

Acknowledgment Financial support by the Deutsche Forschungsgemeinschaft and the Fonds der Chemischen Industrie is gratefully acknowledged.

References

1. Jørgensen CK (1971) Modern aspects of ligand field theory. Elsevier, New York
2. Matsumoto M, Kobayashi H (1985) J Org Chem 50:1766–1768
3. Bica K, Gaertner P (2006) Org Lett 8:733–735
4. Valkenberg MH, deCastro C, Holderich WF (2001) Appl Catal A 215:185–190
5. Berthold HJ, Spiegl HJ (1972) Z Anorg Allg Chem 391:193–202
6. Kauffmann T, Laarmann B, Menges D, Neiteler G (1992) Chem Ber 125:163–169
7. Uchiyama M, Matsumoto Y, Nakamura S, Ohwada T, Kobayashi N, Yamashito N, Matsumiya A, Sakamoto T (2004) J Am Chem Soc 126:8755–8759
8. Hoffmann RW (2005) Angew Chem 117:6433–6435, Angew Chem Int Ed 44:6277–6279
9. Fürstner A, Martin R, Seidel G, Goddard R, Lehmann CW (2008) J Am Chem Soc 130:8773–8787
10. Fürstner A, Krause H, Lehmann CW (2006) Angew Chem 118:454–458, Angew Chem Int Ed 45:440–444
11. Kharash MS, Tawney PO (1941) J Am Chem Soc 63:2308–2315
12. Bazhenova TA, Lobovskaya RM, Shivaeva RP, Shilov AE, Shilova AK, Gruselle M, Leny G, Tchoubar BJ (1983) J Organomet Chem 244:265–272
13. Jonas K (1985) Angew Chem 97:292–307, Angew Chem Int Ed Engl 24:295–311
14. Jonas K, Krüger C (1980) Angew Chem 92:513–531, Angew Chem Int Ed Engl 19:520–537
15. Jonas K (1990) Pure Appl Chem 62:1169–1174
16. Fürstner A, Martin R, Majima K (2005) J Am Chem Soc 127:12236–12237
17. Fürstner A, Majima K, Martin R, Krause H, Kattnig E, Goddard R, Lehmann CW (2008) J Am Chem Soc 130:1992–2004
18. Lei A, He M, Zhang X (2002) J Am Chem Soc 124:8198–8199
19. Oppolzer W, Fürstner A (1993) Helv Chim Acta 76:2329–2337
20. Varela JA, Saá C (2003) Chem Rev 103:3787–3801
21. Chopade PR, Louie J (2006) Adv Synth Catal 348:2307–2327
22. Shima S, Pilak O, Vogt S, Schnick M, Stagni MS, Meyer-Klaucke W, Warkentin E, Thauer RK, Ermler U (2008) Science 321:572–575
23. Frey M (2002) ChemBioChem 3:153–160
24. Darensbourg MY, Lyon EJ, Smee JJ (2000) Coord Chem Rev 206–207:533–561
25. Fontecilla-Camps JC, Ragsdale SW (1999) Inorg Chem 47:283–333
26. Peters JW, Lanzilotta WL, Lemon BJ, Seefeldt LC (1998) Science 282:1853–1858
27. Nicolet Y (2001) J Am Chem Soc 123:1596–1601
28. Nicolet Y, Cavazza C, Fontecilla-Camps JC (2002) J Inorg Biochem 91:1–8

29. Kayal A, Rauchfuss TB (2003) Inorg Chem 42:5046–5048
30. Whaley CW, Rauchfuss TB, Wilson SR (2009) Inorg Chem 48:4462–4469
31. Kubas GJ (2001) Metal dihydrogen and σ-bond complexes. Kluwer/Plenum, New York
32. Chong D, Georgakaki IP, Mejia-Rodriguez R, Sanabria-Chinchilla J, Soriaga MP, Darensbourg MY (2003) Dalton Trans 4158–4163
33. Mejia-Rodriguez R, Chong D, Reibenspies JH, Soriaga MP, Darensbourg MY (2004) J Am Chem Soc 126:12004–12014
34. Munyejabo V, Guillaume P, Postel M (1994) Inorg Chim Acta 221:133–139
35. Li Kam Wah H, Postel M (1989) Inorg Chim Acta 165:215–220
36. Damiano J-P, Munyejabo V, Postel M (1995) Polyhedron 14:1229–1234
37. Guillaume P, Postel M (1995) Inorg Chim Acta 233:109–112
38. Li Kam Wah H, Postel M, Tomi F (1989) Inorg Chem 28:233–238
39. Jonas K, Schieferstein L, Krüger C, Tsay Y-H (1979) Angew Chem 91:590–591, Angew Chem Int Ed Engl 18:550–551
40. Jonas K, Schieferstein L (1979) Angew Chem 91:590, Angew Chem Int Ed Engl 18:549–550
41. Bogdanovic B, Schwickardi M (2000) Angew Chem 112:4788–4790, Angew Chem Int Ed 39:4610–4612
42. Martin R, Fürstner A (2004) Angew Chem 116:4045–4047, Angew Chem Int Ed 43:3955–3957
43. Hieber W, Beutner H (1960) Z Naturforsch Teil B 15:323–324
44. Hieber W, Beutner H (1963) Z Anorg Allg Chem 320:101–111
45. Hieber W, Leutert F (1931) Naturwissenschaften 19:360–361
46. Hieber W, Leutert F (1931) Ber Dtsch Chem Ges 64:2832–2839
47. Ellis JE (1975) J Organomet Chem 86:1–56
48. Ellis JE (1990) Adv Organomet Chem 31:1–51
49. Trost B (1996) M. Van Vranken D L 96:395–422
50. Eberhardt U, Mattern G (1988) Chem Ber 121:1531–1534
51. Davies SG, Smallridge AJ (1990) J Organomet Chem 386:195–201
52. Itoh K, Nakanishi S, Otsuji Y (1991) Bull Chem Soc Jpn 64:2965–2977
53. Itoh K, Nakanishi S, Otsuji Y (1994) J Organomet Chem 473:215–224
54. Itoh K, Otsuji Y, Nakanishi S (1995) Tetrahedron Lett 36:5211–5214
55. Nakanishi S, Yamaguchi H, Okamoto K, Takata T (1996) Tetrahedron: Asymmetry 7:2219–2222
56. Yamaguchi H, Nakanishi S, Okamoto K, Takata T (1997) Synlett 722–724
57. Nakanishi S, Memita S, Takata T, Itoh K (1998) Bull Chem Soc Jpn 71:403–412
58. Nakanishi S, Okamoto K, Yamaguchi H, Takata T (1998) Synthesis 1735–1741
59. Roustan J-L, Abedini M, Baer HH (1979) Tetrahedron Lett 20:3721–3724
60. Roustan J-L, Abedini M, Baer HH (1989) J Organomet Chem 376:C20–C22
61. Xu Y, Zhou B (1987) J Org Chem 52:974–977
62. Zhou B, Xu Y (1988) J Org Chem 53:4419–4421
63. Plietker B (2006) Angew Chem 118:1497–1501, Angew Chem Int Ed 45:1469–1473
64. Plietker B (2006) Angew Chem 118:6200–6203, Angew Chem Int Ed 45:6053–6056
65. Jegelka M, Plietker B (2009) Org Lett 11:3462–3465
66. Glorius F (ed) (2007) N-heterocyclic carbenes in transition metal catalysis. Springer, Hamburg
67. Plietker B, Dieskau A, Möws K, Jatsch A (2007) Angew Chem 120:204–207, (2008) Angew Chem Int Ed 47:198–201
68. Holzwarth M, Dieskau A, Tabassam M, Plietker B (2009) Angew Chem 121:7387–7391, Angew Chem Int Ed 48:7251–7255
69. Trivedi R, Tunge JA (2009) Org Lett 11:5650–6652
70. Magens S, Ertelt M, Jatsch A, Plietker B (2008) Org Lett 10:53–56
71. Kumada M, Kochi JK (1971) J Am Chem Soc 93:1487–1489
72. Kochi JK (1974) Acc Chem Res 7:351–360
73. Kharasch MS, Fields EK (1941) J Am Chem Soc 63:2316–2320

74. Kochi JK, Tamura M (1971) J Organomet Chem 31:289–309
75. Kochi JK, Neumann SM (1975) J Org Chem 40:599–606
76. Bedford RB, Bruce DW, Frost RM, Goodby JW, Hird M (2004) Chem Commun 2822–2823
77. Hatakeyama T, Nakamura M (2007) J Am Chem Soc 129:9844–9845
78. Fürstner A, Leitner A (2003) Angew Chem 115:320–323, Angew Chem Int Ed 42:308–311
79. Korn TJ, Cahiez G, Knochel P (2003) Synlett 12:1892–1894
80. Fürstner A, Brunner H (1996) Tetrahedron Lett 37:7009–7012
81. Dunet G, Knochel P (2006) Synlett 407–410
82. Reddy CK, Knochel P (1996) Angew Chem 108:1812–1813, Angew Chem Int Ed 35:1700–1701
83. Dohle W, Kopp F, Cahiez G, Knochel P (2001) Synlett 1901–1903
84. Hölzer B, Hoffmann RW (2003) Chem Commun 732–733
85. Nakamura M, Matsuo K, Ito S, Nakamura W (2004) J Am Chem Soc 126:3686–3697
86. Bedford RB, Betham M, Bruce DW, Danopoulos AA, Frost RM, Goodby JW, Hird M (2006) J Org Chem 71:1104–1110
87. Fürstner A, Leitner A, Méndez M, Krause H (2002) J Am Chem Soc 124:13856–13863
88. Smith RS, Kochi JK (1976) J Org Chem 41:502–509
89. Allen RB, Lawler RG, Ward HR (1973) J Am Chem Soc 95:1692–1693
90. Czaplik WM, Mayer M, von Wangelin AJ (2009) Angew Chem 121:616–620, Angew Chem Int Ed 48:607–610
91. Czaplik WM, Mayer M, Cvengros J, von Wangelin AJ (2009) ChemSusChem 2:396–417
92. Moreau B, Wu JY, Ritter T (2009) Org Lett 11:337–339
93. Wu JY, Moreau B, Ritter T (2009) J Am Chem Soc 131:12915–12917
94. Driller KM, Klein H, Jackstall R, Beller M (2009) Angew Chem 121:6157–6160, Angew Chem Int Ed 48:6041–6044
95. Periasamy M, Rhameshkumar C, Rhadhakrishnan U (1997) Tetrahedron Lett 38:7229–7232
96. Periasamy M, Rhameshkumar C, Rhadhakrishnan U, Brunet J-J (1998) J Org Chem 63:4930–4935
97. Hieber W, Sonnekalb F (1928) Chem Ber 61:558–565
98. Hieber W, Sonnekalb F (1928) Chem Ber 61:2421–2427
99. Hieber W, Sonnekalb F (1929) Chem Ber 62B:422
100. Hieber W, Sonnekalb F, Becker E (1930) Chem Ber 63:973–986
101. Hieber W, Becker E (1930) Chem Ber 63:1405–1417
102. Hieber W, Nast R, Sedlmeier J (1952) Angew Chem 64:465–470
103. Hieber W, Sedlmeier J, Werner R (1957) Chem Ber 90:278–286
104. Hieber W, Werner R (1957) Chem Ber 90:1116–1120
105. Hieber W, Kahlen N (1958) Chem Ber 91:2223–2233
106. Hieber W, Kahlen N (1958) Chem Ber 91:2234–2238
107. Hieber W, Beck W, Braun G (1960) Angew Chem 72:795–801
108. Reppe W, Vetter H (1953) Liebigs Ann Chem 582:133–161
109. Ladoulis SJ, Nicholas KM (1985) J Organomet Chem 285:C13–C16
110. Silverman GS, Strickland S, Nicholas KM (1986) Organometallics 5:2117–2124
111. Foster MW, Cowman JA (1999) J Am Chem Soc 121:4093–4100
112. Reginato N, McCrory CTC, Pervitsky D, Li L (1999) J Am Chem Soc 121:10217–10218
113. Cooper CE (1999) Biochim Biophys Acta 1411:290–309
114. Drapier JC (1997) Methods 11:319–329
115. Dailey HA, Dailey TA, Wu C-K, Medlock AE, Rose JP, Wang K-F (2000) Cell Mol Life Sci 57:1909–1926
116. Ding H, Demple B (2000) Proc Natl Acad Sci USA 97:5146–5150
117. Yang W, Rogers PA, Ding H (2002) J Biol Chem 277:12868–12873
118. Rogers PA, Ding H (2001) J Biol Chem 276:30980–30986
119. Tsai F-T, Chiou S-J, Tsai M-C, Tsai M-L, Huang H-W, Chiang M-H, Liaw W-F (2005) Inorg Chem 44:5872–5881

120. Harrop TC, Song DT, Lippard SJ (2006) J Am Chem Soc 128:3528–3529
121. Lu T-T, Chiou S-J, Chen C-Y, Liaw W-F (2006) Inorg Chem 45:8799–8806
122. Enemark JH, Feltham RD (1974) Coord Chem Rev 13:339–406
123. Tonzetich ZJ, Do LH, Lippard SJ (2009) J Am Chem Soc 131:7964–7965
124. Hung M-C, Tsai M-C, Lee G-H, Liaw W-F (2006) Inorg Chem 45:6041–6047

Index